The Brewing Science Laboratory

Sean E. Johnson

Brewing Lab Sciences
University of Northern Colorado

Michael D. Mosher

Department of Chemistry and Biochemistry
University of Northern Colorado

Front cover photo: Courtesy Karyl Rice—© American Society of Brewing Chemists.

Chapter-opening photos: Pages 1, 53, 127, 161, 213: Elena Pavlovich/Shutterstock.com. Pages 15, 77, 115, 175, 229: Zoltan Major/Shutterstock.com. Pages 31, 93, 145, 193, 245: IntoTheWorld/Shutterstock.com.

To the extent permitted under applicable law, neither the American Society of Brewing Chemists (ASBC) nor its suppliers and licensors assume responsibility for any damage, injury, loss, or other harm that results from an error or omission in the content of this publication in any of its forms and/or of any of the products derived from it. This publication and its associated products do not address every possible situation, and unforeseen circumstances may make the recommendations they contain inapplicable in some situations. Users of these materials cannot assume that they contain all necessary warning and precautionary measures and that other or additional information or measures may not be required. Users must rely on their own experience, knowledge, and judgment in evaluating or applying any information. Given rapid advances in the sciences and changes in product labeling and government regulations, ASBC recommends that users conduct independent verifications of results and findings.

ASBC reserves the right to change, modify, add, or remove portions of these terms and conditions at its sole discretion at any time and without prior notice. Please check the Terms and Conditions of Use stated on the ASBC website periodically for any modifications.

Reference in this publication to a trademark, proprietary product, or company name is intended for explicit description only and does not imply approval or recommendation to the exclusion of others that may be suitable.

Library of Congress Control Number: 2019940701
International Standard Book Numbers:
Print: 978-1-881696-36-0
eBook: 978-1-881696-37-7

© 2020 American Society of Brewing Chemists

All rights reserved.

No portion of this book may be reproduced in any form, including photocopy, microfilm, information storage and retrieval system, computer database, or software, or by any means, including electronic or mechanical, without written permission from the publisher.

Content from the *ASBC Methods of Analysis* has been adapted by permission of the American Society of Brewing Chemists.

Printed in the United States of America on acid-free paper

American Society of Brewing Chemists
3340 Pilot Knob Road
St. Paul, Minnesota 55121, U.S.A.

Preface

Have you ever had a beer that was just amazing? Maybe it's your go-to beer when you're out on the town. And have you ever considered that the beers you've enjoyed from one time out to the next have come from many different batches? Likely, these beers had the same flavor profile, even though different bottles came from batches brewed years apart.

Don't worry! You're not the only one who hasn't considered this reality. It's all about consistency, which is an often-overlooked aspect of brewing. But it's also likely the most important. How does a brewery achieve that consistency? It does so by implementing quality control and quality assurance principles.

We often inform our graduates looking for positions in the chemical industry to consider companies with names that don't include the word "laboratory." In fact, most if not all companies that manufacture products have laboratories inside their walls. Included are companies in industries that manufacture engine parts, conveyor belts, fertilizers, and pesticides as well as those that produce cheeses and beverages. This means that the brewing company down the street likely contains a laboratory.

What does the laboratory do in one of these businesses? While the lab may focus on confirming or informing the production line about the quality of a product, the efficiency of a manufacturing process, or the remaining lifetime for the coolant in a machine, lab workers are also tasked with protecting consumer safety.

In the case of a brewery, the laboratory serves a vital role in the production of beer. The laboratory checks the quality of the product, performs routine analyses on various stages of the product as it goes through the production line, and confirms that hazards to the safety of the consumer are eliminated. The lab also evaluates beer for consistency and trueness to the specifications laid out by the brewery.

As a brewery grows, so does the laboratory. A mom-and-pop neighborhood brewery might have its laboratory on a workbench, but by the time this brewery is producing beer for the local area or distributing beer regionally, its laboratory has become a room full of exciting equipment, glassware, and instrumentation. The brewery may also have assigned an employee to run that laboratory. Even a large brewery has multiple laboratories located near key points in the production facility and multiple laboratory staff members assigned to perform experiments. The largest breweries may take on research experiments beyond the original scope of their labs to find new technologies, new methods, and new ways to improve production.

That's what *The Brewing Science Laboratory* is about. It covers the major aspects of what a brewery laboratory is and does. Without making any assumptions about what the reader knows, this book provides the information necessary to create and operate a successful brewery laboratory. The text is written for the student interested in entering the brewing industry as a quality control technician, but the information it provides will be useful for the small brewery owner and even the brewer who's interested in increasing his or her own laboratory's output.

The Brewing Science Laboratory presents the information in a conversational tone. Topics that need background information are preceded by sections that dive into the necessary details. Information is provided in a need-to-know-now format. This approach is helpful for the novice, who might not feel comfortable talking about chemistry, biology, physics, or other subjects. It's also useful for the well-versed reader who appreciates reviewing the basics before diving into a subject. In short, the text is written for anyone to read, understand, and apply to his or her own situation.

As you can see from the table of contents, the text begins by providing a background in implementing a quality assurance/quality control program, information on how to do so on a budget, and the science behind the analysis. Then the discussion moves into the laboratory to provide basic information on how to work there. The content focuses on the instrumentation used to provide accurate and precise measurements—a topic addressed in many of the *American Society of Brewing Chemists (ASBC) Methods of Analysis*. Particularly useful are the adapted methods provided in Chapter 15. They include concise directions for preparing reagents, step-by-step instructions on conducting analyses, and data sheets that can be filled out as experiments are performed. The data sheets can be reproduced and are available online (asbcnet.org/BSL).

For readers interested in learning more about brewing laboratories and how they can be leveraged to address potential issues in the brewery, the authors strongly recommend spending time on the ASBC website (asbcnet.org) Additional *ASBC Methods of Analysis*, along with guidelines, fishbone diagrams, and other information, are provided online and can help convert the workbench in any brewery into a powerful and indispensable laboratory.

Acknowledgments

Compiling a text like this takes the work of numerous people. Without their help, we would not have been able to get very far. In particular, we wish to thank Dana Sedin (New Belgium), Mark Eurich (New Belgium), Joe Palausky (Boulevard), Kim Bacigalupo (Sierra Nevada), and Scott Britton (Duvel Moortgat) for their help in reviewing and editing the text. Their suggestions have significantly improved this work. In addition, producing this book would not have been possible without the tireless work and attention of ASBC staff.

Most importantly, we thank our wives and families for their unwavering support. Thank you for allowing us to pursue all of our crazy beer adventures. We would not be here without you.

Sean E. Johnson
Michael D. Mosher

Foreword

Humans have been brewing beer for thousands of years. With only a few simple ingredients, even your neighbor can make beer in his or her garage. Yet those who understand brewing appreciate the challenges and complexities of the brewing process. Scientists continue to expand our understanding of the brewing process and brewing raw materials. With this enhanced knowledge comes new challenges to ensure the quality of beer. *The Brewing Science Laboratory* provides a foundation for designing quality into brewery operations.

With more than 7,000 breweries in the United States alone, ranging in size from small brewpubs to multinational breweries, the quality requirements will vary significantly. In some cases, a brewery will serve only draft beer to the customers who visit it, and in other cases, a brewery will package and ship beer across the United States or to international destinations. Many breweries aim to flavor match brands across multiple brewery locations.

Even with these differences in size and geographic scope, all breweries should have a focus on quality to ensure that their customers' expectations are met each time they enjoy a beer. Because of social media and beer-rating websites, no brewery can expect to hide quality issues, such as off-flavors, from the public. There is a cost to maintaining a quality program, but having an inadequate quality program could result in a brewery going out of business. There are many examples of product recalls and quality issues that could have been avoided with a proper quality assurance (QA) program in place.

In addition, breweries continue to take bigger quality risks while pushing the boundaries using novel fruits, spices, mixed cultures, and other materials in their beers. Even the complexity around dry hopping (such as the impact of hop enzymes) can have a large impact on the quality of finished beer. It can be a daunting task to determine where to start a quality program or how to grow a program to cover all the potential risks in the brewing process.

In *The Brewing Science Laboratory*, authors Sean Johnson and Michael Mosher cover topics critical to starting and growing a brewery laboratory program. The book is relevant to breweries starting QA programs, to breweries growing their programs to include more advanced analyses, and to breweries with strong programs in place that need tools for training new members of their QA teams. Sean and Michael's book guides the reader through the fundamentals of chemistry, sensory, and microbiology in an easy-to-understand format. The book covers a wide range of important topics, from the costs associated with setting up a quality laboratory to guidance on proper laboratory techniques. It also provides information on data management, laboratory safety, and laboratory capabilities, including how and when to grow them.

I highly recommend this book to any brewer looking to start or grow quality into his or her brewery operations.

Dana Sedin, PhD
ASBC President, 2018–2019
New Belgium Brewing Company, Quality Assurance

Contents

CHAPTER 1

Introduction   1

Overview of the Brewing Process   3
Quality   5
  Maintaining Quality   5
  Measuring Quality   10
  Investing in Quality   12

CHAPTER 2

Implementing Safety and Quality   15

What Is HACCP?   15
Implementing a Successful HACCP Program   17
American Society of Brewing Chemists   22
  ASBC Methods of Analysis   24
What's Wrong with My Beer?   25
  Common Off-Flavors   26
  How to Attack Off-Flavors   28
  Using New Ingredients (or the Same Ingredient from Another Supplier)   30

CHAPTER 3

The Brewery Laboratory   31

The Brewery Laboratory   31
  Overview of the Lab   31
  Do You Need a Scientist?   33
  Who Performs QA Procedures?   34

The Design of a Brewery Laboratory 35
  Space Requirements 36
  Supplies and Chemicals 38
  Laboratory Safety 40
  Fire Safety 44
  Common Instruments 45

A Brewery on a Budget 46
  The Essentials Are Based on Your QC Requirements 48
  A Lab for Under $10,000 48
  A Lab for Under $1,000 50

CHAPTER 4

The Chemistry Behind the Analysis 53

Essentials of Chemistry 53
  Atoms and Radioactivity 54
  Drawing Atoms 58
  Ions: Atoms with Charges 59
  Combining Atoms 60
  Writing Chemical Formulas 61
  Collections of Atoms 62
  Calculating Ions in Water 66
  pH 69
  Chemical Reactions 71
  Line Drawings 74
  Organic Molecules in Brewing 75

CHAPTER 5

Essentials of Biology and Biochemistry 77

The Barley Seed 77
The Hop Cone 80
Wort: The Perfect Growing Environment 83
Types of Microorganisms 86
The Yeast Cell 88

Basic Metabolism   89
    Glycolysis   90
    The Krebs Cycle   90
    Oxidative Phosphorylation   91
    Fermentation   92
    Requirements for Metabolism   92

CHAPTER 6

Working in the Lab   93

Laboratory Glassware   93
    Types of Glassware   94
    The Meniscus   97
    Using Graduated Glassware   97
    Using Volumetric Glassware   98
    Filtering Liquids   100
    Distillation   101

Handling Chemicals   102
    Obtaining the Mass of a Solid or Liquid   102
    Obtaining the Volume of a Liquid   104
    Preparing a Solution from a Solid   105
    Preparing a Solution by Dilution   105
    Distilling Liquids   106

Handling Microbes   108
    Sterile Versus Sanitized Versus Clean   108
    Glassware Used in the Micro Lab   110
    Proper Use of the Microscope   111
    Petri Dishes and Agar   112
    Broths and Growing Yeast   114

CHAPTER 7

Instruments and the Laboratory   115

Instruments in the Laboratory   115
    Instruments Versus Equipment   115
    Pros of Using Instruments for Analysis   117
    Cons of Using Instruments for Analysis   118

Laboratory Conditions for Optimal Instrument Operation   120
    Temperature   121
    Electrical Power   122
    Humidity   122
    Dust   123
    Lighting   124

Instrument Upkeep and Maintenance   124

CHAPTER 8

Reporting Your Results   127

Graphing with Microsoft Excel or Another Spreadsheet Program   128
    Entering Results   130
    Standard Curves   131
    Plotting Results, Cumulative Sum Charts, and Trending   138

Tables and Record Keeping   140
    Using Functions   140

Macros and Advanced Excel Functions   141
    Programming 101   142

CHAPTER 9

Sensory Analysis   145

Keeping a Sensory Notebook   146

Appearance: Making a First Impression   147

Flavor: Training the Palate   150

Aroma: A Key Component of Taste   151

The Sensation of Tasting   151

Aroma and Overall Flavor   152
    Sensory Thresholds   154
    Sensory Overload   155

How to Set up Your First Sensory Panel   155
    Trusting Your Panel   157

Types of Sensory Analysis   158
    Triangle Test   158
    Paired Comparison Test   158
    Tetrad Test   159
    Threshold Test   159

Performing Sensory Analysis   160

CHAPTER 10

Colorimetric Analysis   161

Eyes Are the Instrument   163
    People as Instruments   164
    Testing for Presence Versus Testing Against a Standard   165

Making a Reference Chart   167
    Colorimeter Setup   168
    Minimum and Maximum Concentrations   171

Colorimetric Methods   172

CHAPTER 11

Spectrophotometric Analysis   175

What Is a Wavelength?   175
    Infrared, Visible, Ultraviolet, and Beyond   177
    Relationship to Colorimetric Methods   179

General Instrument Design   179
    Fixed or Single Wavelength   182
    Continuous Wave   183
    Fourier Transform   184

Infrared Analysis   186
    Pros and Cons of Infrared Analysis   188
    Using Infrared Analysis in Brewing   189

UV-Vis Analysis   189
    Pros and Cons of UV-Vis Analysis   191
    Using UV-Vis Analysis in Brewing   191

CHAPTER 12

Chromatography   193

Polar Versus Nonpolar   194

London Dispersion Forces, Dipole–Dipole Forces, and Hydrogen Bonding   196

Mobile and Stationary Phases   200

General Instrument Design   202

Thin-Layer Chromatography   203

Gas Chromatography   205

High-Performance Liquid Chromatography   210

CHAPTER 13

Other Instruments in the Brewery Laboratory   213

Density Meters   213

Hydrometers   214

Gravimetric Density Meters   217

Coriolis Density Meters   218

Refractometers   219

Viscometers   221

Olfactometers   225

Foam Measurements   226

CHAPTER 14

Microbes   229

Early Research and Inventions   229

Sterilization Revisited   231

Microbes and the Brewing Process   234

Making Media   235

Broths, Plates, and Slants   236

Growing Microbes   238
    Liquid Media   238
    Solid Media   239
    Aseptic Sampling   241
    Membrane Filtration   241

Identifying Bacteria   242
    Reading Plates   242
    Gram Stainings   242
    Polymerase Chain Reaction   243

CHAPTER 15

Data Sheets for *ASBC Methods of Analysis*   245

    ASBC Malt 4: Extract   247

    ASBC Malt 6A: Diastatic Power   250

    ASBC Malt 6B: Diastatic Power (Rapid Method)   254

    ASBC Malt 15B: Grist—By Manual Sieve   258

    ASBC Hops 6A: α- and β-Acids in Hops and Hop Pellets— α- and β-Acids by Spectrophotometry   260

    ASBC Hops 13: Total Essential Oils in Hops and Hop Pellets by Steam Distillation   263

    ASBC Hops 15: Iso-α-Acids in Isomerized Hop Pellets by High-Performance Liquid Chromatography 1993   265

    ASBC Hops 17: Hop Essential Oils by Capillary Gas Chromatography–Flame Ionization Detection   268

    ASBC Wort 5: Yeast Fermentable Extract   271

    ASBC Wort 12: Free Amino Nitrogen (International Method)   273

    ASBC Wort 22: Wort and Beer Fermentable and Total Carbohydrates by High-Performance Liquid Chromatography (HPLC)   276

    ASBC Wort 23A: Wort Bitterness—Bitterness Units by Spectrophotometry   279

    ASBC Wort 23C: Wort Bitterness—Iso-α-Acids in Wort by High-Performance Liquid Chromatography   281

    ASBC Beer 4A: Alcohol—Beer and Distillate Measured Volumetrically   284

ASBC Beer 4C: Alcohol—Alcohol Determined Refractometrically   287

ASBC Beer 4D: Alcohol—Ethanol Determined by Gas Chromatography   289

ASBC Beer 10A: Color—Spectrophotometric Color Method   292

ASBC Beer 11C: Protein—By Spectrophotometer   294

ASBC Beer 13C: Dissolved Carbon Dioxide—Manometric/Volumetric Method   297

ASBC Beer 22A: Foam Collapse Rate—Sigma Value Method (Modified Carlsberg Method)   299

ASBC Beer 23A: Beer Bitterness—Bitterness Units (International Method)   301

ASBC Beer 23C: Beer Bitterness—Iso-α-Acids by Solid-Phase Extraction and High-Performance Liquid Chromatography   303

ASBC Beer 25D: Diacetyl—Ultraviolet Spectrophotometer Method   307

ASBC Beer 38: Magnesium and Calcium   310

ASBC Yeast 4: Microscopic Yeast Cell Counting   313

ASBC Yeast 6: Yeast Viability by Slide Culture   315

ASBC Yeast 9: Morphology of Giant Yeast Colonies (International Method)   318

ASBC Yeast 11B: Flocculation—Absorbance Method   320

ASBC Sensory Analysis 7: Triangle Test (International Method)   323

ASBC Sensory Analysis 18: Tetrad Test   328

Glossary   333

Index   343

CHAPTER 1

Introduction

An innocuous-sounding piece of legislation referred to as H.R. 1337—An Act to amend the Internal Revenue Code of 1954 brought forth during the 95th Congress prompted a resurgence of breweries across the United States. After some debate, the bill passed both the U.S. House and the Senate, and President Jimmy Carter signed it into law in October 1978. This legislation and its amendment, championed by Senator Alan Cranston of California, removed the tax on beer and wine production for personal use in the home. Specifically, up to 200 gallons were exempted from tax each year if the household contained two adults (100 gallons were exempted for only one adult in the household). At least at the federal level, brewing beer in the home was legal for the first time since Prohibition began in 1920. Because the 21st Amendment, which repealed Prohibition, gave states the right to regulate alcohol production, each state then had to create its own legislation on the issue. While some states followed suit immediately, it took until 2013 for all 50 states to allow beer to be brewed in the home for personal use. It should be noted that because each state had to enact its own rules for regulation, the laws associated with brewing beer in the home are different in individual states.

The effect that this legislation had on the beer industry took a little while to realize, but by the mid-1980s, many people were making beer at home. Most of the beers produced in the United States prior to the 1980s were **lagers**—beers of a clean, crisp, effervescent style that proved difficult for novice brewers to reproduce in a kitchen. Instead, homebrewers branched out into exploring **ale** styles, which were much more forgiving in their flavor profiles. The porter style was rekindled; highly hopped ales became the rage; fruit-flavored ales were tried; and new recipes that didn't follow a set style were created. The result was that by the early 1990s, there were a significant number of beer aficionados who preferred the break from the clean, crisp lager.

Leading up to the 2000s, many entrepreneurs began the process of sharing their love of the multitude of beer styles. New breweries opened up across the country. According to the Brewers Association, a nonprofit trade association that promotes the craft beer industry, there were more than 1,500 breweries in the United States by 2000. In 2019, there were well over 6,000 breweries (including microbreweries, brewpubs, and regional breweries) in the United States, with no signs of slowing growth. For the consumer, this is wonderful. Each brewery produces multiple versions of different styles,

lager Beer fermented with *Saccharomyces pastorianus,* a bottom-fermenting yeast; lagers are fermented at colder temperatures than ales.

ale Beer fermented with *Saccharomyces cerevisiae,* a top-fermenting yeast prone to producing more ester flavors.

ranging from the standard pale ales to the high-end, barrel-aged double imperial stouts. Consumers can choose sours, fruit-flavored ales, and even historical styles that were once lost to time.

Unfortunately, rapid growth such as this can have a significant downside. While closures of microbreweries are currently limited (in 2017, there were 101 closures and 731 openings of microbreweries), closures will most likely increase over time (Fig. 1.1). In 2019, the rate of closure for microbreweries hovered between 1.5% and 2.5%, and growth is continuing in this market. However, consider the restaurant industry, which is fairly saturated across the country. Many studies have focused on restaurant success rates. One such study in the Dallas/Fort Worth area concluded that 23% of new restaurant businesses failed in the first year. Within 3 years, approximately 60% of the new restaurants had failed. Will the craft brewing industry follow suit? Only time will tell.

A lot can be learned from the restaurant industry. How to avoid closure is just one of those lessons. The list of things needed in a successful restaurant mimic those needed in a successful brewery. On that list is the need to maintain consistently high quality in the food that's served. Regular customers will switch to other restaurants if the quality of the food is not consistent. And without regular (read "repeat") customers, the restaurant will struggle to keep customers. The same is true in a brewery: Consistency is essential. Ensuring that the current batch of your flagship India pale ale is the same as a previous batch involves more than just following the same recipe.

Breweries and experienced operations across the globe are well aware of the need for consistency. They invest profits back into the company to ensure that every case of beer they produce is exactly the same as the previous case. This involves employing personnel that are focused on monitoring production at every step in the manufacturing process. It involves buy-in from everyone in the company, from the CEO to the dockworker. It involves consistency.

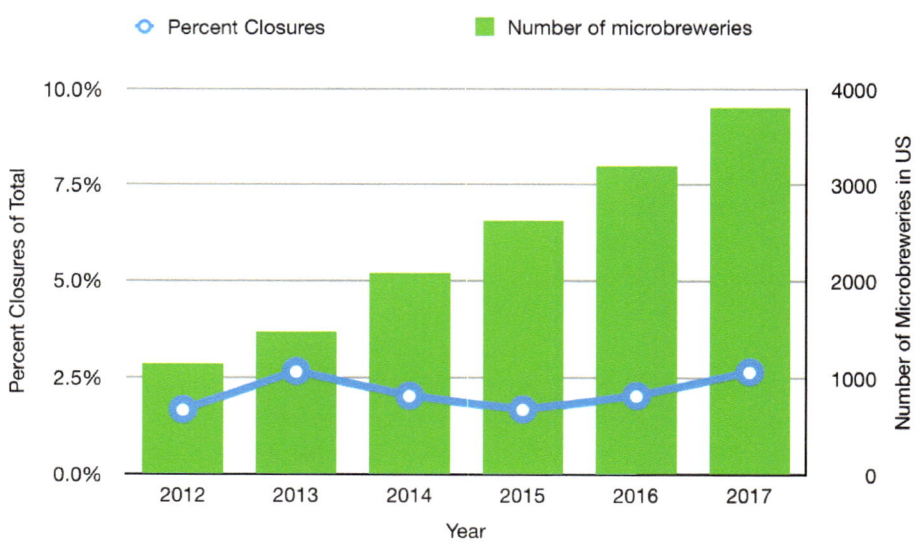

FIG. 1.1. Numbers of microbreweries and rates of closure from 2012 to 2017. (Data from the Brewers Association. Figure courtesy S. E. Johnson and M. D. Mosher—© ASBC.)

Implementing a program that ensures consistency may seem like a daunting and expensive task, but this isn't necessarily the case. If consistency is the goal of the brewery and part of the work ethic for each employee, then implementing such a program is not insurmountable. Yet if funds are limited for the full implementation of a thorough program, even a portion of a full program will provide some adherence to consistent production. This chapter focuses on building consistency in a brewery. And it will introduce you, the reader, to **quality** and how that plays into creating the same high-quality beer again and again.

quality Consistency in a product; can also mean adherence to product specifications, customer satisfaction, lack of off-flavors, or another measure.

Overview of the Brewing Process

To evaluate consistency across the production process within a brewery, we will first examine the brewing process in detail. Starting from the very beginning, we make the transition from farm to glass. Many other sources outline the brewing process in significantly more detail than is described here. The overall process is outlined graphically in Figure 1.2 and discussed in the following paragraphs to verify the terminology used at each step.

Barley, wheat, rye, and oat varieties are grown across the world. Researchers create and evaluate the best new **cultivars** that will fulfill the requirements for the malting and brewing industry. Farmers select those cultivars that will grow best in their area, considering factors such as crop yield, seed size, and profitability. Harvesting the grain reduces the moisture level. Storing the grain removes the dormancy in the seeds. Removing the dormancy increases the number of seeds that are able to sprout or germinate.

At the **maltings**, the barley is sorted to remove metal shards and rocks. Then the grains are steeped in water with aeration to encourage **germination**.

cultivar A variety of hops, barley, or yeast that is in the same genus but has different properties.

malting A location that performs malting; also known as a malthouse.

germination The combination of processes that occur when a seed begins to grow.

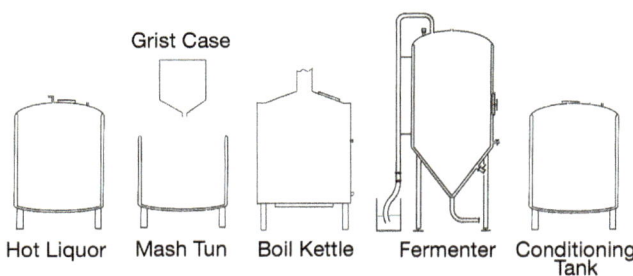

FIG. 1.2. Overview of the brewing process. Malt is milled and placed in the grist case. It is then added to the mash tun with water from the hot liquor tank. After mashing, the sweet wort is moved to the boil kettle and boiled. The bitter wort is then chilled and moved to the fermenter. After fermentation, the green beer is moved to the conditioning tank. Once conditioned, the finished beer can be moved to a serving tank or packaged in bottles, cans, or kegs. (Courtesy S. E. Johnson and M. D. Mosher—© ASBC)

acrospire The part of the seed that becomes the stem of the barley plant.

kilning Heating to remove water from grain; the process stops germination, eliminates some enzymes, and initiates Maillard reactions that result in color and flavor enhancements.

malt Grain that has undergone the steeping, germination, and kilning process.

mill A machine used to crush malt for optimal utilization of starch and enzymes within the seed.

mash tun The vessel in which mashing occurs.

sweet wort The solution containing food and nutrients required for yeast growth obtained after the mashing process has been performed; because of the high concentration of sugars (mostly maltose), the solution is sweet.

lautering A step after mashing that allows the grains to settle to form a filter bed; the sweet wort is drained through this bed.

sparging The process in which additional water is added to the mash to extract additional sugars.

boiling Heating the sweet wort to boil in a separate vessel free of any grain or grain particles.

bitter wort Wort that has been flavored with hops; typically, the liquid obtained after the boiling step.

whirlpool The process of swirling the wort (usually with pumps) as it cools from boiling; this causes solid materials to gather at the center of the vessel.

The maltster removes the water and turns the seeds regularly as they germinate. The barley begins to chit, where the roots become exposed, and the **acrospire** slowly begins growing along the length of the seed under the coat. Changes take place inside the barley seed that result in the activation of enzymes and the breakdown of the starchy endosperm within the seed. Some of the starch within the endosperm is converted into sugars that are transported to the embryo and used as the roots and shoot begin to grow. Once the acrospire has reached the full length of the seed, the process is halted by slowly drying the seeds with warm air in a process known as **kilning**. After kilning, the barley is commonly referred to as **malt**. Kilning continues until the moisture reaches 3–6%: too low and the malt becomes brittle and breaks easily; too high and it loses extract yield and shelf life.

At the brewery, the malt is taken in and stored until use. To start a batch of beer, the brewer **mills** the malt into the **mash tun** and adds hot liquor. The hot liquor is water obtained from various sources. A city supply is often the primary source of the water, which the brewer treats to remove sanitizing agents, increase or reduce alkalinity, and/or add ions such as calcium and sulfate. The mixing of crushed malt and water is accomplished in many different ways; milling (such as wet milling and dry milling), vortex mixing, and mash mixing are all used in the beginning stages of the mashing process. Once the mashing process has started, the brewer selects temperatures to activate enzymatic activity. The result is further decomposition of the barley seed and the production of maltose, glucose, maltotriose, and other fermentable sugars. In addition, the process produces other compounds, including soluble and insoluble proteins, amino acids, polyphenols (tannins), fragments of glucan, and unfermentable sugars such as limit dextrins. This sugar-rich liquid is known as **sweet wort**.

The wort is separated from the spent grains by **lautering** and then **sparging**. The overall process allows the grain to form a filter cake through which the wort passes. The result is the liquid that's used to make beer. This wort contains all the compounds mentioned previously, plus those that are a part of the water used initially to create the mash. Thus, the sweet wort also contains metal ions such as calcium and magnesium, anions such as carbonate and chloride, and any other elements found in the original strike water.

Boiling the wort sterilizes the solution so it can be fermented. This process also halts enzyme activity, concentrates the wort to the correct gravity, and allows the brewer to extract hop oils from hops to produce bitter compounds. Once that boiling process has been completed, the **bitter wort** is **whirlpooled** to separate the bulk of the **trub** (pronounced "trube") from the liquid. **Chilling** occurs by passing the bitter wort through a plate chiller and then adding oxygen to the liquid gas via direct oxygenation or sterile air. In the fermenter, the cooled wort is subsequently mixed with **yeast** in a process referred to as **pitching**. The yeast grows in the presence of the oxygen and then once the oxygen is depleted, begins to **ferment** the sugars in the wort and produce a solution of alcohol, carbon dioxide, and other chemicals. When fermentation is considered complete, the temperature is adjusted to cold crash the solution. The result precipitates the majority of the yeast and clarifies the beer significantly.

The resulting **green beer** moves to the conditioning tank, where it can be evaluated and the flavors, colors, and carbonation can be adjusted if necessary. It is then filtered, carbonated, and packaged. After packaging, the beer is either served from the keg (large pack) or from a can or bottle (small pack). This may involve distributing the beer via a separate company or by the brewery. The process isn't finished until the customer receives the beer in a glass!

Quality

As consumers, we evaluate products in terms of their quality. We award "five stars" to a quality hotel room; "two thumbs up" for a great movie; or some other reward system for products that we think have high quality. But what is quality? Quality is in the eye of the beholder. And in the brewing industry, quality is determined by those who purchase and consume the products of the brewery. The consumer judges the quality of the product.

For the brewery, quality in a product can be determined by recognizing how closely the product adheres to a specific style, how closely it matches some brewery-specific standards for that beer, or how few microbes are found in the finished product. Other measures of quality might also be used at the brewery. While each of these quality measures indicates a standard to which the brewer aspires, the measures are very subjective and represent only quality as the brewer or brewery defines it.

But quality should be thought of as residing with the consumer. Consumers purchase products, and repeated purchases contribute to the success of the business. Even if the brewery can clearly show that its beer matches a particular style, the consumer makes the call on whether to purchase it. A beer may have zero microbes, but that might not matter at all to the consumer. In fact, the consumer may be focused on the attractiveness of the label as the deciding factor to purchase and repurchase the beer.

It can't be stated enough: Quality is in the eye of the consumer. The consumer's judgment of quality is demonstrated by repeat business. If the beer meets or exceeds the consumer's expectations, then it must be a quality product. This sounds like an easy concept to implement in the brewery. However, it is not. Not all consumers have the same ideas of quality, and not all consumers like the same beers. Some prefer the darker, richer flavors found in stouts and porters; others prefer the crisper, lighter flavors of the lager class. Thus, creating a beer that every consumer prefers is nearly impossible. To complicate matters further, a quality product is a moving target. While some consumers may prefer a malty beer today, they may change their preference and go for a dry, hoppy beer tomorrow. To be successful, the brewery must keep a stock of various styles of beers that are well appreciated across a wide range of consumers.

Maintaining Quality

How does meeting the consumer's expectation for quality happen in the brewery? It begins with careful adherence to consistent methods of production. It requires that each batch of beer produced is within brewery specifications

trub A solid material made up of protein, tannins, hop material, and malt material that settles during the whirlpool.

chilling Cooling the beer, postboil; can be done to reach the desired fermentation temperature, to reach the cold storage temperature, or to adjust the temperature for filtration or packaging.

yeast A fungus that ferments different sugars (maltose, glucose, etc.) into ethanol and carbon dioxide during anaerobic conditions.

pitching Adding yeast to the fermenter.

ferment Converting sugars into ethanol and carbon dioxide under anaerobic conditions.

green beer Beer that has finished primary fermentation but still presents off-flavors.

good brewing practices (GBPs) The standards in a brewery that govern quality and consistency in production.

total quality management (TQM) A brewery-wide program that ensures quality production in which all employees constantly work to improve quality.

that are fitted to the consumer's measure of quality. It requires that every employee in the brewery, from the top to the bottom, constantly works to meet those requirements of consistency. It requires that each employee know his or her portion of the process, at the bare minimum, and how that process impacts quality. Those breweries that excel in producing beer with consistent adherence to brewery specifications ensure that their employees know the entire brewing process and are trained on many different lines within the brewery. This allows the brewery staff to work as one. Any issues can be addressed then by multiple people within the brewery.

A brewery can use many methods to improve the chances of meeting those high standards of production. One such method, which was hinted at previously, is the implementation of **good brewing practices (GBPs)**. In this method, every employee has a responsibility to maintain quality. This philosophy is the foundation of a program known as **total quality management (TQM)**. The principles behind the TQM method of quality assurance are listed in Table 1.1. The key components of the method are centered on employee and management investment in the process. Problem solving within the manufacturing process is handled within teams; members from different areas of the entire process meet, consult, and discuss problems as they arise. Then the teams take the actions needed to solve the problems. Each batch of beer provides an opportunity to improve on how the overall process is performed. By constantly improving the manufacturing process, the level of consistency improves. Each production batch meets the specifications better than the previous one. And quality improves.

For a small brewery with a few employees, practicing TQM may require that every employee become involved in the whole brewing process. Distinctions of job responsibilities may be blurred when this happens. Everyone assists with every process and adheres to improving production. Everyone assists with measuring quality parameters for each batch of beer, whether it be sensory analysis, cleaning practices, or collecting and analyzing samples. Every worker, regardless of the size of the brewery, focuses on improving the process. An improvement can be as simple as calibrating the hydrometer weekly or correcting the piping runs to eliminate dead ends. Often, taking large steps is not required to make improvements to the production process. Rather, the repeated implementation of many small steps is what continuously improves the process. That improvement results in consistency.

Other methods are also available to maintain the consistent production of a quality product. Often seen in larger breweries, one such method is known as **hazard analysis and critical control points (HACCP)**. Adhering to this method is required by law. It is used successfully in the food industry and

hazard analysis and critical control points (HACCP) A brewery-wide approach that ensures the consistent production of a product safe for consumption.

TABLE 1.1. Principles of total quality management (TQM)[a]

- Everyone is responsible for quality.
- Processes are the problem—not employees.
- Quality must be measured.
- Improvements to processes must be continuous.

[a] Courtesy S. E. Johnson and M. D. Mosher—© ASBC.

will likely become a requirement for U.S. breweries in the not-too-distant future. While this method is intended to ensure that the production of food products is safe for human consumption, it can also be a method of quality assurance.

Implementing HACCP in a brewery is covered in detail in Chapter 2; however, implementation begins with an analysis of the brewing process, which must be completed prior to any manufacturing. To implement HACCP as both a food safety and a product quality program, employees analyze each step in the production process for potential hazards to the safety and quality of the finished product. Their careful evaluation determines if making a modification at a process point will alleviate the hazard. If the hazard is a risk to the safety of the consumer, the process point is considered a **critical control point (CCP)**. If the hazard is a risk to the quality of the product, it is a **quality control point (QCP)**.

For example, during wort chilling, a hazard to the quality of the product may be identified as microbial contamination resulting from improper cleaning of the chiller. This is a QCP. Stones found in the mash tun would not necessarily be considered a hazard to either the safety of the product or the quality of the product and therefore would not be considered a CCP. Stones may be considered a hazard to the equipment but are not a CCP or QCP. Broken glass that ends up beer during bottling, on the other hand, would be a hazard to the safety of the consumer. The bottle filler would then be a CCP. Similarly, yeast in the mash tun would not be a CCP because any modification at that point would not change the quality of the product. Table 1.2 shows some QCPs and CCPs for breweries to consider. It is important that each brewery do a thorough self-evaluation to make sure that any points that

> **critical control point (CCP)**
> A location in the production line where evaluation of a hazard that risks the safety of the consumer can be reduced or eliminated.
>
> **quality control point (QCP)**
> A location in the production line where evaluation of a hazard that risks the quality of the product can be reduced or eliminated.

TABLE 1.2. Examples of quality control points (QCPs) and critical control points (CCPs)[a]

QCP or CCP?	Task	Frequency	Reason
QCP	Original gravity (OG) measurement	Each brew	The OG measurement can show the level of mash efficiency. Low or high mash efficiency may change the flavor profile of the beer.
CCP	Machine oils	Daily to monthly	Damaged machines can break down or harm employees.
QCP	Water profile	Monthly	Water supplies can change often; speak to local water supply management for information.
QCP	Yeast pitch rate	Each brew	Low pitch rates can cause off-flavors in finished beers.
CCP	Check bottles for breakage	Before bottling	One broken bottle can cause glass shards to fall into other unbroken bottles.
QCP/CCP	Check water pH	Monthly	pH can affect mash enzymes and flavor. High pH values might indicate caustic cleaner in the lines.
QCP	Plate microbes from fermenter	Post cleaning	Microbes can produce off-flavors in finished beer.

[a] Courtesy S. E. Johnson and M. D. Mosher—© ASBC.

need monitoring are being monitored. However, going overboard can waste time and as a result, money.

Each hazard must then be analyzed to set reasonable upper and lower boundaries of acceptable values. This means critically analyzing the hazard and determining what is an acceptable value associated with that hazard. For each hazard the brewery identifies, there should be a range of values that are acceptable to maintain quality production; these values are determined by the brewery after examination of the process. For example, if the sweet wort specification is for the specific gravity to be 1,048 kg/m^3 (1.048 g/mL), then a value of 1,046 kg/m^3 at the low boundary and 1,050 kg/m^3 at the high boundary might be considered acceptable. Charting, as shown in Figure 1.3, illustrates that the batch produced on 3/17/18 had a low sweet wort value. Further evaluation of the process was then carried out, the problem was identified and corrected, and changes to the process were implemented to ensure that the issue didn't occur again. The chart reveals that the correction was successful; the sweet wort gravity was within quality specifications the following day.

The boundaries of the chart are important. For example, if diacetyl is being measured in green beer, a high boundary of 100 parts per million (ppm) could be set as the specification to make sure that the majority of the product passes this QCP. However, that value is much too high given the flavor threshold of 40 parts per billion (ppb) diacetyl in beer. Essentially, every beer that would be produced, whether it tasted like diacetyl or not, would be passed along because it would have a diacetyl concentration of less than 100 ppm. A similar problem would exist if the maximum level of diacetyl was set

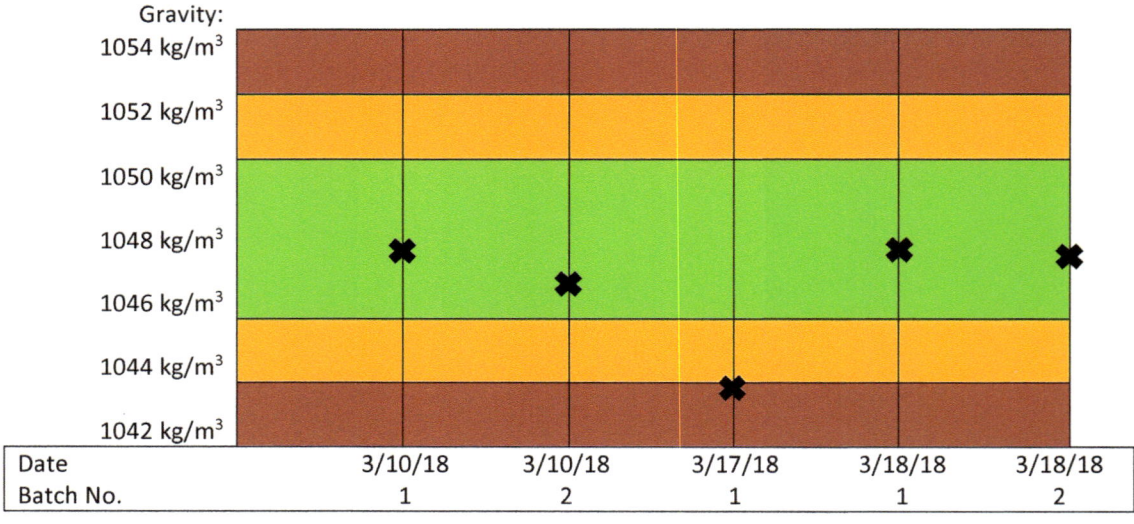

FIG. 1.3. Fictional data illustrating the use of charting in good brewing practices (GBPs). (Courtesy S. E. Johnson and M. D. Mosher—© ASBC)

to 50 parts per trillion (ppt). This level is much too low to set as a value. In fact, it would be pretty difficult to even measure the level of diacetyl at this low limit. If that were the high boundary for diacetyl, essentially none of the beer produced would pass that specification. In other words, significant care and effort must be applied to setting the limits for each hazard. The limits should provide a usable range of values that are acceptable in the product at that point in the production process.

Once the specification details have been mapped across the entire manufacturing process, personnel are assigned to every analysis and CCP or QCP. Those employees are responsible for measuring the hazards to the safety and quality of the product at those points. Employees must make accurate records and report the values to other identified people to assist in the feedback loop for the consistent production of quality product. The HACCP plan also denotes the actions that should be taken at each CCP, should a hazard at that point be identified. Similarly, a quality control program should outline the actions for each QCP. Those actions indicate how to correct the hazard so product quality is maintained and safety issues are avoided.

The basics of these methods require the brewery to identify, measure, record, and react to specific values associated with the beer being produced. The original gravity of the wort must be measured, recorded, and adjusted if it does not match what that beer should be. The carbonation level of the beer should be measured, recorded, and adjusted. And the list goes on. Each of these are examples of measures of the batch of beer being produced and its adherence to the spec sheet for that beer.

These measures can be obtained by the staff within the brewery, or they can be performed by trained members of a dedicated laboratory. By themselves, these measures provide quality control. The values are measured and recorded, and the brewery responds to them by adjusting the product so that the values fall within the range indicated on the spec sheet. When coupled with well-defined procedures for the production process, a quality control program can result in a consistent product. However, as noted earlier, taking a reactionary approach to production is not the ideal way to manufacture a consistent product. There is still a risk of inconsistency in production if only the quality control measures are implemented.

The primary reason for failure here is that constant adjustments may be needed from the start to the finish. Constantly adjusting the batch can change the overall product and result in something that doesn't represent the consistency that the brewer desires. Thus, maintaining production of a quality product in a consistent fashion requires implementation of a system-wide process of quality management. **Quality control** is reactionary if used alone. **Quality assurance** is the systematic analysis of production that utilizes, among other things, quality control measures to ensure a consistent product. In other words, quality control will identify when something went wrong, but that alone doesn't explain what needs to be done to fix the issue. The brewer needs to know what to do to stop the overall production of beer from failing to provide a consistent product.

While there are many different methods of quality assurance, the ones we will explore in this text are relatively easy to implement. However, each does require performing a complete analysis of the manufacturing process. Only if every step in the process has been evaluated and understood in the

quality control A program to ensure quality; reactionary if used alone.

quality assurance The systematic analysis of production that utilizes quality control measures, among other things, to ensure a consistently high-quality product.

context of overall production can issues, problems, and efficiencies be identified. Only then can the quality of beer be analyzed and compared with specifications. That analysis is quality control.

Measuring Quality

As we have noted, measuring values of quality is required during the manufacturing process. At specified points within the process, some specifications may require evaluation. For example, the gravity prior to fermentation may need to be measured. The brewery can perform the measurements by using on-site, in-line, or laboratory modes.

The **on-site analyses** used to measure a process are the most common measurements performed in small breweries. This mode involves taking samples of the batch where the process occurs and performing the analysis at that site. Our example of measuring the gravity of the bitter wort prior to fermentation can be easily accomplished using this mode of action. An employee—usually one responsible for that portion of the process—withdraws a sample from the product stream using sterile techniques. Then the analysis is performed. This mode is relatively quick but has some drawbacks that must be evaluated. The goal is to obtain information about the product stream. If the analysis isn't performed correctly, the on-site analysis can become its own QCP.

Disadvantages of using on-site evaluation include the requirement for sterile sample collection, the duplication of analysis equipment throughout the brewery to perform the analysis, the relatively limited precision in the measurements, and the requirements for adequate training regimens that allow confidence in the result of the analysis. Not all on-site evaluations require sterile sample collection routines. For example, determining the gravity of sweet wort before the boiling process need not follow such strict sampling techniques, although some measure of hygiene is still required. And if different employees are to perform these on-site measurements at different points along the production process, then duplicates of the equipment should be stationed at those points. In our example of measuring the gravity of the wort stream, multiple hydrometers, pycnometers, or automated gravity meters would be required to cover each location where the gravity is analyzed. Without this added expense, the brewery would not be able to perform the analysis. Finally, training employees to accomplish the analyses and refreshing their techniques with regular retraining requires an investment of time and resources that may be cost prohibitive. A small group of employees dedicated to performing analyses could be assigned to perform all the measurements throughout the brewery. This would limit the number of employees that must be trained. Those employees would then form a laboratory unit within the brewery.

Advantages to the use of on-site measurement techniques include the rapid reporting of results, typically within a minute or two. This allows making rapid adjustments to the product stream. For example, if the gravity of the bitter wort exceeds the upper control limit set by the process, water can be added to the bitter wort to dilute the gravity back to be within specifications. Another very strong advantage is the relative cost of performing the on-site analysis. In many cases, the analysis method, such as

---

on-site analysis Analysis of a sample performed at the site of the sample collection; for example, using a hydrometer next to the brewhouse.

determination of the gravity of a wort stream, is less expensive than other modes of analysis.

Laboratory-based analyses are another mode of analysis that can be implemented during the production process. This mode involves a member of the laboratory staff obtaining a sample of the product stream, temporarily packaging it so that it is stable until it can be measured, and then transporting it to the laboratory for analysis. This process has both advantages and disadvantages. In fact, many of the advantages are also the disadvantages of using this mode of quality control. Among the advantages are the accuracy and precision of the measurements, the ability to perform complicated analyses, and the lack of the need to duplicate instrumentation. Making laboratory measurements does require the purchase of expensive instrumentation that often exceeds the cost of a stainless steel vessel. Laboratories must also be staffed with personnel who are trained in the use and maintenance of these instruments. If staff are not trained, the confidence in the values obtained for a particular measurement can be questioned. The biggest disadvantage, however, is the length of time between sampling and reporting the value. For example, in a fermenter tank farm where production utilizes multiple fermenters at the same time, sampling these fermenters to analyze for diacetyl is common. Then the multiple samples are removed to the laboratory for analysis. If the lab requires 30 minutes to process each sample, then the earliest a measurement can be reported will be within about 1 hour (assuming the sterile sampling, packaging, and transportation take no more than a half-hour). How long then will the cellar staff need to wait to find out the diacetyl level in the twentieth tank that was sampled?

With the major disadvantages of time and cost come the tremendous advantages of precision and type of analysis. For example, measuring diacetyl on-site is difficult, but in the lab, the analysis is easily performed. Thus, the laboratory plays a very important role in the brewery but comes with a cost that must be accounted for. If the result of implementing the laboratory is the successful application of a quality assurance program, then the cost of creating the laboratory is justified. In short, having a laboratory reduces the expenses associated with continually addressing quality issues.

In-line analyses can be a useful mode of measurement. These analyses involve placing a measuring device directly into the product stream. For example, a densitometer, which measures the gravity of a solution, can be placed in the piping that runs from the wort chiller to the fermenter. Then as the wort passes the meter, it will report the gravity almost instantaneously. Using this measuring device greatly reduces the time needed to perform the analysis, eliminates the need to withdraw a sample and potentially contaminate the product stream, and allows the employee to react instantaneously to some measure of the quality.

The disadvantages of in-line measurements include the cost, availability, and precision of the analysis. While temperature probes are relatively inexpensive in-line measuring devices, densitometers, dissolved oxygen sensors, and even flow meters can be quite expensive. Moreover, in-line sensors aren't available for every desired analysis. For example, diacetyl measurements still require withdrawal of a sample from the product stream. While in-line analyses can be as precise or more precise than on-site measurements, they are not typically as precise as those performed in the laboratory.

laboratory-based analysis Analysis performed in a laboratory on samples from the production floor.

in-line analysis Analysis conducted by a measuring device placed directly in the product stream.

Investing in Quality

Performing product analyses involves ongoing costs. Conducting even the simplest experiment costs money. For example, simply measuring the temperature during a mash requires buying the thermometer or thermocouple and paying someone to perform the procedure. Brewers understand (or should understand) that the money invested in developing means to measure specifications can be returned to the company in other ways. The advantages to implementing new analyses include the reduction in the number of product recalls, the improvement in the number of batches of product that meet specifications, and the increase in revenue because of an increased customer base. In other words, investing in quality results in improving consistency.

There are disadvantages to making this investment, although they are relatively small. The primary disadvantage is the redirection of funds to on-site, in-line, or laboratory expenses. This involves the training of personnel, the maintenance of sampling and analysis equipment, and the cost of actually performing the analyses. A separate reason that is incorrectly thought of as a disadvantage is allowing personnel other than the brewer or owner of the brewery to make the "go/no go" decision after receiving an unacceptable result. For example, if a particular batch is analyzed and found to be so far out of specification that it can't be corrected, the go/no go decision will rest with the person or department performing the analysis. If an unacceptable level of lactic acid is found in the finished beer and can't be adjusted to be within specifications, the laboratory performing the analysis should have the authority to dictate that the product be destroyed rather than be served. Such decision making should be part of the HACCP program and quality control program and, while expensive, results in the brewery releasing only product that is within specifications.

Every brewery should perform a cost–benefit analysis (Fig. 1.4). Prevention costs increase as more funds are directed to quality assurance. Analysis costs include the funds invested in performing the analyses but also the expense of verifying that the analyses are correct. Failure costs decline as

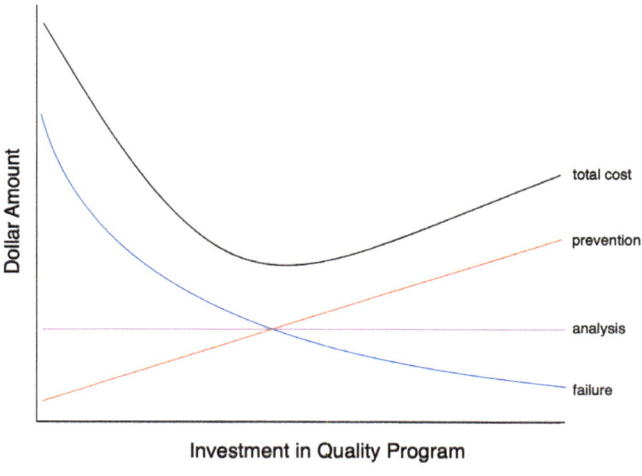

FIG. 1.4. Costs versus return on investment. (Courtesy S. E. Johnson and M. D. Mosher—© ASBC)

funds are funneled into a quality assurance program. These costs are associated with product recalls, spoiled product, and product dumps due to poor product quality.

Most analyses such as these indicate that as funds and resources are directed into a quality assurance program, the consistency of production improves. And as consistency improves, the overall cost to the company is reduced. However, there is a limit on the return on investment in a program. At some point, the addition of new instrumentation, the cost of performing the analyses, or the addition of personnel will have only a limited impact on improving the product quality compared with the cost of new personnel, new methods of training, or the analysis expense itself. Since these costs are based on a given brewery, the point of maximum return on investment is specific to each brewery.

Surprisingly, many breweries do very little to measure product quality and adhere to specifications. Most evaluate the original gravity and finishing gravity of the product because doing so is required by law for tax purposes. Most also measure temperatures during the mashing process, and some even perform yeast viability counts prior to pitching into a fermenter. Fewer breweries perform analyses that accurately report the percentage alcohol by volume, diacetyl concentrations, microbe contamination, or haze measurements. A small investment in instrumentation, equipment, and chemical supplies is all that's needed to perform these evaluations. Such an investment will likely have a dramatic, immediate, and welcomed impact on product consistency.

CHAPTER 2

Implementing Safety and Quality

A smoothly operating brewery should be able to focus its operation on producing a high-quality product that's safe to consume. Each step in the production process should be mapped out with directives on how to handle the product stream. This chapter will focus on rigorously developing a **hazard analysis and critical control points (HACCP)** plan for the brewery. In addition, the chapter will discuss the need for a quality assurance program utilizing **good brewing practices (GBPs)**. We'll also explain the major off-flavors in beer and their use in troubleshooting issues in the brewery. Later chapters will cover the specific analyses in much greater detail.

hazard analysis and critical control points (HACCP) A quality assurance program that ensures consistency in the product.

good brewing practices (GBPs) The standards in a brewery that govern quality and consistency in production.

What Is HACCP?

As noted in the introduction, HACCP is a quality assurance program that ensures consistency in the product. In addition, when carefully implemented, HACCP reduces the number of required analyses only to those that impact the quality of the product. A team of food scientists and engineers from the National Aeronautics and Space Administration, the Natick Research Laboratories, and The Pillsbury Company developed the basic principles of HACCP in the 1950s. The goal of their research was to implement a system of food safety for the production of food for the space program. Three underlying principles were established for HACCP:

1. Assess food production at each step in the process.
2. Identify **hazards** that compromise product quality at each step in the process.
3. Establish monitoring, reporting, and actions associated with each critical control point (CCP).

hazard Anything that harms the safety of the product for consumption.

In the 1960s and 1970s, there were a few reported cases of foodborne illnesses resulting from improperly packaged foods. The U.S. Food and Drug

Administration (FDA) used HACCP principles to develop regulations to protect the public. Through the 1970s and 1980s, more and more food preparation companies began implementing HACCP. Fast-food restaurants recognized food safety as an important business concern. If customers became ill after eating food that was not safe, the result could be significantly detrimental to the business.

So, while regulatory changes are often driven by state and federal government agencies, the food industry took a proactive approach and began implementing strict safety procedures ahead of anticipated regulations. Many businesses also required their suppliers to follow HACCP or other documentable food safety procedures. In addition, the equipment used in the food industry became a focus, as manufacturers examined designs and materials to consider new ways to improve cleaning procedures. Sanitary welding became very important, which resulted in manufacturing equipment to avoid the presence of hidden spaces, hard corners, and small crevices that could hide microbial biofilms.

By the 1990s, interest in and implementation of HACCP (and related modifications of the program) had increased globally across the food industry. In fact, as of January 2006, all food industry operators in the European Union were required to follow Regulation (EC) Number 852/2004 on the hygiene of foodstuffs. This requirement dictated that all food businesses establish, follow, and carefully document their adherence to an HACCP plan. The food preparation industry's proactive approach to food safety was also recognized at the international level. In the early 2000s, the International Standards Organization created standard ISO 22000 on food safety management, which implemented principles of HACCP. Although HACCP continues to evolve (for example, separating verification from validation methods), the basic principles are appropriate for the manufacture of food and food products.

In spite of these regulations, foodborne illnesses and hazards continued. In the United States, readdressing the need for food safety at a national level became important. The U.S. Congress passed the **Food Safety Modernization Act (FSMA)** in 2011, the first large overhaul of food safety laws in about 50 years. In addition to providing regulations for the entire food industry, this legislation includes specific regulations for fresh and processed produce and dairy. Most of the provisions of FSMA, however, do not apply to small businesses. Juices, seafoods, and alcoholic beverages are exempt from some but not all regulations. Since alcoholic beverages are considered foods by the FDA, all manufacturers must comply with at least the section of the FSMA requiring **good manufacturing practices (GMPs)**. For breweries, this regulation is better termed good brewing practices (GBPs).

In short, all breweries must prove that they have identified potential hazards to food safety, have implemented preventive controls, and are evaluating those controls. Small breweries can be exempt from many of the regulations, but they are still required to protect their customers from health hazards. Preventive controls must be in place to ensure that the brewery's product is free of hazards caused by the manufacturing process, correctly labeled for allergens, and properly sanitized. The brewery must also ensure that hazards are controlled throughout its supply chain. Creating and implementing an HACCP plan will fulfill all of these requirements.

Food Safety Modernization Act (FSMA) A law enacted in 2011 that requires the Food and Drug Administration (FDA) to regulate food production and processing.

good manufacturing practices (GMPs) Production standards that govern quality and consistency.

Beer is considered a safe product. The presence of alcohol and hop compounds (particularly the iso-alpha acids), the lower pH, the **reducing environment** (rich in carbon dioxide and devoid of oxygen), and the lack of nutrients make beer a relatively inhospitable environment for most microbes. Human pathogens cannot grow or survive in beer. In fact, the process of manufacturing beer is harmful to most microbes. Unfortunately, with new innovations in craft beer, the boundaries of "safe" are being stretched. Brewers now consider using nuts and milk (both considered allergens), spices (which can be contaminated), and other products such as meat and shellfish (which may carry allergens or be contaminated). Brewers must ensure that they follow GMPs and make their products safe for consumption.

> **reducing environment** An environment that lacks oxygen and as a result doesn't allow oxidation reactions to occur.

Implementing a Successful HACCP Program

Within a brewery, a successful HACCP program requires having a number of important features in place. First, all employees and owners need to understand and embrace the process. They need to understand both how this program will benefit the company and how an HACCP program will protect their product. Running a successful HACCP program takes time and costs money. Without everyone's cooperation, a plan might be put in place but not followed. In that case, not only would the brewery potentially be in violation of food safety regulations, but the risk to product quality and safety would be high. And the risk of losing customers due to lack of consistency would be even greater.

Implementing an HACCP program requires identifying potential hazards to the safety of the product, including steps in which raw materials are added, product is removed for sampling, and no apparent physical modification is made. For example, a hazard could be identified as shellfish in the flavoring agents added to the conditioning tank. Another hazard could be broken glass in the bottled beer. Still another could be cleaning agent residue in the tubing from the boil kettle to the chiller. In short, anything that can affect the safety of the product when it's consumed is considered a hazard, including the main product of the brewery (beer) and other products produced for consumption by humans and/or animals. Hazards can occur in yeast, spent grains, and any other edible or potable product.

HACCP hazards are classified into three categories (Fig. 2.1):

1. Chemical hazards occur when a chemical compound is present at levels that are considered unacceptable. For example, high levels of copper or the presence of caustic cleaners in the product are chemical hazards.
2. Biological hazards involve the presence of microbes that can infect the product stream. While most of the biological hazards in beer focus on the quality of the product, some pose safety concerns for consumers.
3. Physical hazards occur when foreign materials are present in the product stream. Shards of glass, metal shavings, and pieces of plastic in the finished product are all physical hazards.

standard operating procedure (SOP) A detailed explanation of how a task is performed.

Hazards to food safety, as identified by HACCP, differ from the hazards of working in a brewery. Those safety concerns and safety hazards should be carefully addressed via a chemical hygiene plan or worker safety plan and explicitly included in all **standard operating procedures (SOPs)**. Regular training sessions should be held, and safety concerns and safety hazards should be mitigated.

Like a safety program for employees, an HACCP program is built by conducting a series of steps:

1. Conduct a hazard analysis of each step in the production process.
2. Identify CCPs.
3. Establish upper and lower critical limits.
4. Establish monitoring procedures.
5. Establish corrective actions.
6. Establish verification procedures.
7. Implement recordkeeping procedures.

Before implementing these steps, the brewery should assemble an HACCP team that guides the process, creates the program, and verifies that the program is being followed. In addition, that team should evaluate the success of the program and continue to build and strengthen the HACCP program through a reiterative process. The team can also provide feedback to anyone in the brewery on its findings: How many hazards have been identified? How many failures have been found, and how many corrective actions have been successful? By evaluating and constantly revising the overall program, the entire brewery will benefit from smooth operation.

Step 1: Conduct a hazard analysis of each step in the production process. There are many ways to perform a hazard analysis. One of the easiest and likely most thorough involves the brewery staff. Each department in the

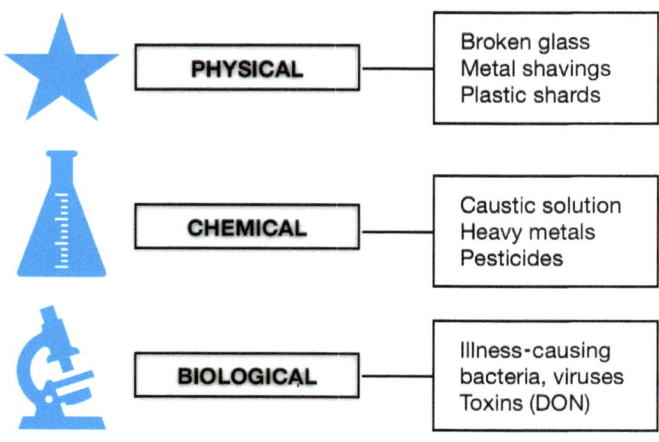

FIG. 2.1. Categories of HACCP hazards. It's important to note that these are hazards to food safety, not to food quality. (Courtesy S. E. Johnson and M. D. Mosher—© ASBC)

brewery is staffed with employees who know the process in detail. Using their understanding of the process can speed the hazard analysis. The best way to do this is to meet and identify each SOP used in that specific department. For example, the cellar staff can meet with the HACCP team to review the SOPs that define cellaring in the brewery. Those SOPs would include processes to transfer wort from the chiller to the fermenting vessel, operational steps associated with the fermenters, process steps to monitor the fermentation, transfer steps that move the green beer to the filtering room, and so on. Each staff member who works daily with the SOPs should be present at the meeting to discuss exactly what he or she does. Not only will such an exercise inform everyone about the SOPs, but it might also identify efficiencies and areas for improvement in the overall process.

Each hazard that's identified is then assigned a risk rating that evaluates the potential for the hazard to occur and the potential for the hazard to cause harm to consumers. Some hazards might have a great potential to occur but little potential to cause harm; others might be just the opposite. Assigning a risk rating allows the HACCP team to evaluate each hazard to determine whether it should be considered a CCP. Using the rating system shown in Table 2.1, the value for the likelihood and the value for the harm are multiplied to obtain the risk rating. Typically, any rating greater than 3 would be elevated to consideration as a CCP.

For example, suppose that the presence of coliform bacteria, such as *Escherichia coli*, in the mash water supply was identified as a hazard. The likelihood of infection was determined to be moderate (2), because it would impact the entire batch if infected water were used. The harm rating was determined to be low (1), because the product of the mash is boiled to sterilize and remove the harm to the product. The risk rating is thus 2 ($2 \times 1 = 2$). In another case, glass shards in the bottled beer was identified as a hazard. The likelihood of this hazard was considered to be severe (3), because lack of control could affect multiple batches of product. The harm was considered to be severe (3), because a consumer could swallow a piece of glass and require hospitalization. The risk rating was thus determined to be 9 ($3 \times 3 = 9$). The HACCP team would consider that the coliform infection was much less a

TABLE 2.1. Likelihood and harm ratings used for risk assessment[a]

Likelihood rating	Description of hazard
1	Low: occurs rarely and/or would affect only a portion of a batch
2	Moderate: occurs occasionally and/or would affect an entire batch
3	Severe: occurs constantly and/or would affect multiple batches
Harm rating	**Description of hazard**
1	Low: consumption is unpleasant but does not cause adverse health issues
2	Moderate: causes mild health issues or adverse health issues after long exposure
3	Severe: causes severe health issues (e.g., hospitalization, death, etc.)

[a] Courtesy S. E. Johnson and M. D. Mosher—© ASBC.

risk to product safety than the presence of broken glass in the bottles. They would definitely consider the glass as a potential CCP and might advance the coliform infection to the same level.

Step 2: Identify CCPs. Working with a group of employees designated as the HACCP team, each brewery worker evaluates the SOPs for which he or she is responsible and identifies steps or processes in which a hazard to the product may exist. The hazards are then evaluated to see if they are CCPs in the production process. The following questions can also be used to determine the answer:

1. Does control of the product exist at the specific step? If control doesn't exist at this step but should exist to ensure product safety, then the process step should be modified so that control is possible. If control already exists, then . . .
2. Does the specific step eliminate or reduce the hazard to an acceptable level? If it does, then the step is a CCP. If it doesn't, then . . .
3. Could contamination of the hazard occur at an unacceptable level? If there's no possibility that the hazard could rise to an unacceptable level, then the step isn't a CCP. Instead, it is only a **control point**. If there is a possibility, then . . .
4. Will a subsequent process step eliminate or reduce the hazard to an acceptable level? If a process step further along in the product stream can eliminate the hazard, then the specific process step isn't a CCP; it's a control point. But if this is the last step in the process, then the step is a CCP.

CCPs are those points at which biological, chemical, and/or physical hazards could pose problems to consumers. These hazards must be monitored so that if they rise above acceptable levels, they can be mitigated. Control points, on the other hand, are points at which monitoring can take place, but there is no risk of a hazard moving beyond acceptable limits. In a brewery, there are many more control points than CCPs.

Step 3: Establish upper and lower critical limits. For each CCP for which control is possible, the acceptable limit of the hazard should be identified. For product safety exclusively, the upper critical limit is that which defines the difference between a safe and an unsafe product. However, if the plan is to monitor product quality, limits can be selected for that purpose. A lower limit should also be selected.

The upper and lower limits should be chosen to represent the limits of detection and linearity within a laboratory (or other) analysis. In other words, setting the lower critical limit to a level that's lower than the smallest value that could be identified in an analysis will result in all measurements being above that limit. Similarly, the upper critical limit should be set at such a point that it can be accurately measured. If no information is available on what the measurement should be, approximately 10 measurements should be obtained and the average and standard deviation of those measurements should be determined. Any values that are ≤2 standard deviations from the

---

control point A specific location along the process stream at which the product specifications can be measured and modified.

average are acceptable. Values that are ≥2 standard deviations from the average are of concern, and values ≥3 standard deviations from the average are major deviations from the norm.

For example, assume that we have identified a quality control point (QCP) for which we are concerned with the level of acetaldehyde—an **off-flavor** in many beer styles. In lagers, the typical **flavor threshold** is 2–3 parts per million (ppm). (This is the lowest level that is typically perceived by humans.) If the acetaldehyde level is to be evaluated, it might make sense to have a lower critical limit of 0.5 ppm and an upper critical limit of 2.0 ppm. Then, an analytical method must be identified to accurately measure the value of acetaldehyde in beer at these levels. It would not make sense to have a lower critical level of 1.0 parts per billion (ppb) and an upper critical level of 10.0 ppb. In addition to choosing levels that are very different from those needed for product quality, it's necessary to identify an analytical method capable of measuring such small quantities. Doing so can be very difficult, because every batch would likely fail the analysis.

For some hazards, especially among the CCPs that are concerned with product safety, the upper critical limit may be set to 0. In the case of broken glass in a bottled, consumable product, the presence of even one piece of glass should be considered unacceptable to the product's safety. Some breweries would consider a cryptosporidium infection in any process waters with the same scrutiny. In these cases, there isn't a lower critical limit, because finding any of the hazard in the product would be unacceptable for product safety.

Step 4: Establish monitoring procedures. Once the CCPs and limits have been identified, the HACCP team establishes protocols for analyzing the hazards at those points, including revising the SOPs and training the relevant personnel to perform the analyses as accurately and precisely as possible. The protocols may establish a process for laboratory personnel to follow. The monitoring procedures should include specific requirements for when product is analyzed, how often it is analyzed, what steps are taken to conduct the analysis, and other specifics. The personnel responsible for evaluating these hazards would be required to obtain samples and to analyze them for the specific hazards.

The best monitoring procedure is one that does not require human intervention. In-line measurements with automatic logging provide the most accurate and precise way to record the information. Using this method is not possible with all CCPs and QCPs, but it should be considered when monitoring can be conducted using instruments. Brewery personnel can make mistakes in performing SOPs. Even with training, an otherwise model employee might accidentally forget to perform an analysis at the prescribed time. Assigning such an analysis to an in-line instrument eliminates the chance for human oversight. In addition to automatically logging results, an in-line instrument can be set to sound an alarm when the hazard it measures rises above the upper critical limit.

Step 5: Establish corrective actions. The HACCP team also ensures that a set of corrective actions is put in place to reduce a hazard to below the upper critical limit. These actions should be written out as part of the HACCP plan and included in the SOPs for specific types of analysis. The corrective actions

off-flavor A flavor in the beer that was not intended or does not fit the style of the beer.

flavor threshold The minimum concentration at which a given flavor can be perceived.

should also dictate what to do with the product between the start of the failure or alarm and the time the corrective action is applied. In every case, the specifics of the process followed should be written down and included in the batch log.

Step 6: Establish verification procedures. Once the HACCP plan has been constructed (steps 1–5), the brewery initiates the program and begins operating under the new product safety procedures. Each monitoring and corrective action should be evaluated regularly to ensure that the overall process is working. Evidence that following the plan is reducing or eliminating hazards should be documented—for example, performing additional product tests to confirm that the hazards don't exist, evaluating consumer comments about the product, and performing in-house audits to verify that all CCPs are being monitored as dictated by the plan.

Step 7: Implement recordkeeping procedures. A HACCP plan may be reviewed periodically. Conducting that review will be very difficult unless records are kept of the actions that have been implemented. In situations in which a state or federal agency is requesting a review, having adequate records of the HACCP plan and associated data is essential.

It's therefore imperative that the HACCP team follow up to evaluate the plan's effectiveness in removing hazards and to ensure that all records of the plan are up to date. This may involve periodically reviewing the entire plan by an in-house audit. External auditing can also be performed to ensure that the HACCP plan is being implemented correctly and producing the desired results.

If records aren't being kept for the HACCP plan or the plan isn't up to date, then it won't be a reliable reference for future issues, problems, and concerns. Only with an up-to-date and well-documented HACCP plan can a brewery confirm that product safety is being addressed. Over time, keeping accurate records also can help to improve the quality of the product.

American Society of Brewing Chemists

American Society of Brewing Chemists (ASBC)
An international organization of scientists and technical professionals that focuses on applying scientific principles to the brewing industry.

The **American Society of Brewing Chemists (ASBC)** is a professional organization of scientists and technical professionals in the brewing, malting, and allied industries. ASBC was founded in 1934 "to improve and bring uniformity to the brewing industry on a technical level," as stated on the organization's website. Today, the international organization is composed of individual and corporate members that represent breweries of all sizes, as well as members from academic institutions, government agencies, and regulatory organizations. ASBC provides its members with these services:

- analytical, scientific process control methods to ensure high quality and safety standards
- science-based approaches and solutions to industry-wide issues
- scientific support to evaluate raw materials for optimum performance
- professional development opportunities

ASBC promotes its vision of being the global community for scientific excellence in brewing by disseminating information to its members in several ways:

- the *Journal of the American Society of Brewing Chemists* (*JASBC*), which reports recent scientific experimental results on issues in brewing and raw materials
- educational outreach through meetings and webinars
- a lab proficiency program, which breweries can use to validate their ability to perform key analyses

In addition, ASBC provides an ever-expanding list of cause-and-effect diagrams (also known as fishbone or Ishikawa diagrams) (Fig. 2.2). These diagrams are based on 5M + E (measurements, materials, methods, manpower,

PROCESS CONTROL FOR YEAST VIABILITY & VITALITY: TYPES OF STRESSES

Physical Stresses
- Temperature
- Hydrostatic pressure
- Osmotic pressure
- Gases
- Shear

CO_2 stimulates growth at low pressures (@0.2 bar), at pressures over 1 bar it reduces yeast growth and fermentation rate.

Temperature (-)...lager yeast cannot grow above 37°C, ale can, *KEX2* gene involved? PAF protein also implicated...a signaling-phospholipid with pleiotrophic effects...expression very temp dependent & significant differences between ale and lager yeast.

[Dissolved carbon dioxide] exceeds saturation limits by 50% during the period of most active fermentation (@ 50-100 hours)...most inhibiting then.

Chemical

Types of Stress: a) temperature, b) pH, c) ethanol, d) starvation, e) hydrostatic pressure, f) osmotic pressure, g) carbon dioxide, h) acid washing and i) global consolidation of the brewing industry for brewers.....its a tough life out there ...not always "happy yeast"!

- Free radicals (if California)
- pH
- [Metal ions]

Effect on Yeast Viability & Vitality

Finland: Immobilized yeast (+)...free cells higher in trehalose than immobilized cells in HGB up to 30°P....immobilized cells have lower mannan/glucans vs. free cells and are more active relative to free in terms of ethanol production and rate of attenuation. Free cells also have higher UFA and ergosterol vs. bound cells.

German lager brewers often can only repitch yeast 3-4 times vs > 10 common in USA. Why? German's have 100% malt, cold, relatively low gravity vs high adjunct, warmer, higher alcohol US processes. Speculate the difference lies in pitching rate practices, where US see 2-4 doublings during fermentation whereas a 12°P pitched at 40-50 million may see < 1? After 2-3 repitchings, need to replace yeast. Super if you are a supplier of yeast propagation systems!

- Tank H:W ratio (-)
- CO_2 saturation (-)
- Temperature (-)
- Stirred fermentations (-)

Synergism of ethanol, temperature and osmotic pressure (-)

Virgin cells slower fermenters.....repitching is okay!!!

Biological Stresses: brewers, brewing scientists, suicide mutations, lethal DNA damage.

Fermentation Stresses: pH, [ethanol], low nutrients, hydrostatic, osmotic, oxidative.

Environmental Stresses **Biological**

ASBC No. IIa (Casey; June/05)

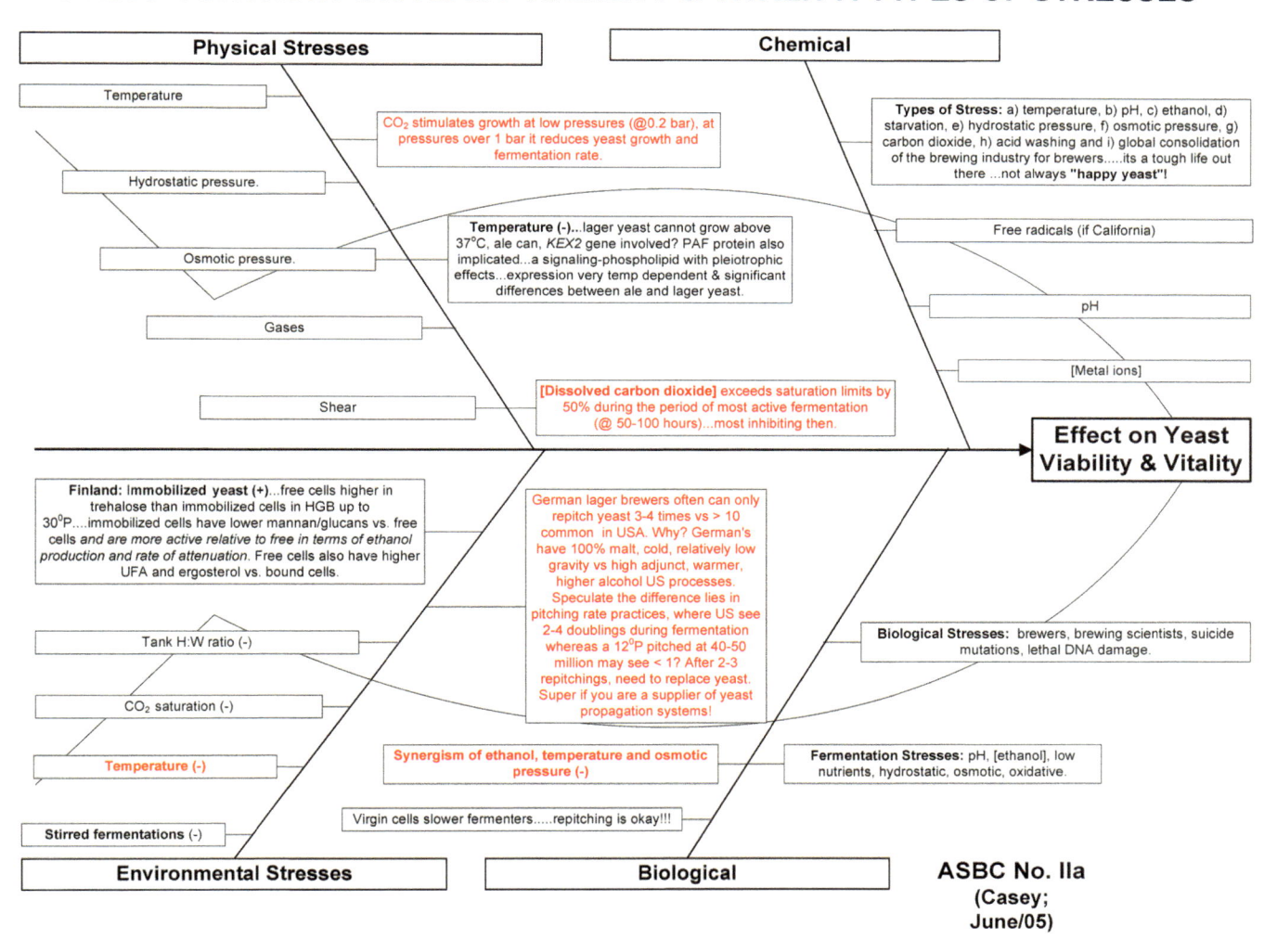

FIG. 2.2. ASBC fishbone diagram on stresses to yeast health (ASBC no. IIa). The diagram illustrates positive and negative impacts of stressors on yeast health. Each fishbone diagram also provides useful information on the topic of the fishbone. (Courtesy ASBC—Reproduced by permission)

ASBC fishbone diagram A cause-and-effect diagram that aids in evaluating issues and determining solutions in the brewing process.

ASBC Methods of Analysis An online collection available to ASBC members that provides numerous analyses that can be performed in the brewing laboratory.

machines, and environment) process improvement diagrams, which show the effects of different specifics in each area mentioned. The **ASBC fishbone diagrams** summarize technical information for specific issues. For example, a cause-and-effect diagram has been created that provides potential causes for gushing in beer. These summaries provide many possibilities of potential causes and are useful when analyzing a specific situation.

ASBC Methods of Analysis

Of particular importance among the publications that ASBC provides to its members is a set of standardized methods of analysis for the major quality points within a brewery. (The methods are also available via subscription to corporations and academic organizations.) The *ASBC Methods of Analysis* provide details on necessary steps to follow to obtain the most accurate and precise results. Some methods require specific scientific instrumentation, while others use relatively common laboratory equipment. All the methods, however, have been tested, verified, and evaluated for their ability to provide consistently accurate results. Following the methods verbatim will increase the likelihood that results are precise and accurate.

Many of the ASBC methods require having some background or experience with laboratory equipment and procedures. While ASBC can provide some training in how to use its methods of analysis, brewers who have limited laboratory experience might find the procedures daunting. In Chapter 15, we will explore some of the analytical methods for determining the concentration of compounds found in beer (assuming that the reader has little experience in the laboratory). This explanation assumes that the reader has some skills using a calculator and spreadsheet and is comfortable working with glassware and other fragile items.

The *ASBC Methods of Analysis* are organized by the substrate being analyzed. For example, there are groups of methods for beer, malt, wort, and so on. Exploring the methods is often the best way to determine if a specific analysis exists for your particular analyte. Again, this book will group analyses by the types of instruments with which they are used. For example, all the methods that use an ultraviolet/visible spectrophotometer are identified, as are all the methods that use a gas chromatograph. Because smaller breweries often have limited equipment, these groupings should allow quick identification of the analyses that can be performed in the small brewery lab. A concordance index and analyte index will help you find the methods that you want.

In Chapter 15, the methods have been rewritten and laid out in a repetitive format for ease of use. At the start of each method, the required materials, supplies, and chemicals are listed. For a reagent that must be prepared in the laboratory, specific directions are provided along with storage information and storage life. (Chapter 6 guides you through the process to measure out and prepare reagents.) Each method then lists the specific steps to follow to perform the analysis. When an instrument is to be used, information is provided on how to use the instrument. Since instruments are produced by many different manufacturers, generic information is provided in the methods. However, Chapters 7 and 10–14 provide significant background information to allow the reader to understand how individual instruments work, why they do what they do, and what it means when they provide results.

The methods also provide data sheets, which the brewer can use to fill in specific numbers as the analysis is performed. These sheets can be photocopied from the book or downloaded from asbcnet.org/BSL and printed out, allowing a permanent record of the analysis to be maintained. In addition, the data from the sheets can be used to provide verification that the analysis was correctly performed; any errors made during the analysis can be found by evaluating the data sheets.

What's Wrong with My Beer?

Suppose that a microbrewer has produced a batch of beer and is unhappy with the results. He or she followed the recipe to the letter, and no safety hazards were detected at any CCP. All the equipment was cleaned in the usual way before use. The target original and final gravities were spot on. Fermentation appeared to proceed normally, but the finished beer is sitting in the conditioning tank and taste tests just don't meet expectations. So, what can the brewer do? Should the batch be repurposed, reflavored to hide the off-taste, blended with other batches, or dumped? The brewer has to make that call.

Dumping a batch of beer is expensive for any size of brewery. But given the tight operating margins of a microbrewery, where a single batch of beer means the difference between making rent or not, dumping a batch of beer can be very upsetting. To avoid dumping the product, the brewer strongly considers ways to mask the off-flavor. Many things can be done to mask flavor, including adding strongly flavored extracts, adding coffee, and dry hopping. If masking is done correctly, a subtle off-flavor can be overpowered so that it's much less perceptible. That means the beer can still be sold, but it isn't good practice and it's definitely not advised if the brewery is attempting to make a consistent product. However, masking an off-flavor is sometimes necessary to keep the brewery financially sound.

Before making that decision, the brewer should take the time to find out what happened to that batch. What caused the off-flavor? Did it result from a particular chemical, a bad process, or an infection with an unwanted microbe? To make that call, the brewer needs to know the origin of the off-flavor. For example, if the off-flavor was caused by a microbial contamination, blending the product with another that lacks the off-flavor will only "feed" the microbes with additional nutrients. The resulting blended product will still have the off-flavor and a much larger batch of product will have quality issues. The situation will likely be worse than if the blending had not been performed.

Of course, identifying the off-flavors in a beer can aid in hunting down the reason the batch failed. In a one- or two-person operation, the brewer may be able to correctly identify each flavor and aroma and know exactly what's wrong. However, running a sensory panel should be considered. (How to create and operate a sensory panel is explored in Chapter 9.) In addition, a quality assurance program should be in place to significantly reduce the chance of a future issue, even if the brewery is staffed with only one or two employees.

Common Off-Flavors

There are a wide variety of flavors in beer. When they occur in a particular brand in a particular concentration on purpose, they are part of the overall flavor of that brand. But when the concentration of the specific flavor is too low or too high or a new flavor appears in the brand, then we say that the beer has an off-flavor. It's important to note that an off-flavor can be either too much of a compound or too little of a compound (which would cause other flavors to predominate). ASBC provides members with many resources to assist with flavor analysis. Specifically, the *Beer Flavor Database* and the *Hop Flavor Database* can be quite helpful.

The most common off-flavors found in beer can be categorized into six main groups, as shown in Table 2.2. These groups are based on the general characteristics of the flavor. Some are fairly pleasant aromas and flavors; others are much less so. Table 2.2 also indicates the descriptive words typically used to classify the flavors. While the description of each flavor may seem straightforward, individuals may perceive the flavor slightly differently. In addition, depending on the level of the compound in the beer and its flavor threshold, different concentrations of the same flavor may be perceived as different flavors. For example, at low levels, the metallic off-flavor is perceived as an iron nail flavor, but at higher concentrations, it may be perceived as a bitter bloodlike flavor. In another example, at very low levels, the oxidation off-flavor will have a musty flavor, but at high levels, it tastes like photocopier paper.

The flavor threshold for each of the common off-flavors is also reported in Table 2.2. These values are reported for the flavor compound(s) in pure water. The lowest reported value of the threshold in lager is also provided. It's important to note that threshold values will change depending on the beer style. We can see this by comparing the threshold value of a compound in water versus lager. For example, the banana flavor is detectable at 3 ppb in water but requires 500 ppb in a lager to be detected. These values range from levels of parts per trillion to parts per million, depending on the compound that causes the flavor. Table 2.2 also lists the chemical compound that may be responsible for a given off-flavor. Note that while a single compound is indicated in the table for each off-flavor, multiple compounds may contribute to the flavors that are noted.

The units for flavor thresholds include parts per million (ppm), parts per billion (ppb), and parts per trillion (ppt) (Fig. 2.3). These units are used in brewing to represent very low concentrations. In fact, they represent fractions of a gram (about the mass of a single raisin) dissolved in a liter (about the same as a quart) of any liquid. Specifically, 1 ppm is equal to 1 milligram (10^{-3} grams) per 1 liter of water. To put that quantity in perspective, 1 ppm is approximately equal to 3 drops of food coloring in a barrel of beer. That's not likely enough to be perceptible to the eye. One ppb is equal to 1 microgram (10^{-6} grams) of compound per 1 liter of water. That's about the same as those same 3 drops dissolved in a tanker truck full of water. One ppt is equal to 1 nanogram (10^{-9} grams) of compound per 1 liter of water. That's about the same as 3 drops of food coloring in a small lake or reservoir. At these concentrations, it would definitely be difficult to see the color from the food coloring.

TABLE 2.2. Common off-flavors in beer[a]

Group	Flavor	Descriptive terms	Flavor threshold (water/lager)[b,c]	Potential causes/issues	Example compound(s)
Fruits	Banana	Banana, pear, nail polish	3 ppb/500 ppb	Yeast stress	3-Methylbutyl acetate
	Green apple	Apple, latex paint, squash	10 ppb/1.1 ppm	Young beer, yeast health	Ethanal (acetaldehyde)
Vegetables	Creamed corn	Corn, cabbage, vegetable	0.33 ppb/1 ppb	Improper boil, pilsner malt	Dimethyl sulfide
	Garlic	Garlic, cooked onion	—/1.2 ppb	Slow boil, pilsner malts, oxygen exposure	Dimethyl trisulfide
	Grainy	Green, wheat, raw grain	0.7 ppb/50 ppb	Improper mash, oversparging	2-Methylpropanal
Old	Papery	Oxidized, paper, sherry	80 ppt/100 ppt	Oxygen exposure, aging	Trans-2-nonenal
	Musty	Moldy, earthy, mushroom	—/0.1 ppm	Mold contamination	2,4,6-trichloroanisole
	Cheesy/Sweaty	Old cheese, goaty, sweaty	—/0.3 ppm	Old hops	3-Methylbutanoic acid
Spoiled	Mercaptan	Rotten vegetables, garbage, sulfury	—/0.2 ppb	Bacterial infection, dead yeast	Ethanethiol
	Lightstruck	Skunky, sulfury	0.2 ppt/1.3 ppt	Light exposure	3-Methyl-2-butene-1-thiol
	Rotten eggs	Rotten eggs, sulfury, sewer, hot springs	—/2 ppb	Infection, stressed yeast	Hydrogen sulfide
	Rancid	Baby vomit, putrid	—/2 ppm	Infection, aerobic mash	Butanoic acid
	Catty	Cat urine, tomato plants	—/15 ppt	Contaminated materials, hop oils	8-Mercapto-p-menthan-3-one
Acidic	Sour/Tart	Sour milk, acidic, citrusy	31 ppm/71 ppm	Natural sources, infection	Acetic acid
	Metallic	Iron, copper, rusty, bitter	—/0.1 ppm	Poor water	Iron ions
	Phenolic	Mousy, barnyard, Band-Aid, medicinal	—/140 ppb	Chlorine sanitizers, Brett contamination	Chlorophenols
Other	Diacetyl	Butter, movie theater popcorn, oily	—/17 ppb	Bacterial infection, poor fermentation	Diacetyl
	Soapy	Goaty, tallow, oily, waxy	—/13 ppm	Poor yeast health, dead yeast	Octanoic acid
	Ethyl acetate	Solventy, nail polish remover	—/20 ppm	Warm fermentation, low oxygen	Ethyl acetate

[a]Courtesy S. E. Johnson and M. D. Mosher—© ASBC.
[b]The flavor threshold is the lowest level reported for each flavor in water and in lager. The flavor threshold may be much greater depending on the style of beer.
[c]ppm = parts per million; ppb = parts per billion; ppt = parts per trillion.

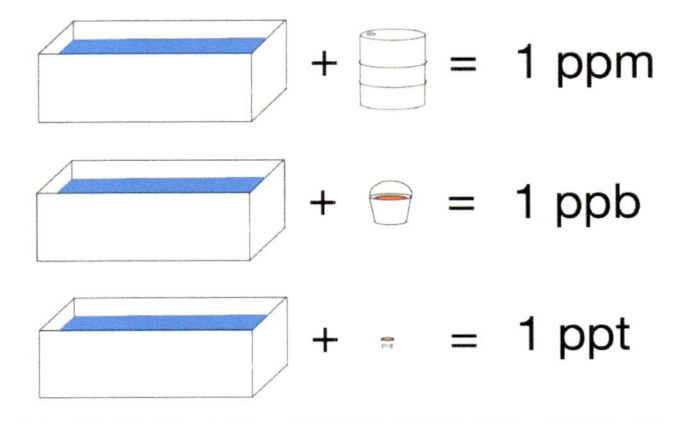

FIG. 2.3. Relative sizes of parts per million (ppm), parts per billion (ppb), and parts per trillion (ppt). One ppm is about the same size as a keg of food coloring in a large Olympic-sized pool. One ppb is about the same size as a small pail of food coloring in the same-sized pool. One ppt is about the same size as a thimble of food coloring in the same pool. (Courtesy S. E. Johnson and M. D. Mosher—© ASBC)

However, most of the flavors we taste are perceptible at these very low levels. Four drops of cinnamaldehyde (the flavoring agent that tastes like cinnamon) in a beer barrel of water is discernible. Four drops equals just a little more than 1 ppm of cinnamaldehyde, which means that even a very small bacterial infection or a slightly unhealthy yeast population could cause significant impacts to the flavor of the beer. It's easy to see how making a small mistake during brewing could be disastrous to the flavor of that batch of beer. This is another illustration of why it's important to implement a thorough quality control and assurance program that monitors the quality of the product via QCPs.

It's possible that each of the off-flavors listed in Table 2.2 could be found in a batch of beer but still be below the flavor threshold. Although no one could taste the off-flavor, there could still be an issue with that batch of beer. The batch could have been slightly oxidized, for instance, but the concentration of trans-2-nonenal could be less than the flavor threshold. It's also possible for the concentration of an off-flavor to be greater than the flavor threshold yet still unperceivable by the person performing the taste analysis. Flavor thresholds vary across tasters. Some people have no problem identifying diacetyl at 17 ppb, others can't perceive the flavor below 100 ppb, and still others can't detect it below 10 ppm. Thus, it's important to carefully select a group of tasters in a panel. (More on this can be found in Chapter 9.)

How to Attack Off-Flavors

Issues with a particular brand of beer can be determined by conducting a careful, thorough analysis of the flavor profile when there are no reported off-flavors and then comparing future batches of the same beer to that profile. If an off-flavor is perceived by a tasting panel and the level of that off-flavor is discerned, then the brewer has the information needed to begin troubleshooting the issue. The process of operating a tasting panel is illustrated in Chapter 9, but for now, let's discuss the steps that a brewer can take to correct the brewing process to avoid that off-flavor.

It can't be stated enough that a thorough quality assurance program will have addressed any issues with the brewing process. Coupled with an HACCP program, a similar quality program designed to reduce issues with product quality will note and correct quality issues early in the process. The chance of a problem existing in the finished product will be greatly diminished.

But if a problem does occur, the brewer should be prepared to go well beyond simply identifying the issue. The first step, identification of an off-flavor using a tasting panel, begins the exploration of the process. Then once the off-flavor has been identified by the panel, a chemical or microbiological search can take place in the laboratory. Specific analyses should be performed that look for the compounds causing the off-flavors. These tests can be done to confirm the presence of the suspect pollutant in the beer and the concentration of that compound so that the brewer can gain insights into how the off-flavor appeared in the finished product.

Correcting issues with off-flavors involves performing these steps:

1. An off-flavor is identified by the tasting panel.
2. The off-flavor is confirmed by either a chemical or microbiological analysis.
3. A laboratory analysis provides the concentration of the off-flavor.
4. The information is applied at the point in the production process that the issue was identified.
5. Further work by the sensory team, with the laboratory, explores what in the production process caused the issue. The ASBC fishbone diagrams are very helpful in this process.
6. Once the issue has been identified, the team updates/edits/modifies SOPs and processes to reduce the chance of recurrence.

As an example of this correction process, suppose that a finished batch of a specific brand of beer has been analyzed by a tasting panel. The beer was removed from the conditioning tank and is now being held until the tasting panel gives the thumbs-up to send the beer to filtration and then packaging. Unfortunately, the tasting panel has found a buttery taste to the beer that wasn't in the previous sample of the brand. The level of that flavor has been identified by the panel as "slight."

Next, the laboratory performs a thorough analysis of the beer. Knowing that the off-flavor is "buttery," the laboratory focuses first on the analysis of diacetyl in the beer; it's found to be 50 ppb, just above the flavor threshold. That information is passed back to the cellar team waiting for the "go/no go" decision on the batch of beer in this beer type. In this example, the cellar team will explore the potential causes of the off-flavor, ranging from issues with the fermentation (the diacetyl rest wasn't completed) to an infection of *Lactobacillus* bacteria. The team might consult the ASBC fishbone diagrams to find potential causes.

Let's assume that the cause of the diacetyl is verified to be a *Lactobacillus* infection that's coming from the carbon dioxide (CO_2) line in the conditioning tank. Verification of that issue would have required additional laboratory work as the source of the diacetyl was explored. The team then cleans the CO_2 line and modifies the SOP for the conditioning tank to require additional scrutiny when cleaning the line. Following this process will identify the issue and then follow up on the corrective actions so they become part of SOPs.

Using New Ingredients (or the Same Ingredient from Another Supplier)

The most common use of a sensory panel is to determine if one batch of beer is the same as the batches that preceded it. This can be an issue when a supplier change is made or the malthouse changes its procedures. Monitoring consistency is also important as a function of time, because barley crops change from season to season. Much of the variation should be caught by the maltings, but minor changes may have large effects on the flavor profile.

It's important for the brewer to know how to implement some form of sensory analysis to ensure that when a supplier provides ingredients, the flavor remains unchanged. Often, making a necessary reduction in the cost of the materials requires changing the supplier and thus the ingredient specifications. The most common way to analyze the sensory aspects of the product is to perform a same/different test. In some cases, however, that may not be the best test to run. For example, during recipe development, the sensory team expects the beers to have different flavors. Only knowing that the beers are different would not be very helpful in this case. Detailed explanations and fill-in-the-blank guides are provided in Chapter 15 for assistance with running different sensory analyses.

CHAPTER 3

The Brewery Laboratory

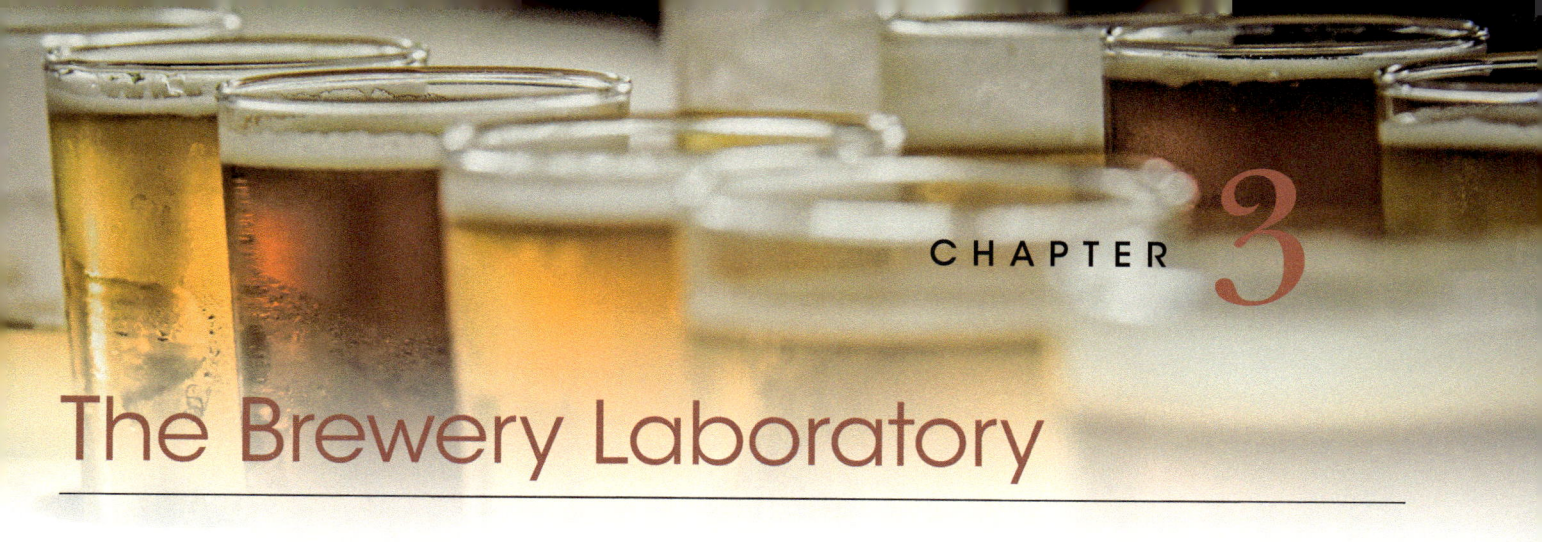

In this chapter, we will look at the brewery laboratory and uncover the important features of that space. This area of the brewery is often overlooked in new brewery startups, but operation of a quality assurance (QA) or quality control (QC) program requires some type of laboratory. And having a brewery laboratory is necessary for a brewery to follow good brewing practices (GBPs), hazard analysis and critical control point (HACCP) principles, and provisions of the Food Safety Modernization Act of 2011 (FSMA).

The Brewery Laboratory

The laboratory in a brewery performs a vital function. As such, the brewery must have a well-designed standard operating procedure (SOP) for each test that it performs. Procedures should be in place to verify that the experiments performed in the lab report values with precision and accuracy. Because the data from the lab are often the basis for making a "go/no go" decision on a given batch, laboratory personnel must be skilled and trained and able to make precise measurements.

Overview of the Lab

The laboratory should occupy a designated space within the brewery and be operated by dedicated employees. The lab contains chemicals, equipment, and instrumentation that are used to perform quality controls and analysis of critical control points. Keeping the lab door closed as a standing policy will limit traffic through the lab.

Limiting unnecessary visitors to the laboratory will help ensure that the space remains as clean as possible. Untrained personnel might accidentally track dust, microbes, and other items into the space. The disruption caused by having to reclean the lab might delay test results or even invalidate them, at the worst.

The work of the brewery laboratory centers on determining specific concentrations of chemicals in a sample. In the laboratory, the chemicals of interest are **analytes** (also called **solutes**). An analyte is anything dissolved in a sample that the lab measures or identifies. For example, calcium, ethanol,

analyte A compound being analyzed within a solution.

solute *See* analyte.

matrix A solution containing the analyte and other compounds that may obscure detection of the analyte.

accuracy The closeness of a measurement to the actual value.

precision The closeness of a measurement to other measurements.

percentage alcohol by volume (% ABV) The volume of ethanol dissolved in 100 mL of beer.

and diacetyl are common analytes in the brewing process. The sample is known as the **matrix** and is best defined as the solution that contains the analytes. Beer is a complex matrix made up of a lot of components. Put another way, beer is a solution of a solvent (water) with hundreds of solutes (or analytes). Unfortunately, the greater the complexity of the matrix, the greater the potential issues in performing an analysis. Since the laboratory must find and report the concentrations of these analytes with confidence, workers in the lab must have good skills and techniques.

In reporting the concentrations of the analytes it measures, the laboratory focuses on two important scientific terms: **accuracy** and **precision**. We have used these words before in this text. Unfortunately, these terms are often misunderstood and misused, but in the scientific world, they have very important meanings. When a measurement is accurate, it represents the actual value of the amount of analyte. This value could be determined by throwing darts at a dartboard and hitting the bullseye. It could also be determined from multiple measurements that miss the bullseye, but the average of all the measurements results in the bullseye (Fig. 3.1). A precise measurement comes from repeated analyses that all arrive at the same answer. That answer doesn't have to represent the actual analyte concentration to be precise. The measurement of any analyte can be accurate, precise, both, or neither.

As an example, let's consider the measurement of **percentage alcohol by volume (% ABV)**, which is the volume of ethanol dissolved in 100 mL of beer. Assume that the actual value of a sample was 4.20% ABV (or 4.20 mL ethanol dissolved in 100 mL beer) and that the laboratory found 4.22% ABV, 4.18% ABV, and 4.20% ABV in three trials of the sample. All the results in this case

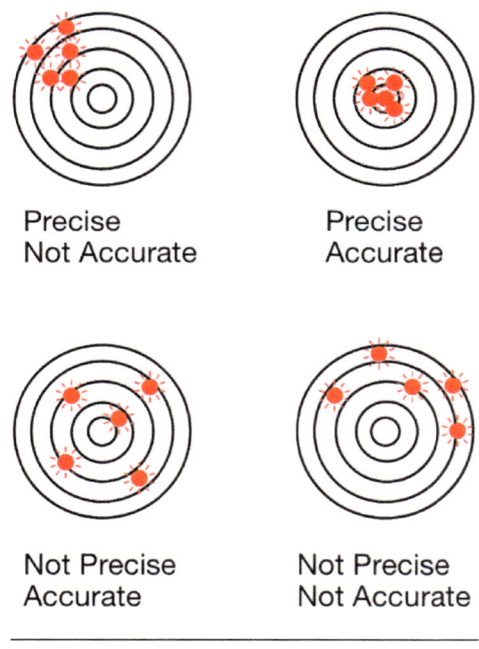

FIG. 3.1. The bullseye analogy of precision and accuracy. (Courtesy S. E. Johnson and M. D. Mosher—© ASBC)

are close together, or precise, and the average is close to the actual value, or accurate. If the laboratory found 4.86% ABV, 4.78% ABV, and 4.82% ABV, the results would be relatively precise but not accurate. If the laboratory found 5.33% ABV, 3.69% ABV, and 7.21% ABV, the results would be neither precise nor accurate.

The specific methods and techniques used to perform the analysis are designed to be as accurate and precise as possible. Every instrument inherently has some error when performing a measurement. The method and the SOP limit that error so it's as small as possible—but the error will always be there. And because the methods are performed by people, human error can affect the measurements, resulting in the possible loss of precision and/or loss of accuracy. With proper training and repeated experience, people will make fewer errors. The ultimate goal of the brewery laboratory is to be both accurate and precise in its measurements.

Do You Need a Scientist?

When we think of a laboratory, we envision workers wearing lab coats and protective goggles and carrying pens and calculators in their pockets. Not surprisingly, the brewery laboratory is staffed by people wearing this very attire. The coats are worn not to distinguish the lab personnel from others but to protect workers' clothing from the harsh, staining chemicals often encountered in the lab. The goggles keep chemicals out of people's eyes, and the pens and calculators? Well, they are used frequently to determine quantities of chemicals to mix for reagents, to calculate test results based on instrument readings, and to look cool!

What about the people that wear the lab coats? Are they required to be scientists? Do they need an advanced degree in chemistry, biology, or physics? The short answer is "No," but the longer answer includes "It is beneficial to have at least one laboratory person with an advanced degree in science." Having access to a scientist who is skilled in laboratory work will improve the accuracy and precision of work done in the lab. A nonscientist can be trained to perform maintenance on the instruments, to perform the methods of analysis, and to operate a laboratory. With a strong interest in science, a laboratory worker can run a successful laboratory. Such a person will need training for instrument maintenance, instrument operation, and laboratory skills. With time, nonscientists can learn to think like scientists. They can learn to pose questions, evaluate systems, find efficiencies, and improve techniques. Their comfort level will be pushed at first, but with trial and error and practice after practice, their skill and comfort working in the lab will improve dramatically. In fact, the authors have extensive experience training nonscientists to perform well in a brewery laboratory. With training, anyone can work well in a brewery laboratory. But having a scientist in that position (or at least one of the positions in the lab) will allow the lab to improve, grow, and flourish.

A scientist knows how to ask the questions needed to improve operating procedures. A scientist knows how to perform analytical procedures. A scientist knows how to run and expand a lab. This chapter and those that follow will provide the information needed to train nonscientists and those who feel out of their comfort zone in operating a full laboratory.

Who Performs QA Procedures?

Analysis of the process stream can be accomplished by many different people. As we noted in Chapter 2, that could involve a dedicated laboratory staff member, an employee who works within a department on the brewery floor, or a specific employee dedicated to sample collection. There are pros and cons to each mode of obtaining a sample for analysis.

Using dedicated laboratory personnel provides the advantages that come with highly trained personnel. These employees understand the reasons that a sample is obtained in a sterile fashion. They know how to ensure that the sample is sterile, and they know how to avoid contaminating the product stream while they obtain a sample. In addition, laboratory personnel have the understanding and the ability to store the sample in a nondamaging environment prior to its analysis. For example, a sterile sample of the liquid in the fermenter can be obtained and then transported to the laboratory to measure diacetyl concentrations. Laboratory personnel know that the sample container must be sealed, kept cold, and analyzed as soon as possible to reduce the rate of yeast metabolism. Following these procedures results in a relatively stable sample that's representative of the time the sample was taken. An untrained worker may not realize the importance of chilling the sample or rushing it to the lab for the analysis. As a result, a mistreated sample might not report the diacetyl level accurately.

An employee working within a given department might be assigned to collect samples for the laboratory. The main advantage of this mode of sample collection is an increase in morale among the workers in that department: They have "ownership" of the samples that are collected. They know exactly when a sample should be collected and can do that more regularly than can laboratory personnel. The disadvantages of this mode of sample collection are the obvious need for regular training of personnel in sampling techniques and a potential reduction in accuracy of results because of relatively unskilled sampling techniques. As noted earlier, some issues with sampling protocol might arise, but those issues can be addressed with adjustments to SOPs.

A specific person can be designated to have the job of collecting all samples in the brewery and providing those samples to the lab. The biggest advantage to this method of sample collection is that one person collects all the samples. After training and regular updates, that person will be able to collect samples in a uniform fashion, and any idiosyncrasies in sample collection will be the same from sample to sample. Thus, the precision of the analyses will improve, because each sample will have been collected in the same way. The disadvantages of having a specific person collect samples are the same as the disadvantages of having employees with the department serve as samplers. The untrained sampler is likely the largest of the disadvantages of this method, although morale in the laboratory (particularly one that isn't very busy) could be reduced.

Although sample collection begins the analysis in the laboratory, we have to remember that there are many ways to perform much of the data collection without using the laboratory. On-site methods still require sampling and still have the issues associated with who performs the analysis. The issues with on-site methods of analysis are alleviated with in-line measurement devices. When those instruments can also keep a data log of the process stream, the values are always available. The best operation is to include the data log within the computer file for a particular batch of beer. The two primary disadvantages of in-line analyses are the cost of the instrumentation and the need to regularly calibrate devices to ensure accuracy and precision. The costs can be significant, because in-line instruments are likely needed in multiple locations within the brewery. If a single in-line instrument costs $5,000, installing 10 instruments within the brewery will cost $50,000. It should also be noted that in-line measurements are not available for every analysis that's needed. Temperature, dissolved oxygen, gravity, and cell counts are all possible in-line analyses. Diacetyl, international bitterness units (IBUs), and food and nutrition (FAN) levels do not have in-line instruments.

Once samples have been taken, they are transported to the lab, and once the samples are there, dedicated personnel take control of them. Following SOPs for the particular analysis, the lab personnel perform the appropriate experiments and record the results. Often, experiments are run in triplicate to provide information on the precision of the measurement. Those results are then recorded and reported. (Chapter 8 outlines the different ways that results can be reported.)

Laboratory personnel are also responsible for keeping the lab in good working order. Doing so includes maintaining the instruments, ordering the chemicals and supplies (such as sample containers), and validating the precision and accuracy of measurements. Validation procedures should be established for every instrument and every SOP, including in-house validations (where standards are prepared and run periodically through the same processes as actual samples) and external validations (such as the *Laboratory Proficiency Program* of the American Society of Brewing Chemists [ASBC]).

There are two ways to perform external validation. In one method, a sample is prepared in the lab, analytically measured, and sent to a verification lab outside the brewery for analysis. The results are then compared with the results produced by the brewery lab to ensure that the methods performed in-house produce the same results as the external lab. It's good practice to send not only an actual sample of the product stream but also a standard sample of the analyte made in the brewery laboratory. Since the lab makes the sample, the concentration of the analyte is well known. In the second external validation method (as is done with the ASBC's *Laboratory Proficiency Program*), samples are sent to the brewery laboratory from the external lab. The brewery lab then performs the analyses and returns the results to the external lab, providing information not only to the brewery about its analysis but also to the external lab about the method of analysis. In other words, external verification provides information about the precision and accuracy of the brewery laboratory itself and the quality of the samples that it prepares.

The Design of a Brewery Laboratory

The brewery laboratory should be included in the design for the entire brewery. Specifically, the lab should be outlined on the blueprints and a budget should be reserved to cover its costs. However, in many cases, newly constructed breweries focus on the brewhouse, cellar, and potential taproom.

The laboratory is often added after the brewery is operational, when the budget is more limited. But if the design is right and careful thought is put into the current and future desired outputs of the lab, a fully functional laboratory can still be created.

Space Requirements

The laboratory in a brewery cannot be a closet! It must be accessible, as noted before, yet isolated from traffic. Some of the major requirements for the space are as follows:

- **A designated space.** The laboratory should be designed so that it's separated with walls from the brewery floor. A bench off to the side will work if a separate space isn't available, but that's definitely not ideal. A better option is to locate the laboratory in a room or wing within the brewery. The space should be able to be locked and accessible only by laboratory staff. Restricting access to the laboratory space will limit the transfer of materials from the laboratory to the brewery and vice versa, significantly reducing the risk of contamination or hazards to the product. Likewise, the risk to contamination and problems with the laboratory testing will be eliminated.
- **A central location.** The laboratory should be centrally located to provide the shortest distance (and thus the quickest response time) to the places that samples will be taken in the product stream. Based on the purpose set for the laboratory, that location may be close to the packaging area, the cellar, or the brewhouse. If the laboratory is to provide analytical measurements across the entire brewery, it should be located so that the trip to and from the brewery floor is as short as possible.
- **Bench space for each instrument.** Each instrument requires a minimum footprint just to include it in the laboratory. Adding additional bench space on each side of the instrument (at least 1 foot) will provide the access needed to address the required setback and the sample preparation. Alternatively, a separate bench for sample preparation can be included in the design. An instrument bench should include access space behind each instrument for electrical, gas, and other services (such as intranet). It's particularly useful to have a small walkway behind the instruments for this purpose and to improve access to perform maintenance on them. If feasible, the laboratory design should provide electrical circuits for the instruments that are completely separate from the other brewery components and even tie-ins for stand-alone electric generators for use during power outages or brownouts. Some instruments require compressed gases, so space needs should be considered for tanks, regulators, and piping to the instruments. Most of the instruments that use compressed gases require infrequent tank replacement, but a few instruments use gases so quickly that the tanks must be replaced weekly. Having easy access to those tanks can be a big benefit. Some instruments also require dedicated ventilation around them for airflow beyond normal ventilation. Such ventilation can be provided with a laboratory-grade hood or other ventilation system that pulls air from the instrument and exhausts it to the outside.

- **Bench space for each piece of equipment.** Like the lab instruments, the individual stir plates, hot plates, balances, and other equipment need space on the bench in order to be used. Each piece of equipment needs at least 1 linear foot of bench, and some require more based on the equipment footprint and required setback. For example, while a balance can fit in 1 linear foot of bench space, it needs at least 3 inches on all sides to have good access. A hot plate may have only a small footprint but require 6 inches of clearance.
- **Bench space to perform laboratory work.** Of course, the laboratory must also have free bench space for personnel to make solutions, prepare samples, manipulate glassware, and so on. Three linear feet of bench makes a nice-sized workspace for one person. This includes adequate space for worksheets, notebooks, computers, and other places to do paperwork (Fig. 3.2).
- **A chemical fume hood and/or laminar flow hood.** The laboratory must also provide designated spaces to perform analyses and obtain samples that are volatile (to avoid releasing fumes into the lab) and to perform microbiological experiments (greatly reducing the risk of contamination). A chemical fume hood draws any vapors out of the space and creates a safer working environment. A laminar flow hood keeps the working environment free of airborne microbes. This is useful when growing and identifying yeast and other microbes.
- **Storage space for chemicals.** The designated lab space should have lockable cabinets, flammable storage cabinets, places to store bottles and boxes of chemicals, and storage locations for waste chemicals.
- **Storage space for glassware.** Cabinets and shelving for glassware are needed within the laboratory. Cabinets can be located above or below benches or be separated from other laboratory benches.
- **Storage space for samples.** The lab will need refrigerators and freezers and cabinets and shelves to store samples at constant temperatures.

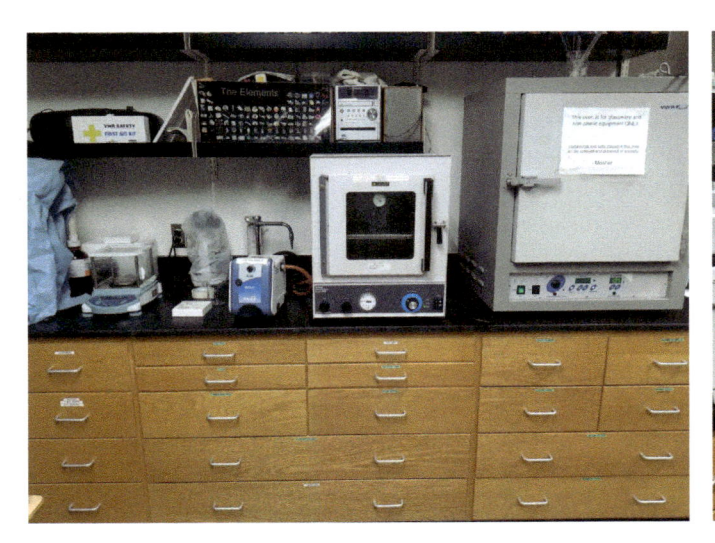

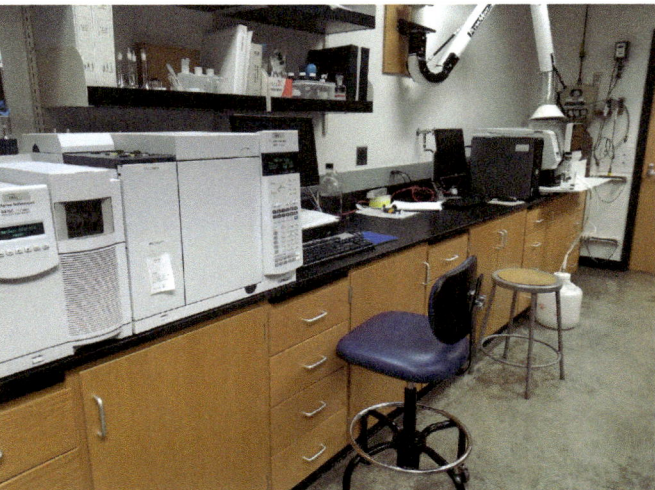

FIG. 3.2. A cramped working area (left) versus an open working area (right). (Courtesy S. E. Johnson and M. D. Mosher—© ASBC)

- **Separate space for yeast supplies and samples.** If the laboratory will work with yeast, ample space should be provided for yeast storage; refrigerators to hold agar slants, petri dishes, and broths; and bench space for incubators and shakers. In addition, the laboratory that handles yeast is best separated from the laboratory that handles the other analyses for the brewery. A sterile environment is needed to handle and grow yeast, so designating a separate yeast space can ensure minimal exposure to microbial infections.
- **Ample electrical and other utility service.** Laboratories require electricity (some instruments require dedicated circuits), water (both tap and distilled), natural gas (for Bunsen burners), and appropriate data service for intranet and internet. Some of these can be supplied (such as bottled distilled water and tanks of natural gas), but it's best to have utilities plumbed into the benches and walls where they are needed.
- **Safety equipment.** A brewery laboratory is required to have safety features such as an eyewash station and a safety shower. In addition, fire extinguishers and other safety equipment are necessary additions to the lab. It's useful to check local and/or state laws for specific safety requirements.

So, what is the minimum space needed for a laboratory? It depends on what types of instruments and how much equipment, glassware, and storage are needed, and that depends entirely on the work done at the laboratory. Will the laboratory be used only for yeast counting and basic analyses such as ABV and IBU, or will it be responsible for storing yeast and providing basic services and advanced analyses such as malt analysis? The smallest lab might be a small room within the brewery or a bench in the back corner of the microbrewery. A larger one might be 2,400 square feet spread across a main lab and many satellite labs throughout the brewery. What's important is not the size of the laboratory space but the presence of the laboratory within the brewery.

Supplies and Chemicals

The laboratory must be stocked with the appropriate equipment, supplies, and chemicals to perform its functions. Laboratory **equipment** includes spatulas and scoops to transfer solids and balances to obtain the mass of chemicals or other ingredients (Fig. 3.3). Magnetic stirrers, vortex mixers, and hot plates are also essential to the laboratory. The electric equipment doesn't specifically determine the concentration of an analyte but is needed to precisely and accurately perform the experiments.

Glassware in the laboratory includes beakers, Erlenmeyer flasks, glass rods, and dishes of all sizes. In addition, the lab should have a set of volumetric glassware, including both volumetric pipettes (used to deliver specific volumes of liquid) and Mohr pipettes (graduated pipettes that can deliver different amounts of liquids), as well as volumetric flasks (used to contain specific amounts of liquid), as shown in Figure 3.4. (Each piece of glassware and its practical application is described in detail in Chapter 6.)

Chemicals are also needed in the laboratory. In addition to a good supply of deionized or distilled water, chemicals include all the reagents

---

equipment Simple laboratory tools, such as spatulas, balances, hot plates, mixers, and so on.

glassware Laboratory equipment that includes glass beakers, flasks, rods, dishes, and pipettes.

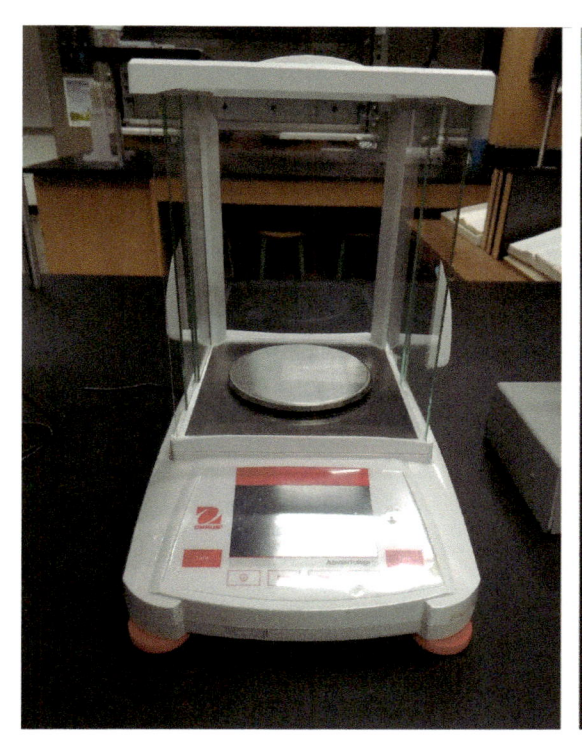

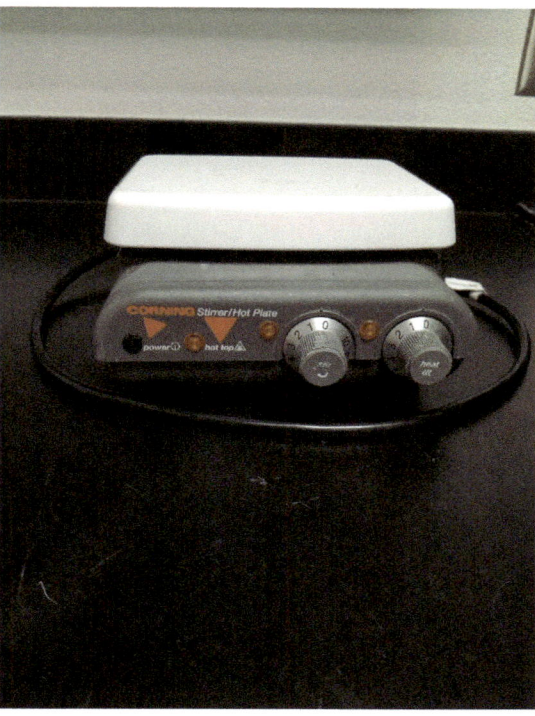

FIG. 3.3. Common laboratory equipment. An analytical balance (left) and a magnetic stirrer (right) are useful in the lab. (Left, Image reproduced by permission of OHAUS. Right, Image reproduced by permission of Corning Life Sciences. Photos courtesy S. E. Johnson and M. D. Mosher—© ASBC.)

FIG. 3.4. Selected laboratory glassware. (Courtesy S. E. Johnson and M. D. Mosher—© ASBC)

required for the different tests the lab will perform. Chemicals may include everything from salt to ninhydrin, depending on how many analyses the lab performs. Some chemicals, such as salt and distilled water, can be purchased at a grocery or hardware store. Others, such as ninhydrin and diacetyl, must be purchased from a chemical supply company. Setting up an account with a supply company can often provide discounts in the costs of these supplies.

Don't forget about storage for all of these materials. While glassware can be stored in drawers or on shelves, chemicals have specific storage requirements that must be followed. The most important rule is to make sure to store chemicals together that have similar reactivities. An oxidizer should never be stored next to a reducer, for example, and flammable liquids must be stored together in a vented flammable cabinet. In fact, there are many different storage protocols. The U.S. **Occupational Health and Safety Administration (OSHA)**, the **Environmental Protection Agency (EPA)**, and the **Centers for Disease Control and Prevention (CDC)** all provide specific guidelines for the safe storage of chemicals. In fact, the website for the CDC provides a particularly useful set of guidelines that are relatively easy to follow (www.ehso.com/ChemicalStorageGuidelines.htm).

Rules for chemical handling, storage, and safety should be recorded and kept on file in the lab. They should be followed to the letter, including the federally mandated requirement that a **Safety Data Sheet (SDS)** be maintained for every chemical in the laboratory. (The term SDS has replaced **MSDS [Material Safety Data Sheet]**, but the use of MSDS is still common.) The SDS contains specific information about the chemical, including its physical properties, toxicity, handling and storage requirements, first-aid measures, and fire-fighting measures. There should be an SDS for the bottle of toluene in the flammable cabinet and separate SDSs for the window cleaner, detergent, and other chemicals stored under the sink. Every worker in the laboratory should know where SDSs are kept and how to quickly find the SDS for each compound he or she uses. In fact, reviewing SDSs during the first few days of a new hire's employment can be beneficial for employee safety. Doing so can also help a new worker find his or her way around the lab or brewery. SDSs are freely distributed when chemicals are purchased and can be stored in a three-ring binder on a shelf in the laboratory for easy access.

Handling, safety, and storage information should also be provided on every chemical container—even containers of water. There are also many different codes that can be used to help relay safety information to users visually, such as the U.S. **Hazardous Materials Identification System** (Fig. 3.5). This numerical rating system incorporates colored labels to identify health, flammability, and reactivity hazards and to indicate the appropriate **personal protective equipment (PPE)** that should be used. The number beside the hazard indicates the severity from 0 (minimal) to 4 (severe). The appropriate personal protective equipment is designated by a letter and a visual guide that shows what equipment should be worn.

Laboratory Safety

Ensuring a safe laboratory environment is the combined responsibility of laboratory personnel and brewery management. However, the primary responsibility lies with the individuals performing the work. While federal, state,

---

Occupational Health and Safety Administration (OSHA) A U.S. federal organization that provides standards for specific work activities and workplaces (such as breweries) to ensure safe and healthy working conditions.

Environmental Protection Agency (EPA) A U.S. federal organization that provides regulations for the protection of the environment.

Centers for Disease Control and Prevention (CDC) A U.S. federal organization that conducts and supports health promotion, prevention, and preparedness with the goal of improving public health.

Safety Data Sheet (SDS) A sheet containing data about a particular chemical, including its properties, handling, storage, and safety.

Material Safety Data Sheet (MSDS) An old term for Safety Data Sheet (SDS) that is still commonly used.

Hazardous Materials Identification System A numerical rating method that incorporates colored labels to identify types of safety hazards.

personal protective equipment (PPE) Apparel that is required to provide safety and health protection to an employee (such as goggles, face shields, gloves, and aprons).

and local laws and regulations establish legal requirements for safety, following those safety rules is prudent for financial reasons and for the health of the laboratory and brewery personnel. Within the laboratory, the head of that group of workers holds particular responsibilities to ensure that all safety rules are followed to the letter, that all workers are trained regularly, and that a **culture of safety** exists within the laboratory.

Maintaining a culture of safety is not simply following the rules. It starts at the top. Owners and all leaders must commit to safety as the most important priority. A culture of safety requires that all SOPs incorporate safe practices. In addition, every worker should think about safety prior to performing an experiment. Every meeting of laboratory workers should involve discussions about safety and procedures. A culture of safety exists when workers don't need to be reminded to wear PPE (such as goggles, lab coats, and gloves) and to work in a hood.

Instituting a culture of safety in a laboratory can be difficult. Some workers may rebel against the strictness of the rules and steps required for safety. Writing and implementing new SOPs that include those safety procedures requires both time and effort. Workers should be consulted, at the least, and included in the implementation process to increase their buy-in and ownership of safety practices. Once everyone prioritizes safety before doing his or her job, the culture of safety will be an integral part of the daily routine. At that point, maintaining the culture among the workers can be easily accomplished through weekly meetings.

culture of safety The core values and behaviors that underlie a commitment to safety within the brewery by workers and leaders.

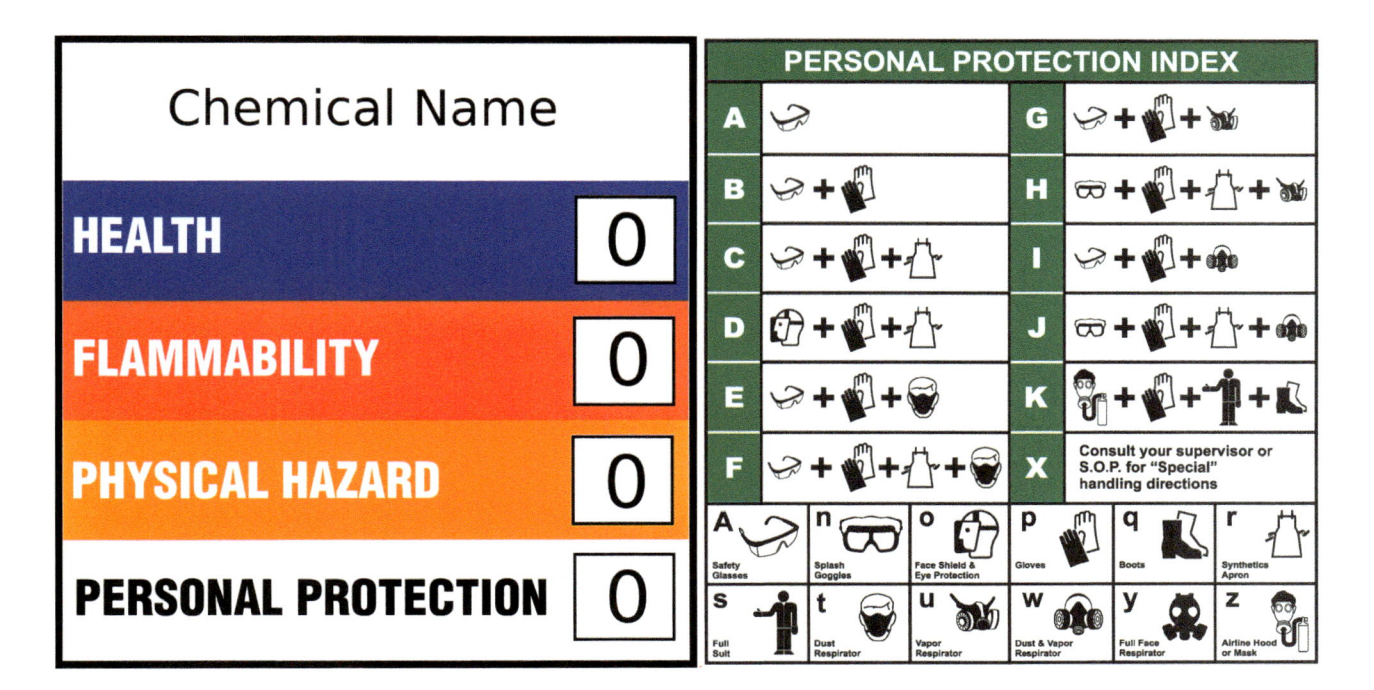

FIG. 3.5. The U.S. Hazardous Materials Identification System. Left, Health, flammability, and reactivity hazards are denoted with numbers indicating severity. Right, Personal protection is denoted with a letter that indicates the appropriate protective equipment to be worn. (Left, Reproduced from Wikipedia. Right, Courtesy SafetySign.com/Brimar Industries—Reproduced by permission.)

American National Standards Institute (ANSI) A private organization that sets standards for products, processes, and people.

Goggles. Safety begins with the simple introduction of eye protection. Approved goggles should be provided to all laboratory employees. The best option for goggles is a pair of indirectly vented or unventilated chemical splash goggles that meet the standard set by **American National Standards Institute (ANSI)** Z87.1-2003 (Fig. 3.6). Based on federal, state, or local laws, safety glasses or another form of eye protection may be substituted for splash goggles. Once issued to an employee, goggles must be worn in the laboratory any time that any chemical is being used—even if the chemical is being used across the laboratory from where the employee is located. Most accidents involving chemical spills on a person result from a spill that someone else made. These rules also apply to all visitors to the laboratory. (There should not be many visitors.) Rules such as wearing goggles in the lab are part of the culture of safety, and putting on goggles should be second nature to anyone who enters the lab.

Goggles work, keeping a chemical that's splashed into someone's face from contacting his or her eyes. Unfortunately, vapors from a chemical and sometimes the actual chemical do come in contact with the eyes, especially if the goggles are not fitted correctly. Workers who wear contact lenses can have significant damage to their eyes from chemical splashes. Capillary action pulls the chemical into the space between the eye and the contact lens, whether a liquid chemical or a vapor. Thus, workers who wear contact lenses are at much greater risk of eye damage even in the absence of splashes. Good safety practices should prohibit workers from wearing contact lenses. Goggles often fit over existing eyewear.

Gloves. Gloves provide an extra level of protection for workers using chemicals, but with this added protection comes the possibility of further safety and health concerns. Latex allergies may require the use of an alternative material. Simple checks can assure that the appropriate gloves are being

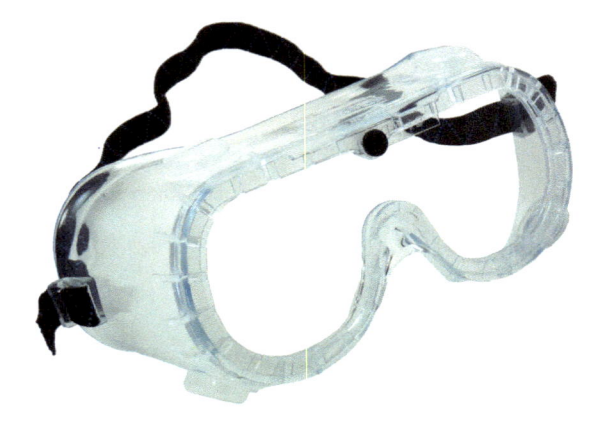

FIG. 3.6. Splash goggles offer excellent protection of the eyes against splashes and spills of chemicals in the laboratory. (Goggles produced by Long Dar Plastic. Photo courtesy Home Science Tools—Reproduced by permission.)

used. Will the glove material protect workers from the chemicals being used? In some cases, wearing gloves can lead to accidental chemical exposure to the wearer and others. Chemicals can be easily spread if gloves are not removed immediately after use and cleaned or disposed of properly. This concern can be reduced by requiring workers to remove their gloves before leaving the lab, opening drawers, handling pens, using computers, or handling instruments.

Even if the appropriate gloves are used, certain chemicals can diffuse through the glove material and be held against the skin. While it may appear that the gloves are providing protection, they aren't if a spill has resulted in chemicals being on the gloves. Personal injury can also occur if a glove wearer touches his or her hair, clothing, or exposed skin.

Fire Extinguishers. In addition to protecting workers from chemical exposure with PPE, a laboratory must be equipped with safety devices, including appropriate fire extinguishers mounted by exit doors. All fire extinguishers are not the same. Each fire extinguisher has a code that assists in determining what types of fires it is capable of fighting. The categories of fire extinguishers are listed in Table 3.1.

Some fire extinguishers are rated to fight multiple types of burning materials. The most common fire extinguisher that will address the most common fires in a brewery laboratory is categorized with an "ABC" code. This means the fire extinguishers can handle burning paper, wood, and liquids, as well as electrical fires.

Each employee in the laboratory should be trained to use a fire extinguisher, not only learning the process but also handling the fire extinguisher and fighting an actual fire. In some cases, the best option is to extinguish

TABLE 3.1. Fire extinguisher codes[a]

Category	Types of materials	Color code	Examples of fire retardants
A	Paper, wood, trash	green triangle (A)	Foam, water
B	Flammable liquids and gases	red square (B)	Carbon dioxide, foam
C	Live electrical equipment	blue circle (C)	Dry chemical, carbon dioxide
D	Flammable metals	yellow star (D)	Dry chemical
K	Grease, oils	black hexagon (K)	Wet chemical

[a] Courtesy S. E. Johnson and M. D. Mosher—© ASBC.

the flames. In other cases, the best option is to use the extinguisher to aid in everyone's escape from the laboratory. Most fire departments offer some training in using a fire extinguisher that will help make that determination. Those services should be used.

Eyewash Station. Another necessary piece of safety equipment in the brewery laboratory is an eyewash station. It allows chemical splashes and foreign matter in the eyes to be easily addressed. There are many styles of eyewash stations, from the kind that look like a drinking fountain to the kind that simply provide a squeeze bottle of sterile saline. The eyewash stations should adhere to safety standards established by local and state laws and be clearly marked and accessible at all times.

Safety Shower. A safety shower should also be included in the laboratory. It can be quite useful if a splash or spill occurs on a person where the affected area can't be rinsed off in the sink. So even though a safety shower wouldn't be necessary for a spill on the forearm, it would be necessary for a spill that soaks a worker on the chest, head, torso, or legs. The correct operation of the shower involves simply standing under the shower head and pulling a cord (often attached to a metal triangle).

Unfortunately, many acid and alkali spills still cause burns on people who have used a safety shower. The most affected areas of the body are those where elastic or tight-fitting clothes touch the body; waistbands, bras, socks, and shoes can trap chemicals near the body. Even when the victim starts the flow of water in a safety shower, a chemical spilled on his or her upper torso can still cause burns where the victim's clothes contain elastic. If the shower washes the chemical off the upper torso, it migrates down to the pants and shoes. Thus, the SOP for use of the safety shower should include removing all clothing.

Fire Safety

Many chemicals used in the brewery lab are in aqueous solutions and therefore not typically fire hazards. However, some analyses require the use of flammable solvents or materials. As stated previously, it's important to store flammable materials together in a fire-resistant cabinet. Chemical storage should follow OSHA standards and also conform to fire codes and any federal, state, and local regulations.

There is a chance, however small, that a fire could start in the laboratory. If the fire is small and contained, it can often be smothered with a piece of glassware or put out with the fire extinguisher. But if lab personnel don't feel comfortable fighting the fire or the fire is larger than the size of a hand, the fire department should be notified immediately. In such a case, firefighters will arrive and try to extinguish the flames. Fighting a normal fire is hazardous enough, but when a room full of chemicals is on fire, firefighters have even more significant risks. The risks posed by the chemicals in a brewery should be discussed with the fire marshal before firefighters are ever needed, and each room within the laboratory should be placarded under the guidelines of the **National Fire Protection Agency (NFPA)** (Fig. 3.7).

National Fire Protection Agency (NFPA) An international nonprofit organization devoted to eliminating risks and losses from fire and related hazards; developed a hazard identification system for emergency responders.

The specific diamonds in the placard outline the health hazard (blue), flammability hazard (red), instability hazard (yellow), and special hazard (white). The numbers in the blue, red, and yellow diamonds rank the hazard from 0 (minimal) to 4 (extreme). For example, the health hazard rating is minimal (0) in a setting where no precautions are needed; slight (1) where a breathing apparatus may be worn; moderate (2) where a breathing apparatus and face mask must be worn; serious (3) where a full protective suit with a breathing apparatus must be worn; and extreme (4) where even short-term exposure to a fire in the room will be fatal.

It's important to note that the placard doesn't say anything about the hazards of using chemicals in the lab; the placard notes the hazards to firefighters when the room is on fire. Special codes include *OXY* for oxidizers, *ACID* for strong acids, *ALK* for strong alkali, *COR* for corrosive materials, and *W* for water-reactive compounds, plus a radiation symbol for radioactive material.

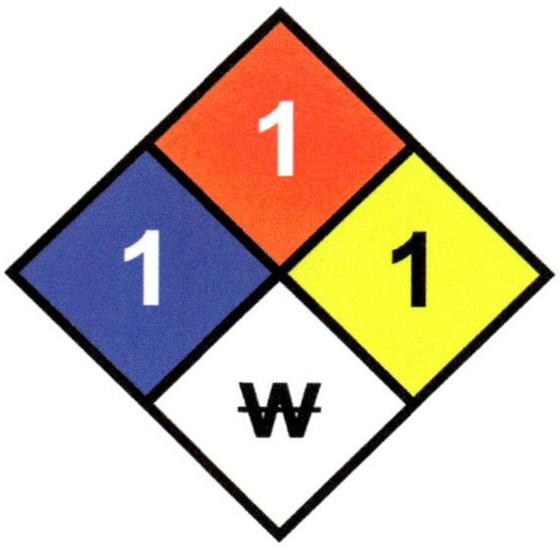

FIG. 3.7. NFPA placards contain fire-fighting information. (Courtesy S. E. Johnson and M. D. Mosher—© ASBC)

Common Instruments

Some **instruments** are commonly found in the brewery laboratory and are different from standard laboratory equipment. Remember that laboratory equipment includes stir plates, ring stands, clamps, balances, and other physical equipment that are used to perform analyses but do not provide the concentration of the analyte. Instruments, on the other hand, are expensive, sensitive machines that include precisely machined parts. They often include onboard computers that communicate with laptops or desktop computers that are dedicated to those instruments. Some produce printouts directly via a screen or onboard printer. Others provide an electrical signal to the dedicated computer, and that signal is then translated into data that can be read by laboratory personnel.

Common instruments in the laboratory can be categorized based on how they work (Table 3.2):

- Colorimeters are not often thought of as instruments, but they work using principles that are very similar to instruments. The main difference is that the user acts as the integrating circuit to determine the concentration of an analyte. (Colorimeters are described in detail in Chapter 10.)
- Spectrophotometers are instruments that use the interaction of light with the analyte of interest or a derivative of the analyte that has a color. These instruments include an ultraviolet-visible spectrophotometer and a fluorimeter. (More information about how these instruments work is provided in Chapter 11.)
- Another class of instrument physically separates all the compounds in the sample and detects each compound individually. This class includes the gas chromatograph (GC) and the high-performance liquid chromatograph (HPLC). (This class is discussed in Chapter 12.)
- Other instruments focus on the physical properties of the bulk sample, rather than a specific analyte. These instruments include the hydrometer, refractometer, and viscometer. (They are outlined in detail in Chapter 13.)

instruments Sensitive machines with precisely machined parts that measure the concentration or identity of an analyte; often include on-board computers that communicate with laptop and desktop computers dedicated to given instruments.

TABLE 3.2. Categories of laboratory instruments[a]

Category	Mode of operation	Examples
Colorimeters	Uses visual comparison of a sample to a set of standards	Comparitor, SRM Color Wheel
Spectrophotometers	Measures the light absorbed or emitted by a sample	Ultraviolet-visible (UV-vis), infrared (IR), or near-infrared (NIR) spectrometer, fluorometer
Chromatographs	Separates individual analytes and determine amounts	Gas chromatograph (GC), high-performance liquid chromatograph (HPLC), thin-layer chromatograph (TLC)
Physical properties	Measures characteristics of a bulk sample rather than a specific analyte	Viscometer, hydrometer

[a] Courtesy S. E. Johnson and M. D. Mosher—© ASBC.

A laboratory may have all of these instruments. A laboratory may also include instruments not discussed in this text, such as an XRF, ion chromatograph, or electron spin resonance spectrometer. Most of these instruments perform high-level analyses that are beyond the scope of the majority of the methods in the *ASBC Methods of Analysis*. Costs for these instruments are often quite prohibitive for a small or even a midsized microbrewery.

All instruments should have well-designed SOPs, and all the laboratory workers who use the instruments should be well trained in how to prepare samples for the instrument, perform the analysis using the instrument, and understand and apply the data obtained from the instrument. Implementing such training will improve the quality of the data that the laboratory produces by increasing precision and accuracy. It is also useful for laboratory workers to have some training in basic maintenance and basic standardization procedures. Doing so will reduce the downtime of the instruments in the laboratory due to maintenance issues.

A Brewery on a Budget

Quite a lot of information has been presented in this chapter yet not enough to build a brewery laboratory from scratch. If you purchased everything discussed thus far, the total cost would be quite high. In fact, the cost could run into the high $100,000s if not well over $1 million. While the lab would be very well outfitted, the cost could "break the bank."

This section outlines some of the important things you should know to plan a great laboratory without taking a significant hit to the budget. The authors strongly recommend the ASBC *Grow Your Own Lab* guide (located on asbcnet.org) (Fig. 3.8). This quick-reference guide outlines the types of equipment that a laboratory should obtain as a function of the production of the brewery.

The Brewery Laboratory   47

A Guideline to Growing Your Quality Laboratory

An Instrument Guide for purchasing lab equipment and expanding your quality program by using the official *Methods of Analysis* of the American Society of Brewing Chemists (ASBC). A subscription to the *ASBC Methods of Analysis* is free with your ASBC membership.

Legend: Recommended to be purchased / Optional Purchase

General

Volume produced (bbls x 1000) per year	ASBC Method of Analysis & Method Number
Thermometer	Temperature control
Packaged Beer Archive Shelving	Shelf stability testing
Refrigerator / Cooler	Sample storage, reagent storage
Waterbath	Attemperate samples and media, organoleptic diacetyl testing
Lab Informatics System	Advanced process control software

Raw Materials and Packaging

Equipment	ASBC Method of Analysis & Method Number
Crimp Gauge*	Crimp Determination Test-Crowns (Bottle Closures-6)
Torque Meter*	Removal Torque Procedures for Crown (Bottle Closures-5B)
Double-seam Gauges*	Can double-seam inspection
Double-seam Cross-section Imager*	Can double-seam inspection
Analytical Balance or Top-Loading Scale	Total Contents of Bottles & Cans By Calculation from Measured Net Weight (Fills-1), Grist (Malt-15), media and reagent preparation
Grist Sieves	Malt Modification by Friability (Malt-12), Grist by Standard Sieve (Malt-15A), Grist by Manual Sieve (Malt-15B)
Sieve Shaker	Grist by Standard Sieve (Malt-15A)
Friability Meter	Malt Modification by Friability (Malt-12)
Drying Oven	Moisture (Brewers' Grains-3 & Malt-3), Preparation of Sample (Brewers' Grains-2), Total Contents of Bottles & Cans (Fills-1)
Mash Bath	High-Dried, Caramel, and Black Malts (Malt-9), Soluble Extract (Brewers' Grains-5), Extract (Malt-4)
Universal Lab Disk Mill	Preparation of Sample (Brewers' Grains-2), High-Dried, Caramel, and Black Malts (Malt-9), Extract (Malt-4)

*package testing equipment is recommended to be purchased with associated packaging equipment

Chemistry

Equipment	ASBC Method of Analysis & Method Number
Hydrometer	Apparent Extract (Beer-3), Apparent Extract by Hydrometer (Wort-4), Extract (Wort-3), Soluble Extract (Brewers' Grains-5), Total Contents of Bottles & Cans By Calculation from Measured Net Weight (Fills-1), Yeast Fermentable Extract (Wort-5)
pH Meter	Beer pH (Beer-9), pH of Water Suspension (Filter Aids-2), Total Acidity (Beer-8), Wort pH (Wort-8)
CO_2 Meter	Dissolved Carbon Dioxide (Beer-13)
Low-Range Oxygen Meter (ppb)	Dissolved Oxygen for brite/packaged beer (Beer-34)
Digital Density Meter	Extract (Wort-3), Malt Extract (Malt-4), Real Extract (Beer-5), Soluble Extract (Brewers' Grains-5), Specific Gravity by Digital Density Meter (Beer-2B), Total Contents of Bottles & Cans By Calculation from Measured Net Weight (Fills-1)
Distillation Equipment	Alcohol (Beer-4A), Diacetyl (Beer-25B) *Note:* Recommend Gas Chromatograph for Diacetyl above 90K bbls/yr
Alcohol Meter	Alcohol (Beer-4)
UV-Vis Spectrophotometer	Beer Bitterness (Beer-23), Beer Color (Beer-10), Diacetyl (Beer-25B), FAN (Wort-12), Iron (Beer-18A, C), Total Polyphenols (Beer-35), Wort Color (Wort-4), Alpha and Beta Acids in Hops (Hops-6), Hop Storage Index (Hops-12), Thiobarbituric Acid Index (Wort-21)

Chemistry (continued)

Equipment	ASBC Method of Analysis & Method Number
Centrifuge	Beer Bitterness (Beer-23), Color (Beer-10), Yeast Solids % by Spin-down (Yeast-5B)
Shaker Table and/or Wrist Shaker	Beer Bitterness (Beer-23), Diacetyl (Beer-25B), Beer Decarbonation by Rotary Shaker (Beer-1D), Headspace equilibration for Total Package Oxygen
Fumehood	Chemical preparation (various analytical methods)
Titration Burette	Total Acidity (Beer-8)
Turbidimeter or Haze Meter	Physical Stability (Beer-27)
Gas Chromatograph	Alcohol Determined by GC (Beer-4D), Diacetyl (Beer-25F), Lower Boiling Volatiles in Beer or Ale (Beer-29)
Foam Meter	Foam Collapse Rate-Sigma Value (Beer-22)
High-Range Oxygen Meter (ppm)	Dissolved Oxygen for wort

Microbiology

Equipment	ASBC Method of Analysis & Method Number
Microscope	Dead Yeast Cell Stain (Yeast-3A), Differential Staining (Microbiological Control-3), Microscopic Yeast Cell Counting (Yeast-4), Presence of Bacteria (Yeast-2B), Yeast Morphology (Yeast-2A), Yeast Viability by Slide Culture (Yeast-6)
Hemocytometer	Dead Yeast Cell Stain (Yeast-3A), Microscopic Yeast Cell Counting (Yeast-4)
ATP Luminometer	Swab Surface Hygiene Using ATP Bioluminescence (Microbiological Control-1)
Autoclave or Pressure Cooker	Sterilization, Culture Media (Microbiological Control-4 & 5)
Incubator with CO_2 Packs, or Anaerobic Incubator	Anaerobic growth conditions, (Microbiological Control-5)
Vacuum Apparatus	Yeast Fermentable Extract (Beer-16, Wort-5), Detection of Microorganisms by Membrane Filtration (Microbiological Control-2C)
Stir plate / Hot Plate	Microbiological media preparation, sample homogenization, beer degassing, chemical preparation
Laminar Flow Hood	Sterile environment for microbiological applications
Thermocycler & PCR-related equipment	Real-time PCR for spoiler identification

Sensory

Equipment	ASBC Method of Analysis & Method Number
Quality Assurance Analysis	True-to-brand testing
Sensory Training	Sensory webinar series, Flavor Terminology and Reference Standards (Sensory-12), Flavor Standard Spiking Calculator (Tools)
Descriptive Analysis	Descriptive Analysis (Sensory-10)
Difference Testing	Choice of Method (Sensory-3), Paired Comparison Test (Sensory-6), Triangular Test (Sensory-7), Duo-Trio Test (Sensory-8)
Threshold Sensory	Threshold of Added Substances—Ascending Method of Limits Test (Sensory-9)
Dedicated Tasting Area	Test Room, Equipment, Conduct of Test (Sensory Analysis-2)

Other Useful Tools Located Online in the *ASBC Methods of Analysis*

Standards and other useful calculators, lab basics instructional videos, training presentations, fishbones and identification guides for all types of troubleshooting.

Getting Started—Suggested Reading Visit asbcnet.org

- *Brewing Microbiology, Third Edition* - F. G. Priest and Iain Campbell
- *Control Charting Guidelines for Quality Control* in the ASBC Methods of Analysis
- *Hops: The Practical Guide to Aroma, Bitterness, and the Culture of Hops* - Stan Hieronymus
- *Malt: A Practical Guide from Field to Brewhouse* - John Mallett
- *Water: A Comprehensive Guide for Brewers* - John Palmer and Colin Kaminski
- *Yeast: The Practical Guide to Beer Fermentation* - Chris White

FIG. 3.8. The ASBC *Grow Your Own Lab* guide provides excellent suggestions for the equipment that should be considered as the brewery increases production. (Courtesy ASBC—Reproduced by permission)

The Essentials Are Based on Your QC Requirements

The best way to plan the brewery laboratory is to determine the specific needs of the brewery: Which tests does the laboratory need to perform? Which tests does the HACCP plan require? Which tests will provide the information about the product stream that will be helpful in producing consistently high-quality beer? When these questions are answered, two lists will likely be created: a list of bare-bones analyses that the lab must perform and a wish list. These lists will dictate what equipment, instrumentation, space, and other features are needed in a well-designed brewery laboratory.

If the sky is the limit on budget, then the laboratory should include at least five spaces to handle the main tasks for evaluating the product stream. A wastewater treatment facility should also be considered for wastes, which may need its own lab.

The first space is a **propagation lab** that deals exclusively with yeast storage, yeast health, and growing samples of pitchable yeast for the yeast propagation system. Second is a **microbiology lab** that focuses solely on measuring the microbial levels in the product stream and identifying any microbes that are found. Third is a separate **packaging lab** that handles packaging requirements, such as bottle quality, quality of the can seams, labeling, carbon dioxide (CO_2) package levels, and other more physical activities in the packaging realm. The **analytical lab** handles analysis of everything from raw materials to finished product. It focuses on analyte determination in the product stream, such as diacetyl, dissolved oxygen (O_2) levels, ABV values, IBU levels, and sulfur dioxide (SO_2) levels. An important addition to the brewery is a **sensory lab** that uses sensory analysis methods to evaluate off-flavors, trueness to style, and other parameters of the beers that are produced.

Most of the cost for a laboratory with a "sky-is-the-limit" design lies in the cost of the major equipment and instrumentation. Even a relatively simple piece of equipment can cost $600; a good balance for weighing solids can cost more than $6,000. Instrumentation is often considerably more expensive; purchasing used instrumentation at auctions or from decommissioned laboratories will still result in a large expense. It would not be unrealistic to spend up to $1 million for the entire laboratory, including equipment and instruments, glassware, benches and ventilation system, and storage space.

A Lab for Under $10,000

If a brewery is under a tight budget, it's possible to put together a lab for under $10,000. The laboratory will still need benches, a refrigerator, and a ventilation system to handle volatile chemicals. These items can be found on used-equipment websites, purchased from other laboratories, and bought at auctions. Colleges and universities have equipment auctions, and purchasing items from these sources can help reduce costs and keep the entire laboratory under the $10,000 limit.

Even though a small microbrewery often has a walk-in refrigerator, it should not be used as the laboratory refrigerator. Toxic chemicals and other materials are often stored in the refrigerator that shouldn't be stored in the same location as consumable materials.

---

propagation lab A lab that deals exclusively with yeast storage, yeast health, and growing samples of pitchable yeast for the yeast propagation system.

microbiology lab A lab that focuses solely on measuring and/or identifying the microbial levels in the product stream.

packaging lab A lab that handles adherence to packaging specifications, such as bottle quality, quality of the can seams, labeling, and CO_2 package levels.

analytical lab A lab that handles analysis of everything from raw materials to finished product.

sensory lab A lab that uses sensory analysis methods to evaluate off-flavors, trueness to style, and other parameters of the beers produced.

Specific items needed in the laboratory are listed in Table 3.3, along with the purpose and approximate cost (as of 2018) for each item. This list does not include the benches, a refrigerator, and a ventilation system. Of course, the costs shown in the table are approximate and should be verified before considering any purchase. In addition, the costs are the minimum costs to obtain the equipment and instrumentation. As such, they don't add up to $10,000, but with some careful buying, all the items could be purchased for that amount. For instance, if a brewery splurges on a really nice balance instrument, then other items might have to be found at an auction house.

Note, as well, that the table doesn't list the costs of supplies such as filter papers, distilled water, and chemicals and reagents. However, as was noted before, many of these items can be purchased at a hardware or grocery store. In addition, while chemicals may be expensive, a small bottle usually goes a long way.

TABLE 3.3. Basic items needed in a brewing laboratory[a]

Item	Mode of operation	Approximate cost (2018 dollars)
Hot plate	To heat samples	$150
Stir plate	To stir samples	$150
Miscellaneous glassware	Volumetric pipettes, volumetric flasks, beakers, Erlenmeyer flasks, stir bars, glass rods, test tubes, test tube racks, graduated cylinders, burettes	$600
Ultraviolet-visible (UV-vis) spectrophotometer	To measure the interaction of light with chemical samples	$350 (used); $3,500 (new)
Laminar flow hood	To reduce contamination when working with microbiological samples	$500–$5,000 (depending on size)
Balance (analytical)	To measure masses to tenths of milligrams	$1,000 (used); $6,000 (new)
Centrifuge	To remove solids from lab samples	$100
Microscope (including hemocytometer)	To count yeast and viability	$200–$500
O_2 meter	To check oxygen levels in the package stream	$200–$500
CO_2 volume meter	To check CO_2 volumes prior to packaging	$200–$500
Hydrometers	Multiple hydrometers to measure gravity	$300
pH meter	To measure the pH of prepared solutions and samples	$200–$1,000

[a] Courtesy S. E. Johnson and M. D. Mosher—© ASBC.

A Lab for Under $1,000

A basic lab can be set up for a lot less than $10,000, producing a bare-bones lab that will be effective for a small microbrewery. The equipment and instruments will need to be purchased as used, and not all the glassware will be covered. Many of the items will need to be repurposed from other uses; for example, glass canning jars can be used in place of beakers.

Which items should be first on the purchase list? Most important is the microscope and hemocytometer. Because the microbrewery will pitch yeast into a fermenter, the brewer needs to be assured that he or she is not under-pitching or overpitching the yeast. Luckily, the microscope and hemocytometer can be obtained from a department store or online from individual sellers. On sale, the pair can run from $150 to $200 (in 2018 dollars). The hemocytometer is rather fragile (it's made of glass), so having a second one is a good option.

Next is a test tube rack and a collection of test tubes. The 20 mL test tubes are very useful and will be necessary for yeast counting. They also work for many of the chemical analyses, so purchasing them with the microscope is part of planning for the future expansion of the lab.

The list should also include a couple of glass pipettes and a bulb. These are designed to deliver accurate quantities of liquid, if used correctly. They will definitely be used in the yeast-counting experiments and also will be necessary for other experiments using other equipment.

Hydrometers should be next on the list of equipment to purchase. A set of three that span the high-gravity range and the low-gravity range will provide the best accuracy and precision when determining the density of the wort, fermenting wort, or finished beer. Hydrometers can be found on sale occasionally. A recent search by the authors found a set of hydrometers for $50, and a storage rack was included. The hydrometers could be supplemented with a hand-held refractometer, which runs about $25 at the inexpensive range of the scale.

A balance is also quite useful in a startup lab. A really nice one that goes to tenths of milligrams might not be affordable if the brewery is on a tight budget, but a balance that reads to milligrams can be found for an affordable price. A digital electronic balance priced at about $300 will do the job quite well.

The next most useful piece of equipment is the ultraviolet-visible (UV-vis) spectrometer, which is covered in detail in Chapter 11 along with other spectrophotometers. Unfortunately, a new one likely can't be found for the $400 remaining in the $1,000 lab budget. The low end of the cost range for a single-wavelength UV-vis spectrometer is about $2,500 new. A colorimeter may be substituted for the spectrophotometer. However, a colorimeter measures only at preselected specific wavelengths and cannot select a different wavelength. That means that if you are trying to measure the absorbance of a sample at 530 nm and the closest wavelength your colorimeter can select is 560 nm, the measurement will not be as responsive (if responsive at all), and accuracy and precision will suffer significantly. In addition, the colorimeter typically is only a visible colorimeter; it often isn't capable of reading wavelengths in the ultraviolet region. That doesn't seem like a serious issue until you try to measure IBUs using the instrument. In sum, although a new colorimeter can be purchased for under $200, the much better option is the spectrophotometer.

Used spectrophotometers are available on the secondary market. Some are refurbished back to full operating state. Analog UV-vis spectrophotometers can be obtained online. While they can be hard to find for under $400, the authors did find some online for sale; in fact, a few were listed in the $250–$350 range. It should be noted that assuming the spectrophotometer works, it will likely need to have replacement lamps for the UV range and for the visible range. These can be expensive, as they often have nonstandard plug-ins, which means you have to buy the one that fits the instrument. But if these can also be found, the $300 UV-vis spectrophotometer from 1980 will be a welcome, useful, accurate, and precise addition to the startup lab, even if it is a little out of date.

If costs have been closely monitored and savings made where possible, then about $100–$150 should be left for glassware. That amount can purchase a set of Erlenmeyer flasks; a set of volumetric flasks (10 mL, 25 mL, 100 mL, 500 mL, and 1 L); and a few graduated cylinders (10 mL, 25 mL, and 100 mL). Funds may even be left for another test tube rack and a larger box of 20 mL test tubes. As we noted before, it's possible to find glassware at auctions and from laboratories that are closing or upgrading, including college, university, and even some industrial labs.

Then, as additional funds become available for the laboratory, other pieces of equipment and instrumentation can be added. The next big purchase will likely be an instrument to analyze diacetyl and ethanol concentrations. Low-cost used GCs and alcolyzers can be found; however, their maintenance and running costs are significantly higher than those of a UV-vis spectrophotometer.

Overall, a brewery laboratory isn't an expensive addition to a brewery with an existing space that can be used for a lab. Although it should have been included in the initial brewery plan, it can be added after the brewery is running. Based on the costs to get the brewery running, even a small microbrewery can set up a brewery laboratory for very little investment.

CHAPTER 4

The Chemistry Behind the Analysis

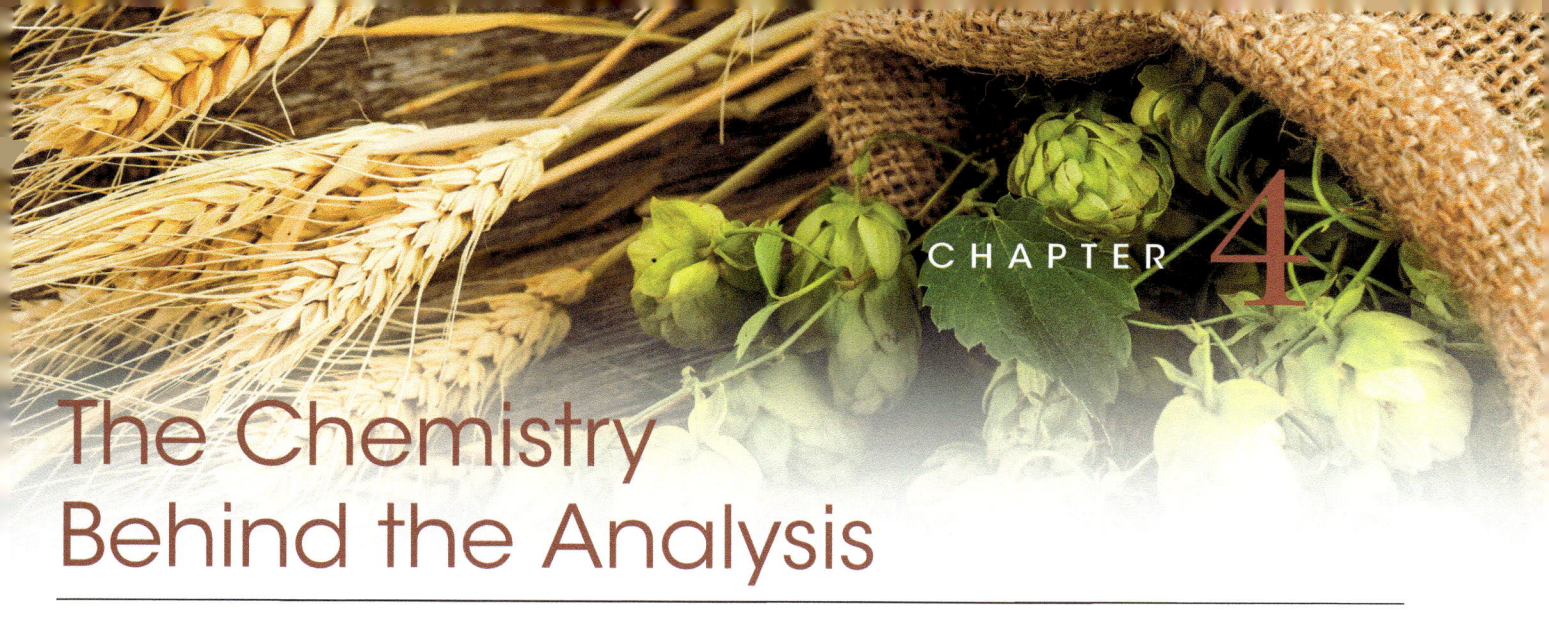

Brewing is an art. Designing a recipe and crafting a quality beer can take years to master. Brewing is also a science. Brewers and all brewery workers should understand that science to improve their ability to produce high-quality beer. That includes knowing some engineering principles about how liquids move, filter, and are pumped. It also includes understanding some of the basic chemistry behind the brewing process.

This chapter focuses on the chemistry behind many of the laboratory analyses used in the brewery lab. Other sciences are useful in the brewery laboratory, as well. We will start with a brief primer on basic or general chemistry in this chapter. Chapter 5 will cover how this information can be applied to living systems as it covers barley, hops, and yeast.

Essentials of Chemistry

Chemistry is often thought of as the application of physics to the matter that makes up our world. This chapter explores and tries to define and describe chemistry: how particles, ions, and molecules (i.e., matter) interact, the changes they undergo, and the energy associated with those changes. Chemistry is a very broad field, and aspects of physics and mathematics are embedded within the very core of the science. Chemistry is also one of the subjects that many high school and college students try to avoid. Unfortunately, because the science of chemistry pervades everything within our daily lives, students who avoid taking chemistry lack some basic understanding of the world around them.

For example, without chemistry, there would be no plastic. We use plastic in almost all aspects of our everyday lives: The clothes we wear are made of fabrics blended with plastic, such as a 50:50 cotton–polyester blend. The soda we drink is stored in containers made from polyethylene terephthalate. The rope we use, the cars we drive, and even the seats we sit on all have some plastic in them. Yes, it's true that we don't need to understand the basics of chemistry to use these items, but without that understanding, we

chemistry The science that studies matter, its changes, and the energy associated with those changes.

Atoms and Radioactivity

All things in this world are made of **matter**. The air we breathe, the water we drink, and the cell phone we can't seem to put down are all made of matter. Matter is everywhere. Matter is everything, except for light (energy), heat (energy), and the deep vacuum of space. There are many different types of matter, ranging from atoms to ions to molecules to compounds.

An **atom** is the smallest indivisible unit of matter. Atoms are very small and definitely not visible to the human eye, but they combine to make up everything around us. Chemists have identified 118 different types of atoms that exist in nature or can be made in the lab. These types of atoms are listed in a table known as the **periodic table** (Fig. 4.1). The periodic table is an organized presentation of all the known **elements**—collections of the same types of atoms. The organization of the periodic table is the main reason

matter Anything that occupies space and has a mass.

atom The smallest indivisible unit of matter.

periodic table An organized table of all the known elements in the universe, whether natural or produced by humans.

element The smallest collection of atoms of the same type.

The Periodic Table of the Elements

1/IA																	18/VIIIA
1 H hydrogen 1.01	2/IIA											13/IIIA	14/IVA	15/VA	16/VIA	17/VIIA	2 He helium 4.00
3 Li lithium 6.94	4 Be beryllium 9.01											5 B boron 10.81	6 C carbon 12.01	7 N nitrogen 14.01	8 O oxygen 16.00	9 F fluorine 19.00	10 Ne neon 20.18
11 Na sodium 22.99	12 Mg magnesium 24.31	3/IIIB	4/IVB	5/VB	6/VIB	7/VIIB	8/VIIIB	9/VIIIB	10/VIIIB	11/IB	12/IIB	13 Al aluminum 26.98	14 Si silicon 28.09	15 P phosphorus 30.97	16 S sulfur 32.06	17 Cl chlorine 35.45	18 Ar argon 39.95
19 K potassium 39.10	20 Ca calcium 40.08	21 Sc scandium 44.96	22 Ti titanium 47.88	23 V vanadium 50.94	24 Cr chromium 51.99	25 Mn manganese 54.94	26 Fe iron 55.93	27 Co cobalt 58.93	28 Ni nickel 58.69	29 Cu copper 63.55	30 Zn zinc 65.39	31 Ga gallium 69.73	32 Ge germanium 72.61	33 As arsenic 74.92	34 Se selenium 78.09	35 Br bromine 79.90	36 Kr krypton 84.80
37 Rb rubidium 84.49	38 Sr strontium 87.62	39 Y yttrium 88.91	40 Zr zirconium 91.22	41 Nb niobium 92.91	42 Mo molybdenum 95.94	43 Tc technetium 98.91	44 Ru ruthenium 101.07	45 Rh rhodium 102.91	46 Pd palladium 106.42	47 Ag silver 107.87	48 Cd cadmium 112.41	49 In indium 114.82	50 Sn tin 118.71	51 Sb antimony 121.76	52 Te tellurium 127.6	53 I iodine 126.90	54 Xe xenon 131.29
55 Cs cesium 132.91	56 Ba barium 137.29	57-71 lanthanides	72 Hf hafnium 178.49	73 Ta tantalum 180.95	74 W tungsten 183.85	75 Re rhenium 186.21	76 Os osmium 190.23	77 Ir iridium 192.22	78 Pt platinum 195.08	79 Au gold 196.97	80 Hg mercury 200.59	81 Tl thallium 204.38	82 Pb lead 207.20	83 Bi bismuth 208.98	84 Po polonium [208.98]	85 At astatine 209.98	86 Rn radon 222.02
87 Fr francium 223.02	88 Ra radium 226.03	89-103 actinides	104 Rf rutherfordium [261]	105 Db dubnium [262]	106 Sg seaborgium [266]	107 Bh bohrium [264]	108 Hs hassium [269]	109 Mt meitnerium [278]	110 Ds darmstadtium [281]	111 Rg roentgenium [282]	112 Cn copernicium [285]	113 Nh nihonium [286]	114 Fl flerovium [289]	115 Mc moscovium [289]	116 Lv livermorium [293]	117 Ts tennessine [294]	118 Og oganesson [294]

lanthanides	57 La lanthanum 138.91	58 Ce cerium 140.12	59 Pr praseodymium 140.91	60 Nd neodymium 144.24	61 Pm promethium [145]	62 Sm samarium 150.36	63 Eu europium 151.96	64 Gd gadolinium 157.25	65 Tb terbium 158.93	66 Dy dysprosium 162.50	67 Ho holmium 164.93	68 Er erbium 167.26	69 Tm thulium 168.93	70 Yb ytterbium 173.05	71 Lu lutetium 174.97
actinides	89 Ac actinium [227]	90 Th thorium 232.04	91 Pa protactinium 231.04	92 U uranium 238.03	93 Np neptunium [237]	94 Pu plutonium [244]	95 Am americium [243]	96 Cm curium [247]	97 Bk berkelium [247]	98 Cf californium [251]	99 Es einsteinium [252]	100 Fm fermium [257]	101 Md mendelevium [258]	102 No nobelium [259]	103 Lr lawrencium [266]

FIG. 4.1. The periodic table lists all the known elements. Each box contains a one- or two-letter element symbol and name. The number in the upper-left corner of the box is the atomic number. The number under the element name is the average atomic mass; the number in parentheses is the mass of the most stable isotope. At the top of each column in the table is a number used to group the elements within that column into a family of similar elements. Two numbering systems are often used: The first is the newest and indicates the columns are numbered from **1** to **18**. The second uses roman numerals and letters. (Courtesy S. E. Johnson and M. D. Mosher—© ASBC)

that this single table is so important in science. Each column in the periodic table is composed of elements that have similar properties. For example, sodium and lithium are in the same column, and both react violently when added to water.

In their natural or most stable states, most elements are found as individual atoms. Some, however, occur as discrete pairs of atoms or even larger collections. For example, oxygen is an element made up of 2 oxygen atoms that are physically attached. In nature, oxygen exists as a pair of atoms. Iron, on the other hand, is made up of single atoms.

To see another type of organization within the periodic table, we have to examine the atom up close. At the center of each atom is a **nucleus**. The nucleus contains two kinds of subatomic particles: neutrons and protons. **Neutrons** are particles that lack an electric charge (hence their name). They can exist within a nucleus in almost any whole number. If two atoms differ only in the number of neutrons in their nuclei, they are **isotopes**. Isotopes have similar properties but are different, as we'll see shortly.

The other type of subatomic particle found in the nucleus is a **proton**. A proton has a positive electric charge and a mass that's approximately the same as that of a neutron. Together, the proton and neutron make up the bulk of the mass of the atom. The number of neutrons indicates the specific isotope of the atom, and the number of protons dictates the specific type of atom. For example, all hydrogen atoms have only 1 proton. If an atom had 2 protons, it would be helium; if an atom had 3 protons, it would be lithium; and so on. Take another look at Figure 4.1. Note the number in the small box in the upper-left corner of each element box. That's the **atomic number**: the number of protons in the nuclei of the atoms that make up that specific element.

Just remember that it's possible for hydrogen atoms to have different numbers of neutrons. In other words, not all hydrogen atoms have the same overall mass. The most abundant or common hydrogen atom has no neutron in its nucleus. It can be denoted as "hydrogen-1"; the "1" indicates that the sum of protons and neutrons in the atom is 1. This number is the **mass number** of the element. A less common type of hydrogen atom, known technically as "hydrogen-2" and commonly as "deuterium," has 1 proton and 1 neutron in its nucleus. A very uncommon type of hydrogen, known as "tritium" or "hydrogen-3," has 1 proton and 2 neutrons in its nucleus.

Let's consider another example. All types of the element carbon are made up of atoms that have 6 protons. (The atomic number of carbon is 6.) The three most common isotopes of carbon differ in the number of neutrons in their nuclei and their mass number changes; carbon-12 has 6 neutrons, carbon-13 has 7, and carbon-14 has 8.

A scoop of carbon contains all the different isotopes. The abundance of each isotope of carbon is shown in Table 4.1. The abundance is the typical ratio of isotopes in any given sample of carbon. Because each isotope has a different mass number, the average mass of the atoms in a sample of carbon is the weighted average of the masses of the isotopes. In other words, 98.9% of all carbon atoms are carbon-12, 1.1% are carbon-13, and only a trace are carbon-14. The result of a weighted average gives a sample of carbon with an average mass of 12.01 {$(0.989 \times 12) + (0.011 \times 13) + (0.00 \times 14) = 12.01$}. Calculating each average mass is relatively straightforward using this process and assumes that you know the isotope mass. Unfortunately, because

nucleus The center of an atom; made up of neutrons and protons.

neutron A neutrally charged particle found in the nucleus of an atom.

isotopes Atoms that differ only in the number of neutrons in their nuclei.

proton A positively charged particle found in the nucleus of an atom; determines the name of the element.

atomic number The number of protons in an atom.

mass number The number of protons and neutrons in the nucleus of an atom.

of reasons beyond the scope of this text, isotope masses differ slightly from isotope mass numbers as atoms get heavier. This doesn't pose an issue to the brewer who has a periodic table containing the average atomic masses.

The units for the average mass of an atom are in atomic mass units (amu). Mathematically, the average mass turns out to be equivalent to the molar mass, which is the number of grams (g) of that atom per mole (mol) of atoms. Scientifically, we say that carbon has an atomic mass of 12.01 amu and a molar mass of 12.01 g/mol.

So, what's a **mole**? In science, a mole is a counting unit similar to a dozen, a gross, or a pair (not a small, furry animal!). Whereas a dozen eggs is 12 eggs, a gross of bottle rockets is 144 bottle rockets, and a pair of pears is 2 pears, a mole of marbles is equivalent to 6.02×10^{23} marbles. That's a HUGE number! In fact, if you were born with a mole of dollars ($\$6.02 \times 10^{23}$) and spent it at a rate of $1 billion per second of your life, by the time you were 75, you wouldn't have spent even 1% of your money. Yes, that's a large number, but it needs to be that large to count enough atoms in a sample you can see. It takes exactly 6.02×10^{23} atoms of carbon-12 to produce a pile with a mass of 12.00 g. A mole of carbon (assuming the standard abundance ratio) has a mass of 12.01 g. Put another way, atoms are so tiny that you need a terribly large number to even see them.

It's also important to remember that grams are not the only unit of measurement for mass. Kilograms, pounds, ounces, and many more units of mass are commonly used in the brewing industry, particularly in the United States. Worldwide, the SI system is common. It's based on powers of 10, which means units are always related by multiplying or dividing them by 10 a certain number of times. For example, 1 kilogram (kg) equals 1,000 g, or $1 \times 10 \times 10 \times 10$ g. We can report this by saying that 1 kg has a mass that's 1,000 times as much as 1 g.

For an extremely small or large value, writing out the entire number can be tedious. Using a method of writing numbers known as **scientific notation** saves time by reporting the value of a measurement multiplied by 10 to a power. That power indicates whether we are talking about kilograms, milligrams, or micrograms (Table 4.2). For example, 1 kg can be written as 1×10^3 g. As another example, the off-flavor of lightstruck beer can be detected in quantities as small as 5 nanograms per liter (equivalent to 0.000000005 g). Using scientific notation, we write that value as 5×10^{-9} g.

In summary, the periodic table lists the average atomic mass for each element, so we don't have to know the abundance of each isotope and do the calculation ourselves. And the number for the average atomic mass can be used directly as the number of grams per mole. In other words, a mole

mole A number equivalent to 6.02×10^{23}.

scientific notation A method of representing numbers using a factor of 10.

TABLE 4.1. Isotopes of carbon[a]

Isotope number	Number of protons	Number of neutrons	Isotope mass number	Abundance (%)
Carbon-12	6	6	12	98.9
Carbon-13	6	7	13	1.1
Carbon-14	6	8	14	Trace

[a]Courtesy S. E. Johnson and M. D. Mosher—© ASBC.

(6.02×10^{23} atoms) of gold has a mass of 196.97 g. (To verify this, check the periodic table.) We should be cautioned to remember that while most elements are a single atom, some elements are a pair of atoms. A mole of hydrogen atoms has a mass of 1.008 g, but a mole of hydrogen gas has a mass of 2.016 g, because the element hydrogen occurs as 2 hydrogen atoms. Oxygen, nitrogen, and all the elements in group 17 (group VIIA) occur as pairs of atoms.

Let's go back to the structure of an atom, because there's still more to see. Encircling the nucleus is a cloud of **electrons**: minute subatomic particles that contribute very little to the mass of the atom. Electrons move around the nucleus of the atom, like satellites, at about 2,200 kilometers per second (4.9 million miles per hour). That's very fast, but it's still less than 1% of the speed of light. Each electron has a negative electric charge that's essentially the same quantity of charge possessed by the proton. Because an atom has no net charge, the number of electrons in an atom must equal the number of protons, so the quantity of positive electric charge equals the quantity of negative electric charge. Table 4.3 lists some common atoms and the numbers of subatomic particles they contain.

electrons Negatively charged particles that surround the nucleus of an atom.

TABLE 4.2. Names and conversion factors for common SI units[a]

Name[b]	Conversion (from grams)	Conversion
Mega-, M	# g × 10^6	1 Mg = 1,000,000 g
Kilo-, k	# g × 10^3	1 kg = 1,000 g
Milli-, m	# g × 10^{-3}	1 mg = 0.001 g
Micro-, µ	# g × 10^{-6}	1 µg = 0.000001 g
Nano-, n	# g × 10^{-9}	1 ng = 0.000000001 g

[a]Courtesy S. E. Johnson and M. D. Mosher—© ASBC.
[b]The prefixes in this column apply to grams, meters, liters, and other SI units but are written here in grams.

TABLE 4.3. Common atoms and numbers of subatomic particles[a]

Atomic number	Name	Number of protons	Number of electrons	Mass	Number of neutrons in most abundant isotope
1	Hydrogen	1	1	1	0
2	Helium	2	2	4	2
6	Carbon	6	6	12	6
7	Nitrogen	7	7	14	7
8	Oxygen	8	8	16	8
12	Magnesium	12	12	24	12
20	Calcium	20	20	40	20
26	Iron	26	26	56	30
30	Zinc	30	30	64	34

[a]Courtesy S. E. Johnson and M. D. Mosher—© ASBC.

radioactive An element that's relatively unstable and undergoes reactions within its nucleus.

half-life The time required for a reaction to lose 50% of its current reactants.

From Table 4.3, we can see that the atomic number (i.e., the number of protons) and the number of electrons is the same. For all the neutral atoms, this is always true. But the number of neutrons in the most abundant isotope deviates as atoms get larger. For example, the most abundant isotope of oxygen contains the same number of neutrons as protons and electrons, but the most abundant isotope of iron has more neutrons than protons. And when there is an imbalance in the number of neutrons and protons in the nucleus, there is a chance that the element will be radioactive.

Radioactive elements spontaneously undergo reactions to become stable. For instance, uranium-238 is a radioactive isotope of uranium. The atom has 92 protons and electrons but 146 neutrons. When it decays, it forms thorium-234, which further decays to other elements that are radioactive. Finally, the series of radioactive decays arrives at lead-206, which is a stable element that isn't radioactive.

Each step in the decay process takes time. In some cases, the **half-life** of the process is minutes, and in others, it's thousands or even millions of years. The half-life is the amount of time it takes for half of the sample to undergo the reaction. For example, radon-222, one of the decay products of uranium, has a half-life of 3.82 days; every 3.82 days, the amount of radon-222 drops by half. So if on day 1 there are 120 atoms of radon-222 in your basement, 3.82 days later, there will be 60 atoms left. After another 3.82 days, there will be only 30 atoms, and after another 3.82 days, only 15 atoms. (Here, we are using the term "half-life" to refer to the radioactive decay rate, but it can also be used to describe the speed of any reaction.)

Drawing Atoms

We have already discovered one way to write the name of a specific atom. We can write carbon-12, hydrogen-2, or oxygen-16 to illustrate the specific isotope of the atom we wish to describe. However, a simpler way involves writing the atomic symbol for the element. The atomic symbols are also provided on the periodic table (see Fig. 4.1). They are one- and two-letter abbreviations of the original element names, essentially. Always capitalize the first letter of an atomic symbol when you write it.

We can add numbers at the four corners of the atomic symbol, as shown in Figure 4.2, to improve or modify the information about the atom. These numbers have specific meanings and allow us to place key information on

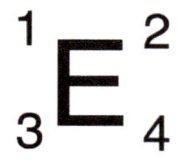

$${}^{1}_{3}E^{2}_{4}$$

FIG. 4.2. The structure of the element symbol. The four numbers indicate the positions of the modifiers for the element. Position 1 is the mass number, position 2 is the charge on the atom, position 3 is the atomic number, and position 4 is the number of that particular atom in a formula. (Courtesy S. E. Johnson and M. D. Mosher—© ASBC)

$${}^{1}_{1}H \quad {}^{14}_{7}N \quad {}^{40}_{20}Ca \quad {}^{238}_{92}U \quad {}^{12}C$$

FIG. 4.3. Examples of atomic symbols. Because the atomic number and the element symbol mean the same thing, the atomic number can be omitted from the symbol, as shown in the symbol for carbon (far right). (Courtesy S. E. Johnson and M. D. Mosher—© ASBC)

the symbol quickly (Fig. 4.3). The bottom-left corner is reserved for the atomic number. As shown in the examples in Figure 4.3 (starting from the left), 1 is the atomic number of hydrogen, 7 is the atomic number of nitrogen, and 20 is the atomic number of calcium. The atomic number is often omitted, because it has the same meaning as the atomic symbol. The top-left corner contains the mass number (i.e., the isotope mass produced by adding the number of protons and the number of neutrons). By subtracting the atomic number from the mass number, we can quickly determine the number of neutrons in the isotope we are talking about. If the mass number is omitted, then we can assume that all the isotopes of that element are being considered, not just a specific isotope. The number in the top-right corner indicates the overall charge on the isotope. For atoms, the charge is 0, which is often omitted. When the number isn't 0, we must include it (as we will see in the next section). Finally, the lower-right corner indicates the number of atoms physically attached to each other. If we are writing the element oxygen, we will use a "2" to indicate that there are 2 oxygen atoms in the element. Omitting this number implies that only 1 atom is present.

Ions: Atoms with Charges

Changing the number of neutrons in an atom changes the mass of the atom and results in a different isotope of that element. Changing the number of protons in an atom changes the identity of the atom. For instance, changing the number of protons in a carbon atom from 6 to 7 results in an atom of nitrogen. Changing the number of electrons in an atom doesn't change the identity of the atom or the mass of the atom. Instead, it changes the charge on the atom. If an electron is added to the atom, there will be more negative charges than positive charges, and the atom will become an **ion** with a charge of –1. Remember that a negative charge comes from a count of all the electrons, and a positive charge comes from a count of all the protons.

A negatively charged atom is called an **anion**, and its name changes slightly by adding "-ide" at the end. Adding an electron to a "chlorine" atom results in a "chloride" anion. Similarly, removing an electron from an atom results in a **cation**: an ion with a charge of +1. Removing an electron from a "sodium" atom results in a "sodium" cation. While an anion typically has a name change, a cation does not. Removing an electron from a calcium atom results in a calcium +1 cation. Removing another electron from the calcium +1 cation results in a calcium +2 cation. Figure 4.4 shows how symbols are written to represent ions. If a specific isotope is being considered, the atomic number and mass number can be included. However, we are usually interested only in the type of element, rather than a specific isotope, so we omit the atomic and mass numbers from our symbol.

ion An atom or collection of atoms with a net charge caused by the addition or elimination of an electron.

anion A negatively charged atom or collection of atoms.

cation A positively charged atom or collection of atoms.

$$^{40}_{20}Ca^{+2} \quad Ca^{+2} \quad O^{-2} \quad Cl^{-}$$

FIG. 4.4. Ion symbols. (Courtesy S. E. Johnson and M. D. Mosher—© ASBC)

$$Ca \longrightarrow Ca^{+2} + 2e^-$$

$$Cl + e^- \longrightarrow Cl^-$$

FIG. 4.5. Forming anions and cations from atoms. Atoms that lose electrons become cations; atoms that gain electrons become anions. (Courtesy S. E. Johnson and M. D. Mosher—© ASBC)

How many electrons can be removed or added to an atom? The answer is based on the location of the atom in the periodic table. Atoms on the left side of the stair-step line in the periodic table tend to lose electrons to form cations. In short, an atom can lose a maximum number of electrons equal to the number of the column in which it's located in the periodic table. Similarly, anions form by adding the number of electrons equal to the number of columns from the right side of the periodic table. Figure 4.5 illustrates this process for some simple anions and cations.

How far the column is from the edge of the periodic table indicates the maximum number of electrons that can be added or removed. However, the stability of the ion that's formed is what determines the most likely number of electrons that are removed or added. The atoms at the edges of the periodic table are very predictable. All the elements in column 1 prefer to have a +1 charge (loss of 1 electron), and all the elements in column 2 prefer to have a +2 charge (loss of 2 electrons). The elements in the last column (group 18, group VIIIA) don't lose or gain electrons; oxygen and sulfur prefer a −2 charge (gain of 2 electrons); and all the elements in group 17 (group VIIA) prefer a −1 charge (gain of 1 electron). Some elements are much less predictable. The elements in groups 3 through 12 (groups IB through VIIB) can lose different numbers of electrons. For example, iron forms two different ions; it prefers a +3 charge in oxidizing environments but a +2 charge in reducing environments.

Combining Atoms

When anions and cations exist in the same solution, they associate with each other. In other words, the positive charge of the cation and the negative charge of the anion attract one another. The cation and anion come together to produce a neutrally charged species known as an **ionic compound**. In scientific slang, ionic compounds are also known as **salts**. Because the anion and cation associate only with each other, they can be separated by dissolving the ionic compound in a solvent. No chemical reaction occurs; all that happens is that the solvent separates the pair of ions. For example, table salt is an ionic compound. When added to water, the ionic compound first dissolves in the water and then separates into the ions that make up table salt.

If the cation in an ionic compound has a +2 charge, then it will attract two −1 charged anions or one −2 charged anion. The result will be a neutral

ionic compound The combination of an anion and cation, such that the overall charge is 0.

salts A common synonym for "ionic compounds."

$$Ca^{+2} + 2Cl^{-} \longrightarrow CaCl_2$$

$$Ca^{+2} + O^{-2} \longrightarrow CaO$$

FIG. 4.6. Combining cations and anions. When these compounds dissociate after dissolving, the reverse of this process occurs and the ions are released into solution. (Courtesy S. E. Johnson and M. D. Mosher— © ASBC)

species. In other words, when sodium (a +1 cation) combines with the –1 charged anion of chloride, one sodium and one chloride come together. When dissolved in water, one sodium chloride (table salt) compound dissociates (i.e., separates) into one sodium +1 cation and one chlorine –1 anion. If the cation is calcium with a +2 charge, it will combine with two –1 chloride anions or one –2 oxide anion. Similarly, scandium has a +3 charge and will combine with three –1 charged chlorides (Fig. 4.6). Note how the symbols are combined to create the formula for the ionic compound. The formula lists the number of each ion in the compound using subscripts. If there isn't a subscript, we assume that the subscript is a 1.

One small issue with ionic compounds is that not all of them dissolve in water. And if the ionic compound isn't soluble in water, then it won't dissociate into the individual cations and anions that make up its formula. Knowing which compounds do and don't dissolve requires knowing the formula. If the formula contains any element from group 1, group 2, or group 17, the ionic compound is soluble in water. Other rules can be applied to determine which ionic compounds are soluble, but it's often easier to look up the compound's solubility using the internet or another source.

Writing Chemical Formulas

A formula is often used to identify numbers and types of atoms in a specific compound. A **formula** is simply the element symbol (or symbols) followed by the number (or numbers) of atoms. Based on our previous discussion of an atom using the atomic symbol, the bottom-right corner is where to add the number of atoms. If there is no number after the atomic symbol, then it's assumed that there is only 1 of that atom. For example, the formula NaCl indicates that there is 1 sodium atom and 1 chlorine atom in the compound. The formula $CaCl_2$ indicates that there is 1 calcium and 2 chlorine atoms (Table 4.4).

In addition to including the number of each ion so that the total charge is 0, the convention requires writing the cations before the anions. Consider the formula for table salt, NaCl. The formula means that sodium is the cation and chloride is the anion. It would never be written as ClNa. A formula containing oxygen and magnesium is best written as MgO, not OMg. (Remember that magnesium is a +2 cation and oxygen is a –2 anion.) If none of the atoms are cations or anions, the convention is to write the elements from the left

formula A list of the elements and the number of each found in a compound.

side of the periodic table before those from the right side (as we will see later in this discussion). Sulfur dioxide, SO_2, is an example of this type of formula.

Ionic compounds occur when a metallic element is combined with a nonmetallic element. Just as cations and anions must combine (i.e., opposites attract), a metal and a nonmetal (opposites) must combine to make an ionic compound. Metals are elements that are hard, malleable, and ductile and that conduct heat and energy. Metals prefer to lose electrons and form cations. Nonmetals, in contrast, are soft and brittle and act as insulators. Nonmetals prefer to gain electrons and form anions.

Instead of trying to remember which elements are metals and which are nonmetals, look at the periodic table (see Fig. 4.1). The bold, stair-step line in the table separates the metals on the left from the nonmetals on the right. Just remember that an ionic compound is made from a metal on the left side of the periodic table and a nonmetal on the right side of the periodic table.

Collections of Atoms

In nature, we can find copper metal. The atoms that make up this metal are all neutral—as they are for all metals. A bar of gold is a collection of gold atoms. An iron nail is made up of iron atoms. But this isn't how most compounds exist. Instead, metals can combine with nonmetals to make ionic compounds by forming cations and anions. It's also possible for nonmetals and nonmetals to combine in a completely different fashion. In fact, this combination is the most common type of chemical.

The combination of nonmetals with nonmetals results in a tight association of the atoms. The atoms don't form cations or anions when they associate in this way. Instead, the electrons on the atoms delocalize and encircle all the atoms in the collection. The result is a **molecule**. The number of atoms depends on many different factors, some of which we will cover briefly later in this chapter. For an ionic compound, where there is a specific charge on the cation and on the anion, only one possible compound can result. This isn't true for molecules. For example, carbon and oxygen are both nonmetals. They can combine in different ways and in different numbers to make a series of different molecules (Table 4.5). The number of atoms that combine and how the atoms are attached within the molecule determines the name of the compound. Individual compounds have different properties and different reactivities, even though they are made from the same atoms.

molecule A combination of atoms that doesn't have a net charge.

TABLE 4.4. Some common ionic compounds[a]

Chemical name	Chemical formula	Common name
Sodium chloride	NaCl	Salt
Calcium oxide	CaO	Quicklime
Magnesium nitride	Mg_3N_2	Magnesium nitride
Potassium iodide	KI	Potassium iodide
Calcium chloride	$CaCl_2$	Calcium chloride

[a]Courtesy S. E. Johnson and M. D. Mosher—© ASBC.

Molecules can be soluble in water. Not all of them are, but the rules for their solubility are much less obvious than those for ionic compounds. However, even if a molecule isn't soluble, the atoms within the molecule won't dissociate into ions. In fact, the atoms in a molecule aren't easily separated. Instead, a chemical reaction is required. Some chemical reactions require very little energy to separate the atoms within the molecule; others require significant amounts of energy. The atoms of carbon dioxide don't separate when CO_2 dissolves in water. As another example, water is made up of the combination of 2 hydrogen atoms and 1 oxygen atom. We write its formula as H_2O. We can boil water and produce steam, but it's still water with the 3 atoms attached to each other. Without a chemical reaction, the hydrogen and oxygen atoms remain attached.

It's also possible to have more than two different types of atoms form a compound, as can occur in ionic compounds—for instance, in the formula CaClF. A similar reaction also occurs in molecules, as shown in Table 4.6. For a molecule that contains a very small number of atoms, the formula often determines the identity of the molecule. For example, H_2O and CO_2 can each be arranged in only one way, and their formulas can be used to identify water and carbon dioxide. Unfortunately, because nonmetals can combine with other nonmetals in so many ways, when the number of atoms in the

TABLE 4.5. Molecules produced by combining carbon and oxygen[a]

Chemical name	Formula	Structure[b]
Carbon monoxide	CO	C≡O
Carbon dioxide	CO_2	O=C=O

[a] Courtesy S. E. Johnson and M. D. Mosher—© ASBC.
[b] Note the different numbers of attachments between the atoms in the two molecules.

TABLE 4.6. Molecules produced by combining carbon, hydrogen, and oxygen[a]

Chemical name	Formula	Structure
Ethanol	C_2H_6O	H-C(H)(H)-C(H)(H)-O-H
Dimethyl ether	C_2H_6O	H-C(H)(H)-O-C(H)(H)-H
Glucose	$C_6H_{12}O_6$	HO-C(H)(H)-C(H)(OH)-C(H)(OH)-C(OH)(H)-C(H)(OH)-C(=O)-H
Fructose	$C_6H_{12}O_6$	HO-C(H)(H)-C(H)(OH)-C(H)(OH)-C(OH)(H)-C(=O)-C(OH)(H)-H

[a] Courtesy S. E. Johnson and M. D. Mosher—© ASBC.

isomers Compounds that have the same number of atoms but differ in how the atoms are attached.

formula increases, we can't identify what the molecule is called. For example, the relatively simple formula C_2H_6O refers to a molecule with three different types of atoms. Unfortunately, two different molecules, ethanol and dimethyl ether, have the same formula, as shown in Table 4.6. These are known as **isomers**: compounds that are different in everything but the formula. As the number of atoms increases, the structure become more complicated and the number of different molecules with the same formula increases.

How the atoms come together to make different compounds is beyond the scope of this text. However, some basics can be gleaned from the structures shown in Table 4.6. Specifically, carbon always prefers four attachments within the molecule, oxygen requires two attachments, and hydrogen requires only one. The periodic table also implies that nitrogen requires three attachments and that the atoms in group 17 (group VIIA) prefer only one attachment.

We noted earlier that atoms are relatively difficult to remove from a molecule. However, there is a class of reactions for which removal is relatively straightforward. When added to water, some molecules lose a hydrogen cation (i.e., a proton) and others gain a hydrogen cation. This reaction doesn't happen with all molecules, but it does occur in certain cases, as described in greater detail later. The result is that the molecule becomes an anion or a cation. For example, ammonia is a molecule with the formula NH_3. When added to water, it exchanges a proton with water (Fig. 4.7). The result is the formation of a cation known as "ammonium" (NH_4^+) and an anion known as "hydroxide" (OH^-).

polyatomic ion An ion made up of multiple atoms.

counterion An ion with the opposite charge; an anion is the counterion to a cation.

These **polyatomic ions** are capable of forming ionic compounds, just like metals and nonmetals. The charge is associated with the entire cation or anion, not just one atom of the set. Thus, when added to water, a polyatomic ion will separate from the **counterion**, but it will not separate into the individual atoms. Note that the number of atoms attached in a polyatomic ion doesn't match with the rules for atoms from Table 4.6. In Figure 4.7, nitrogen has four attachments instead of three, and so it has a + charge. Oxygen has only one attachment to the hydroxide ion instead of two, and so it has a – charge.

Table 4.7 illustrates some common polyatomic ions found in ionic compounds. Polyatomic ions are relatively stable and very common in chemical structures. In fact, a brewer is more likely to run into an ionic compound that's made from a polyatomic ion than one that's not. Unfortunately, the only way to know the names of polyatomic ions is to memorize them and their charges.

FIG. 4.7. Polyatomic ions: ammonium and hydroxide. (Courtesy S. E. Johnson and M. D. Mosher—© ASBC)

Let's see how the names come together for ionic compounds containing polyatomic ions. For example, beer stone can be a problem during brewing. The chemical name for beer stone is "calcium oxalate." Based on the chemical name, the formula is CaC_2O_4, because calcium prefers a +2 charge and the oxalate anion prefers a −2 charge. Epsom salt (magnesium sulfate) has the formula $MgSO_4$, and baking soda (sodium carbonate) has the formula Na_2CO_3. Calcium carbonate has the formula $CaCO_3$.

When there's more than 1 polyatomic ion in the formula, we use parentheses to separate it from the rest of the formula. The result is easier to read. For example, ammonium carbonate has the formula $(NH_4)_2CO_3$. Without the parentheses, the formula is $NH_{42}CO_3$, which doesn't look right. Basically, the name of an ionic compound is formed by saying the name of the first atom or ion in the formula (the cation) and then saying the name of the anion, regardless of whether it's a polyatomic ion. For example, $Ca(OH)_2$ is calcium hydroxide, K_3PO_4 is potassium phosphate, and $NaHCO_3$ is sodium bicarbonate. Table 4.8 identifies some important ions in water and the functions they can serve.

TABLE 4.7. Common polyatomic ions[a]

Name	Formula (and charge)
Hydroxide	OH^-
Nitrite	NO_2^-
Acetate	CH_3COO^-
Carbonate	CO_3^{2-}
Sulfate	SO_4^{2-}
Phosphate	PO_4^{3-}
Ammonium	NH_4^+
Bicarbonate	HCO_3^-
Nitrate	NO_3^-
Chlorate	ClO_3^-
Oxalate	$C_2O_4^{2-}$
Sulfite	SO_3^{2-}

[a] Courtesy S. E. Johnson and M. D. Mosher—© ASBC.

TABLE 4.8. Common ions in the chemistry of brewing water[a]

Ion	Effect(s) on brewing
Ca^{2+}	Acidifies the mash; coagulates proteins; improves yeast health
Mg^{2+}	Acidifies the mash; helps with IBU reactions
Cl^- to SO_4^{2-} ratio	Enhances perceived malt character or bitterness
H_2CO_3 and HCO_3^-	Affects residual alkalinity

[a] Courtesy S. E. Johnson and M. D. Mosher—© ASBC.

Calculating Ions in Water

When ions or molecules dissolve in water, their amount in the water becomes an important value. We could use parts per million (ppm), parts per billion (ppb), and parts per trillion (ppt) measurements to describe the concentrations. However, these units are much too small for some concentrations. In those cases, we use scientific units known as "molarity" and "normality."

Molarity (M) is the number of moles of solute (analyte) per liter of solution. M can be easily calculated if we know the number of moles added to a flask and the volume to which the flask was filled. The number of moles is easily determined if we know the mass of a solid solute in grams or the volume of a liquid solute in milliliters (Fig. 4.8).

To perform the calculation, we first need to know some information about our solute. The most important piece of information is the formula. Working with the formula, we use the periodic table and add the atomic mass for each atom to determine the compound's **molar mass**. For example, if the solute is sodium chloride, NaCl, we use the periodic table to add the mass of 1 sodium atom (22.99 amu) and 1 chlorine atom (35.45 amu). The result is 58.44 amu, equivalent to 58.44 g/mol for the molar mass of NaCl; said another way, 1 mole of NaCl has a mass of 58.44 g.

As a side note, look up the masses of the neutral atoms in the periodic table. Because the mass of the electron contributes essentially nothing to the mass of the atom, there's no difference between the mass of an ion and the mass of an atom. So we can use the masses of the atoms in the periodic table to determine the molar mass of the formula, even when we should be using the masses of the ions that make up the formula.

When a formula gets more complex, the calculation doesn't. We just make sure to add the molar mass of each atom in the formula. When the formula contains parentheses, we make sure to multiply all the atoms within the parentheses by the number. For ammonium oxalate, $(NH_4)_2C_2O_4$, the total number of individual atoms is 2 nitrogens, 8 hydrogens, 2 carbons, and 4 oxygens. Adding the masses from the periodic table for all of these atoms results in a molar mass of 124.10 g/mol. In other words, 1 mole of ammonium oxalate has a mass of 124.10 g.

To continue making our solution and determining its molarity, we weigh the solute on a balance to obtain the mass that we will use. Even a liquid solute can be treated this way if it's transferred into a small vial to be massed. Alternatively, if we measure the volume of a liquid solute and know its density, we can calculate the mass of that volume. Then we add the solute to the flask and dilute it with the solvent to the volume that we desire. We can then calculate the molarity.

> **molarity (M)** A unit of concentration determined by the number of moles of solute (analyte) per liter of solution.

> **molar mass** The mass of 1 mole of a substance.

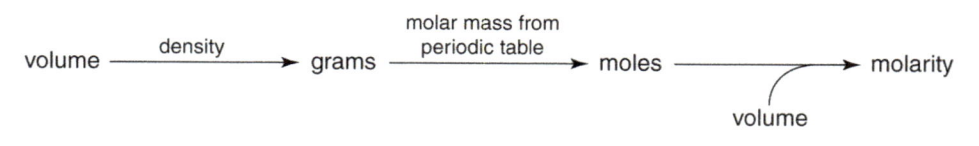

volume —(density)→ grams —(molar mass from periodic table)→ moles —(volume)→ molarity

FIG. 4.8. The method for determining molarity. (Courtesy S. E. Johnson and M. D. Mosher—© ASBC)

For example, if 1.254 g of sodium chloride (NaCl) is diluted to 100 milliliters (mL) in a volumetric flask, we can perform the calculation as illustrated in the equations below. We start by determining the number of moles of NaCl by using the molar mass as a fraction known as an **equivalence**. (The molar mass equals 58.44 g/mol, so we can write 1 mol NaCl/58.44 g NaCl.) Equivalences can be flipped (as in 58.44 g NaCl/1 mol NaCl) without changing the meaning of the fraction. Specifically, this equivalence converts a mass into the number of moles or vice versa, depending on how the fraction is written. Then we do the same thing to determine the number of liters (L) of solution. Here, however, we use an equivalence (1,000 mL/1 L) to convert the volume in milliliters to the volume in liters, because 1,000 mL is equivalent to 1 L:

equivalence A mathematical construct in which the numerator and the denominator are equivalent.

$$\text{moles of NaCl} = 1.254 \text{ g} \times \frac{1 \text{ mol}}{58.44 \text{ g}} = 0.0215 \text{ mol NaCl}$$

$$\text{volume of solution} = 100 \text{ mL} \times \frac{1 \text{ L}}{1,000 \text{ mL}} = 0.100 \text{ L solution}$$

$$\text{molarity} = \frac{0.0215 \text{ mol NaCl}}{0.100 \text{ L solution}} = 0.215 \text{ M NaCl}$$

Finally, we divide the moles of NaCl by the volume in liters to obtain the molarity. Note that the units, mol/L, are the units for molarity (M).

Equivalences in scientific calculations are easy to use and very helpful. They allow us to convert from one set of units to another. In addition, using equivalences, we can tell if we have written the fraction correctly as we do our calculation. For example, we could use an equivalence to determine the number of ounces of hops in a bag of Cascade that weighs 11.0 pounds (lb). In this case, we would use an equivalence that relates the number of ounces (oz) in a pound: 16 oz equals 1 lb, or written as an equivalence, 16 oz/1 lb. Equivalences do not impact the precision or accuracy of the calculated result. They are considered infinitely precise.

The equation is set up as:

$$11.0 \text{ lb Cascade} \times \frac{16 \text{ oz}}{1 \text{ lb}} = 176 \text{ oz Cascade}$$

So how did we know to put "16 oz" on the top and "1 lb" on the bottom? We did this because we wanted to cancel the pound units from the equation. If we had inverted this fraction, none of the units would have canceled. Here's what we would have obtained:

$$11.0 \text{ lb Cascade} \times \frac{1 \text{ lb}}{16 \text{ oz}} = \frac{0.688 \text{ lb}^2}{\text{oz}} \text{ Cascade}$$

This equation does not produce an incorrect answer; however, the units (lb²/oz) make the value rather difficult to understand. And as such, the value isn't really that useful. The best thing to do is to arrange the fraction so the units cancel and the result becomes meaningful.

Molarities can also be used in calculations once they are known. The molarity of a solution is most often used this way: The brewer measures out a volume of a solution to measure a known amount of a solute or analyte. Doing so is particularly useful when you need to precisely measure out a very small amount of an analyte that wouldn't be easy to measure on a balance. Let's assume that we have a 0.0250 M solution of glucose ($C_6H_{12}O_6$). If we obtain 10.0 mL of that solution, how many grams of glucose are in that volume? Our calculation here uses the equivalences and arrives at the answer in one long equation:

$$10.0 \text{ mL soln} \times \frac{1 \text{ L}}{1{,}000 \text{ mL}} \times \frac{0.0250 \text{ mol glucose}}{1 \text{ L soln}} \times \frac{180.16 \text{ g glucose}}{1 \text{ mol}}$$

$$= 0.0450 \text{ g glucose}$$

Did you note the use of the equivalences as we converted from milliliters to liters and the use of the molar mass of glucose that we calculated from the periodic table? The other equivalence in the equation is the molarity of the solution. It's written as 0.0250 moles per 1 L. Note that in this calculation, all the units cancel, so we're left with only grams of glucose. In other words, 10 mL of the 0.025 M glucose solution is the same as 0.045 g of glucose.

Normality (N) is another concentration unit that's frequently used in the brewery lab. Normality is the number of molar equivalents of solute per liter of solution. And what is a "molar equivalent"? Essentially, it's the number of reactions that the solute can perform per mole of solute in an experiment. In most cases, the normality of a solution is equivalent to the molarity. Often, normality is used when we are working with acids or bases, but it's a general term that can be used with any solute. For acids, the number of equivalents is equal to the number of protons (H^+) that are donated in the reaction. These protons are easily identified in the formula, because they are often written first. For example, hydrochloric acid (HCl) and nitric acid (HNO_3) donate 1 proton in a reaction, whereas sulfuric acid (H_2SO_4) donates 2 protons. For bases, it's much harder to tell, but the number of equivalents is often the number of hydroxide ions (OH^-) in the formula.

Thus, a 1.0 N solution of nitric acid donates 1.0 mole of protons during a reaction. Because the formula contains only 1 proton, a 1.0 N nitric acid solution is equivalent to a 1.0 M nitric acid solution. However, a 1.0 N solution of sulfuric acid is equivalent to a 0.5 M sulfuric acid solution, because sulfuric acid donates 2 protons in a reaction. (The normality is twice the value of the molarity.)

A calculation reveals how to incorporate the number of equivalents into the value. Let's assume that we have 10.0 g of sulfuric acid (H_2SO_4) and dilute it to 1.0 L of solution. What is the normality of that solution?

$$10.0 \text{ g } H_2SO_4 \times \frac{1 \text{ mol}}{98.08 \text{ g } H_2SO_4} \times \frac{2 \text{ mol eq } H^+}{1 \text{ mol } H_2SO_4} = 0.204 \text{ mol eq}$$

$$\frac{0.204 \text{ mol eq } H^+}{1 \text{ L solution}} = 0.204 \text{ N } H_2SO_4$$

> **normality (N)** A unit of concentration determined by the number of equivalents of analyte per liter of solution.

Note that we use an equivalence that relates the number of equivalents of H⁺ to the number of moles of H_2SO_4. In this case, 1 mole of H_2SO_4 has 2 equivalents of H⁺ (1 mol/2 mol eq). What is the molarity of this solution? It's half the normality, because 2 equivalents of H⁺ are generated by the sulfuric acid.

As we noted at the start of this section, we could use the volume of a liquid analyte to determine the number of moles of that analyte (using the density of the liquid). Then once the volume is known, we would add the density into the calculation as if it were an equivalence. The calculation gets one step longer, but it still arrives at the correct value. The only issue is that measuring small volumes can be difficult unless specific equipment is used that can remain accurate and precise at those volumes.

pH

People talk about the **pH** levels of their hot tubs and swimming pools and even mention pH as a value of their hot liquor in brewing. Technically, the pH of a solution is the negative logarithm of the molarity of hydrogen cations in the solution. More simply, the pH of a solution is just a measure of the concentration of protons (H⁺) in solution. If there are more protons in solution than occur in pure distilled water, we say that the solution is **acidic**. If there are fewer protons in solution than occur in pure distilled water, we say that the solution is **alkaline** or basic. Mathematically, if the concentration of protons is greater than 1.0×10^{-7} M, then the solution is acidic. Alkaline solutions are basic and occur when the concentration of protons is less than 1.0×10^{-7} M. That's a really small number of protons, but it doesn't take much to make something acidic or basic.

pH A measure of the concentration of protons (H⁺) in a solution.

acidic A solution with a pH less than 7.

alkaline A solution with a pH greater than 7.

Can we calculate the pH if we know how many protons are in the solution? The equation to do this calculation is quite straightforward. For example, if we have a 0.025 M solution of protons, we can calculate the pH as 1.60:

$$pH = -\log[H^+]$$

$$pH = -\log(0.025) = 1.60$$

Similarly, if we know the pH, we can determine the concentration of protons in solution. Assuming we have a pH of 4.5, we can determine that the concentration of protons is 3.2×10^{-5} M:

$$[H^+] = 10^{-pH}$$

$$[H^+] = 10^{-4.5} = 3.2 \times 10^{-5} \text{ M}$$

Note that a pH value doesn't have any units and is reported simply as a number. That's the opposite of a concentration, which is reported in units of molarity.

One question remains about pH: How do we know how many protons are in the solution so we can perform the calculation? The answer depends on the acid or base that's being used. For HCl, H_2SO_4, and HNO_3 (three of the seven strong acids), the concentration of protons is equal to the concentration of the acid. For example, a 1.5×10^{-3} M solution of HCl contains

1.5×10^{-3} M protons, but a 1.5×10^{-3} M solution of H_2SO_4 contains 3.0×10^{-3} M protons. (Remember that sulfuric acid donates 2 protons for each molecule of H_2SO_4.) Unfortunately, the calculation for the other acids—such as acetic acid, phosphoric acid, and carbonic acid—is beyond the scope of this text.

For an alkaline solution, the pH is easily determined by noting the concentration of hydroxide ions (OH^-) that are present. A 1.0 M NaOH solution contains 1.0 M OH^- ions, and a 0.20 M $Ca(OH)_2$ solution contains 0.40 M OH^- ions. Once the concentration of hydroxide is known, we can calculate the pH:

$$pH = 14 - pOH$$

where

$$pOH = -\log[OH^-]$$

Let's assume we have a 0.0085 M solution of potassium hydroxide. Its pH can be calculated in two steps:

$$pOH = -\log(0.0085) = 2.07$$

$$pH = 14 - 2.07 = 11.93$$

A good rule of thumb is to assume that any compound with a hydroxide ion can provide that hydroxide ion when dissolved in solution. This assumption isn't exactly true, but it works about 80–90% of the time. The reason for its failure the rest of the time is that some ionic compounds contain hydroxide ions that just aren't soluble in water. And when that happens, hydroxide ions won't be floating free in solution. For example, $Al(OH)_3$ isn't soluble in water and doesn't contribute any hydroxide to determining the pH of the water. However, for most simple hydroxide compounds—such as NaOH, KOH, LiOH, $Ca(OH)_2$, and $Mg(OH)_2$—the rule holds true. More detailed rules for which compounds are and are not soluble are provided later in this chapter.

The pH of a solution can be easily determined using one of several methods. The easiest method is to use a pH paper strip. This type of test strip is coated with molecules that change color as protons are made available in solution. When a drop of solution is added to the strip, the strip changes color based on the pH of the solution. By comparing the color to a chart provided with the test strip product, the user can determine the pH of the solution. Although this method is simple, using pH paper strips is often much less accurate than other methods.

Using a pH meter is a better method for determining the pH of the solution. This device is a little more expensive in the long run than the cost of continually buying pH strips, but the results it produces are much more accurate and precise. Before each use, the pH meter is standardized against solutions with known pH values. Then, by inserting the probe into the solution to be evaluated, the pH can be read off the display. Hand-held models tend to give pH readings to the tenths decimal place, and tabletop models can read to two or three decimal places reproducibly. Of course, the more decimal places you want, the more expensive the pH meter.

Chemical Reactions

When the brewery laboratory performs an analysis, one of the methods usually involves performing a reaction to make the analyte of interest change color. The intensity of that color in the solution determines the concentration of that analyte. (We will talk much more about the relationship between intensity of color and concentration in a Chapter 8.) The reactions are quite varied and can be difficult to understand, but the general principles of reactions are straightforward.

There are a number of classifications of reactions. Of most importance to the brewer and brewery lab are acid-base reactions, precipitation reactions, and oxidation-reduction reactions. Chemists write all of these reactions using the same conventions. The reactants (or starting materials) are placed on the left side of the equation with plus signs (+) between them. Then an arrow read as "yields" is drawn pointing to the right. Next, all the products of the reaction are shown. Let's look at some examples as we talk about the three main types of reactions.

Acid-Base Reactions. In an **acid-base reaction**, an acid and a base are the two reactants. The products of all acid-base reactions are an ionic compound (a salt) plus a small molecule (usually, water). Consider, for example, the reaction of hydrochloric acid and sodium hydroxide. The reaction is written as follows:

$$HCl + NaOH \rightarrow NaCl + HOH$$

We expanded the water molecule and wrote it here as HOH so that it's easier to see where it came from in the reaction. Note how the "H" from HCl and the "OH" from NaOH came together to make water (HOH or H_2O). We read this as "1 molecule of HCl and 1 molecule of NaOH come together to yield 1 molecule of sodium chloride and 1 molecule of water."

What will happen if we mix hydrochloric acid with calcium hydroxide? Here's that equation:

$$2\ HCl + Ca(OH)_2 \rightarrow CaCl_2 + 2\ H_2O$$

Note that in this case, it takes 2 HCl molecules to donate enough "H"s to match the number of "OH"s in the calcium hydroxide. We denote that by adding a "2" in front of the HCl. We also denote that we made 2 water molecules by placing a "2" in front of the H_2O. Now we read the equation as "2 hydrochloric acids and 1 calcium hydroxide yield 1 calcium chloride and two waters."

The interesting thing about the equation is that we can use the ratio of the compounds in the equation (known as the **stoichiometric ratio**) in many ways. For example, instead of saying "1 molecule of calcium hydroxide reacts with 2 molecules of hydrochloric acid," we can say "1 mole of molecules of calcium hydroxide reacts with 2 moles of molecules of hydrochloric acid." Extending this allows us to perform calculations using the equations. In other words, the numbers in front of the compounds in an equation can be used to write equivalences. For the reaction above, we could write 2 mol HCl/1 mol $Ca(OH)_2$, 2 mol HCl/2 mol H_2O, and so on.

acid-base reaction The reaction between an acid and a base; the products of the reaction are an ionic compound (salt) and a small molecule (usually, water).

stoichiometric ratio The relationship between compounds in a reaction.

Then we can insert these equivalences into calculations to determine tons of information about our reaction. For example, how many grams of water should be formed when 1.5 g of $Ca(OH)_2$ react with hydrochloric acid? The answer is easily determined by setting up an equation. Let's start with the 1.5 g of $Ca(OH)_2$ and apply the equivalences to convert grams of $Ca(OH)_2$ into grams of water:

$$1.5 \text{ g Ca(OH)}_2 \times \frac{1 \text{ mol Ca(OH)}_2}{74.10 \text{ g Ca(OH)}_2} \times \frac{2 \text{ mol H}_2\text{O}}{1 \text{ mol Ca(OH)}_2} \times \frac{18.00 \text{ g H}_2\text{O}}{1 \text{ mol H}_2\text{O}}$$

$$= 0.73 \text{ g H}_2\text{O}$$

In fact, the equation can be used to calculate anything we want. Note the use of the equation in the equivalence known as a **stoichiometric factor**:

$$\frac{2 \text{ mol H}_2\text{O}}{1 \text{ mol Ca(OH)}_2}$$

stoichiometric factor The relationship between compounds in a reaction written as an equivalence.

By changing this factor based on the equation, we can determine any of the other compounds in the equation.

Precipitation Reactions. Precipitation reactions are less common in the brewery laboratory than the other two types of reactions but still quite useful for determining some analytes of interest. These reactions involve the production of a solid material. After the solid is formed in the reaction, it falls out of the solution. This result is written in the reaction by including a subscript "s" in parentheses after the compound that forms the precipitate.

precipitation reaction A reaction that forms a solid as the product.

So, which compounds form precipitates? Those that do must be insoluble in water. Luckily, we can use a set of **solubility rules** to quickly determine whether an ionic compound is soluble in water (Table 4.9). We just need to know the names of the cation and anion and then consider them against the rules. As we noted before, solubility rules are useful only for ionic compounds. Molecules (made from nonmetals combining with other nonmetals) don't always adhere to a set of rules concerning solubility.

solubility rules A set of common statements that help determine whether an ionic compound is soluble in water.

TABLE 4.9. Solubility rules for ionic compounds[a]

Soluble compounds	Precipitates
Group 1 elements	Carbonates (except Group 1 and NH_4^+ salts)
Acetates, chlorates, and nitrates	Phosphates (except Group 1 and NH_4^+ salts)
Chlorides, bromides, and iodides (except those of silver, mercury, and lead)	Sulfides (except Group 1 and NH_4^+ salts)
Sulfates (except calcium, strontium, barium, silver, mercury, and lead)	Hydroxides (except Group 1 and calcium, strontium, barium, and NH_4^+ salts)

[a] Courtesy S. E. Johnson and M. D. Mosher—© ASBC.

Let's consider some compounds to verify the use of the rules. Brewers often add calcium carbonate to the water to treat it and raise the calcium content. However, the solubility rules say that calcium carbonate should not be soluble in water. And the rules are correct: Calcium isn't very soluble. Better ways to raise the calcium content are to use calcium chloride (soluble), calcium sulfate (soluble), and calcium hydroxide (also soluble.) Of course, we would have to consider the impact of the chloride, sulfate, or hydroxide on the brewing water.

Let's examine the reaction of silver nitrate and sodium chloride using an equation. The equation below shows that that one of the products of the reaction is insoluble. In the lab, mixing silver nitrate and sodium chloride forms a precipitate. Parenthetically, the amount of chloride in an aqueous solution can be determined by adding silver nitrate and determining the amount of turbidity (cloudiness) that results. When compared with solutions made with specific amounts of chloride, the concentration of chloride in the water can be rapidly determined:

$$AgNO_3 + NaCl \rightarrow NaNO_3 + AgCl_{(s)}$$

How did we know that 1 sodium ion combined with only 1 nitrate polyatomic ion or that 1 silver ion combined with only 1 chloride ion? Remember our charges. Nitrate is a polyatomic ion with a −1 charge, so silver must have a +1 charge. Sodium is in group 1, so it must be +1, and chloride is in group 17 (group VIIA), so it must be −1. Combining the ions in the products reveals the correct formula for each.

Oxidation-Reduction Reactions. **Oxidation-reduction reactions** (also known as "redox reactions") involve the transfer of electrons during the reaction. The result is that the charges on the atoms change. For the reaction to involve both oxidation and reduction, one of the charges must increase and the other must decrease. So which is which? **Oxidation** involves the loss of electrons and a corresponding increase in the charge on the species. **Reduction** involves the gain of electrons and a decrease in the charge.

Consider the reaction represented by the equation below. If we break it down into the individual elements and ions, we see that the copper starts out in its elemental state with a charge of 0 but then becomes +2 charged when it combines with chloride to make copper chloride. The silver starts out as +1 in silver chloride but then becomes elemental silver with a charge of 0 in the product:

$$Cu + 2\ AgNO_3 \rightarrow Cu(NO_3)_2 + 2\ Ag$$

Because the copper goes from 0 to +2, it is oxidized. Similarly, the silver goes from +1 to 0, so it is reduced.

Redox reactions occur frequently during the brewing process. In fact, the many steps in yeast metabolism involve oxidation-reduction reactions. For example, the final step in the production of ethanol from acetaldehyde is a redox reaction. The acetaldehyde is reduced to ethanol. One of the reagents in the metabolic process must then be oxidized for the reaction to take place. (We will cover that in much more detail in Chapter 5.)

oxidation-reduction reaction A reaction that involves the oxidation of one compound and the reduction of another.

oxidation A loss of electrons from a species.

reduction A gain of electrons to a species.

Line Drawings

The molecules found in brewing science often get so complex that they are difficult to draw using the methods we have discussed in this chapter. For these cases, chemists have devised a method that greatly simplifies the process: the use of line drawings. A line drawing represents all the carbon atoms and any hydrogen atoms on those carbons as lines. All the other atoms are shown in the drawing.

Figure 4.9 shows three molecules that have been drawn both as line drawings (on the left) and as drawings showing all the atoms (on the right). Note that as the molecules get more complex, the line drawings become easier to draw and easier to comprehend in terms of seeing the different parts of the molecules. Even the simpler molecule of butyric acid, which is produced by some spoilage bacteria and has a particularly rotten smell, can be quickly drawn using a line drawing.

Let's walk through drawing a butyric acid molecule. First, we place the pen on the paper to represent the carbon on the left side of the molecule. Then we draw a line to the next carbon atom. We then change directions and

Acetaldehyde

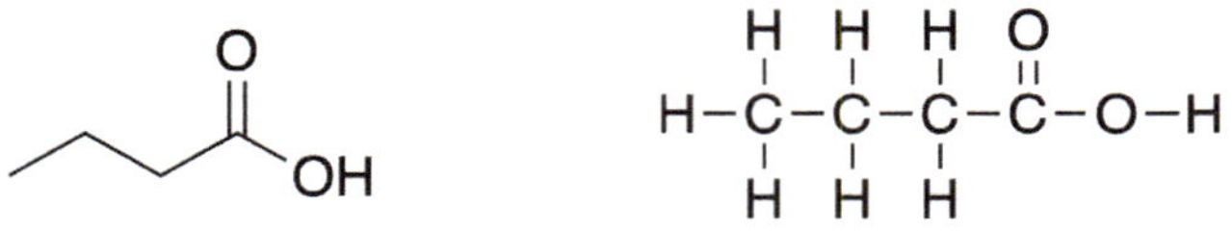

Butyric acid

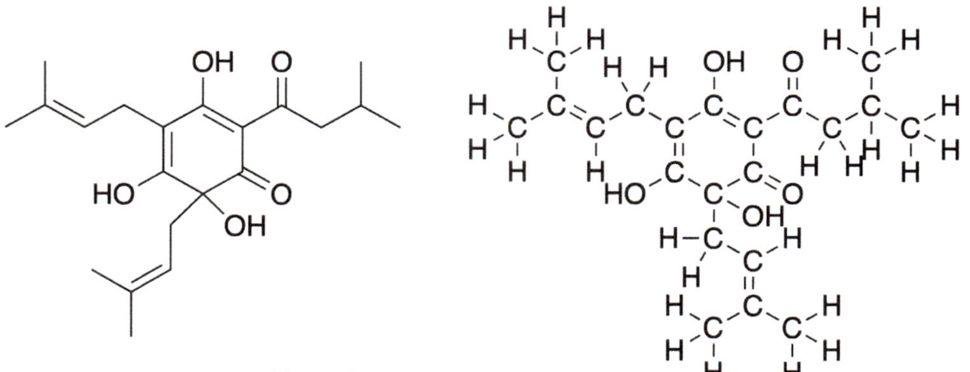

Humulone

FIG. 4.9. Line drawings of molecules (on the left) versus drawings that show all the atoms (on the right). (Courtesy S. E. Johnson and M. D. Mosher—© ASBC)

draw a line to the next carbon and finally change directions again as we draw a line to the last carbon atom. The oxygen is connected with a double line (a double bond) to the last carbon, and the OH is connected with a single line. Why is the hydrogen atom included in the drawing? Because it's attached to an atom other than carbon, it must be included.

Note that in a line drawing, the number of hydrogen atoms on each carbon atom can be quickly determined by counting the number of lines that emanate from each bend in the drawing. The number of hydrogens on each carbon is equal to 4 minus the number of lines. Figure 4.10 shows a line drawing of limonene, a molecule found in hop oils. The circled atom is a carbon atom with 3 hydrogens, the squared atom contains only 2 hydrogen atoms, and the triangular atom contains a carbon with no hydrogens. Once you get the hang of drawing and counting, a line drawing provides a quick way to represent a molecule.

Organic Molecules in Brewing

Reactions that are written regularly in brewing science often include line drawings of molecules, rather than formulas. This graphic representation facilitates being able to count the number of hydrogen atoms in a line drawing. We need a formula when we want to know the molar mass of the compound for our calculations. In Figure 4.10, the formula for limonene can be determined by counting the ends of the lines as carbons and then adding up the hydrogens based on the fact that carbon always makes four bonds ($C_{10}H_{16}$).

In addition to using line drawings, we can take other shortcuts. For example, we noted earlier that acetaldehyde can be reduced to ethanol during fermentation under anaerobic conditions (Fig. 4.11). The equation that represents this reaction may not show some important information. First, the stoichiometric ratio is often omitted, so we may be forced to examine the reaction in detail to determine the ratio. Second, in a redox reaction, sometimes only one portion of the reaction is written. For example, if the

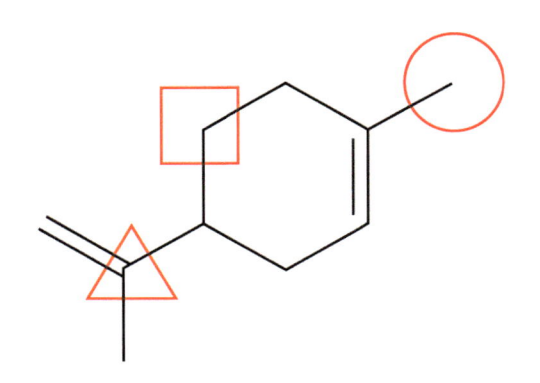

FIG. 4.10. Limonene. The circle, square, and triangle denote carbon atoms with 3, 2, and 0 hydrogen atoms, respectively. Can you find the 2 carbon atoms with only 1 hydrogen? (They are the carbons at 4 o'clock and 8 o'clock in the six-membered ring.) (Courtesy S. E. Johnson and M. D. Mosher—© ASBC)

FIG. 4.11. Shorthand equation showing the reduction of acetaldehyde to ethanol. (Courtesy S. E. Johnson and M. D. Mosher—© ASBC)

focus is on the reduction, only the reduction is written. It's then understood that the oxidation is still included but has been omitted from the equation. Third, some products of the reaction are omitted if they aren't important to the discussion.

In short, the reaction equation becomes a quick shorthand way to relate the information you want others to know. We have to do the legwork ourselves if we want to figure out the rest of the reaction information.

CHAPTER 5

Essentials of Biology and Biochemistry

We explored how chemistry impacts brewing science in Chapter 4. That discussion provided the basic information needed to understand the science behind how barley, hops, and yeast work. This chapter will cover those areas of brewing science and provide insights into the specific analyses performed during beer production. We will start with barley and the malting process and then explore the hop plant and its importance in brewing. We will finish the chapter with a discussion about microbes—in particular, the yeast cell and its metabolism during the fermentation stage of brewing.

The Barley Seed

Barley is a member of the grass family of organisms and was likely first cultivated by prehistoric peoples in Mesopotamia between 10,000 and 8,000 BCE. Scientists have discovered 32 species that belong to the genus *Hordeum*, and *Hordeum vulgare* is the scientific name for all the species of barley.

There are two-row and six-row versions of the barley plant because of a single genetic mutation. Prior to about 1940, two-row and six-row barley plants were considered separate species: *H. vulgare* was assigned to the six-row variety with nonshattering spikes, and *H. distichum* was assigned to the two-row version with nonshattering spikes. In any case, the two types of barley look quite different in the field, and the seeds they produce have somewhat different makeups.

The two-row variety is so named because the seeds align in two columns opposite each other on the spike. The two-row variety contains three pairs of columns on the spike. Two-row barley has been predominantly used in the brewing industry, because its seeds have greater starch content, less protein, more husk material, and slightly lower enzyme concentration. However, the use of **adjuncts** and the advances in barley hybrids have increased the use of six-row barley. In fact, if a brewer has a large adjunct contribution to the

barley The most common grain used in brewing; a member of the grasses family of organisms.

Hordeum vulgare The scientific name for all species of barley.

adjunct A nonbarley source of starch or sugar used in brewing.

Fusarium graminearum The scientific name for a fungus that grows on barley and produces vomitoxin.

vomitoxin A compound that causes vomiting in humans and can result in "gushing" of packaged beer.

deoxynivalenol (DON) The chemical name of vomitoxin, a toxic compound.

steep To submerge barley in water during the malting process to increase the hydration level of the barleycorn before germination.

germination The combination of processes that occur when a seed begins to grow.

acrospire Created by germinating barley; would become the stem of the barley plant if germination was not stopped by kilning.

kilning Heating to remove water from grain; the process stops germination, eliminates some enzymes, and initiates Maillard reactions that result in color and flavor enhancements.

Maillard reactions A series of complex organic chemistry reactions that involve the combination of sugars and proteins and the loss of water.

beer, the use of slightly cheaper six-row barley with greater enzyme and husk contributions will provide greater extract during mashing.

The process of converting a barley seed into malt involves using barley's own mechanics. While the particular details of this process are beyond the scope of this text, a brief review is worthwhile. After harvest, barley is dried to a moisture content of 13–18%, sorted and cleaned of any unwanted material (such as rocks and metal), and then stored until its dormancy period has passed. Before the dormancy period, barley seed will not grow. The barley is then transferred to the maltster to be converted into barley malt via the process outlined in Figure 5.1. The seeds should be closely inspected and analyzed for the presence of fungi, pests, disease, and other damage.

In particular, the maltster is concerned with contamination by *Fusarium graminearum* and other *Fusarium* species that cause head blight. Growth of these fungi is favored by wet periods, especially those that occur late in the growing season. The byproduct of head blight is the production of a toxic compound, **vomitoxin** (chemically known as **deoxynivalenol**, or **DON**). Some brewers reject any malt that contains a measurable amount of vomitoxin. Others place a limit of 0.5 parts per million (ppm) or even 1.0 ppm on the maximum concentration of this compound in the malt. Not only is vomitoxin toxic to humans, but small amounts can cause gushing in packaged beer, such that when the bottle is opened, all the content is ejected.

Once the barley passes inspection, it's then **steeped** in water and oxygenated. After the barley is appropriately hydrated, the water is removed and the grains are slowly stirred until they begin to grow. This step is **germination**. The maltster monitors the process by noting the growth of the roots at one end of the seed and the growth of the **acrospire**, which will eventually become the shoot. When the acrospire has grown to the length of the seed, the malt is considered to be fully modified. It's then dried by circulating air across the seeds. Slowly, the temperature of the air is increased until the percentage of moisture in the malt is reduced to 3–6%, depending on the specifications for the malt.

Applying warm air during the drying process, known as **kilning**, causes **Maillard reactions** to take place (Maillard is pronounced "MY-yar"). This series of complex organic chemistry reactions, named after Louis-Camille Maillard's study of them in 1912, involves the combination of sugars and proteins and the loss of water. The resulting molecules have flavors such as

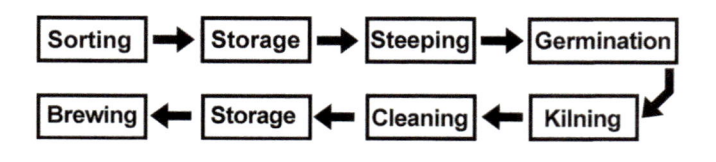

Sorting → Storage → Steeping → Germination → Kilning → Cleaning → Storage → Brewing

FIG. 5.1. The process of converting harvested barley to malt. (Courtesy S. E. Johnson and M. D. Mosher—© ASBC)

caramel, umami, toast, coffee, and burned foods. If the malt is heated only a little, the Maillard reactions make more of the caramel and umami flavors. If the heat is applied too long or the malt gets too hot during the drying process, the flavors fall more to the burned end of the spectrum. Maillard reactions are not the same as the caramelization reactions that occur when sugar is heated, although many of the flavors and the appearance of the product are similar. After malting, the product is then sent to the brewer to be mashed during the brewing process.

What happens inside the barley seed during germination? The most important reactions include the activation of malt enzymes (large molecules that facilitate reactions in living systems); the production of **gibberellic acid** (a molecule that triggers enzyme production and release); and the modification of the **starchy endosperm** (the largest portion of the seed, made up of starch storage cells) (Fig. 5.2). This modification includes the degradation of the protein and cell walls that surround the starch granules in the endosperm. While some of the starch is converted into maltose and other sugars and used by the seed to grow, most of the starch remains to be used by the brewer. During the process, **enzymes** are activated—for example, β- (beta-) amylase, α- (alpha-) amylase, β-glucanase, protease, and phytase. Their action releases the starch along with the phosphate and other nutrients that will be used in the mashing process and also helps to degrade some of the protein within the seeds. Unfortunately, many of the enzymes are degraded during kilning. The remaining enzymatic activity, as it relates to converting starch into sugar, dictates the usefulness of the malt in brewing, is the **diastatic power**. Whereas fully modified malts that are only lightly kilned have sufficient diastatic power to convert all the starch, partially modified malts, highly kilned malts, and adjuncts have little, if any, diastatic power.

gibberellic acid A molecule that triggers enzyme production and release within barley.

starchy endosperm The largest portion of the seed; made up of starch storage cells.

enzymes Molecules that help reactions occur; enzymes are temperature and pH sensitive.

diastatic power The enzymatic activity as it relates to converting starch into sugar.

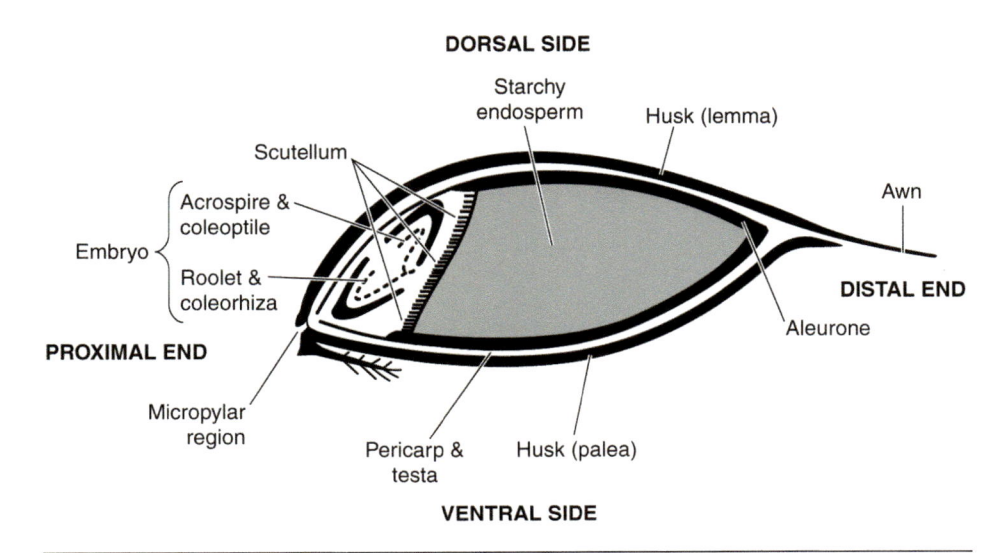

FIG. 5.2. Key parts of barleycorn. (Courtesy S. E. Johnson and M. D. Mosher—© ASBC)

The Hop Cone

Humulus lupulus The scientific name for hops.

cultivar A variety of hops, barley, or yeast that is in the same genus but has different properties.

rhizome A perennial root-like structure; a part of the hop plant that grows year-round.

bine An annual vine-like structure that dies off at the end of each season; wraps clockwise around objects as it grows toward light and touch.

dioecious A plant species that has male and female versions.

strobili The flowers of hop plants.

The hops used in the brewery are the species *Humulus lupulus*. In fact, this plant species was written about by Pliny the Elder (23–79 CE), but its use in beer was first recorded in a set of rules governing the operation of a monastery written by Abbot Adalhardus in 822 CE. It has been suggested that hops may have been used much earlier in other northern European countries. Whenever and wherever they were first used, it's clear that they were used regularly in brewing by the 1300s.

Hops were a wild plant found throughout many of the world's temperate regions. Finding hops in the wild was not difficult, but having a reproducible supply was important for brewers. Hops began to be cultivated shortly after they became used in beer and then domesticated. Today, researchers have developed many different hybrids of the hop plant. Each is known as a variety or **cultivar**.

The hop plant consists of a **rhizome** (a perennial root-like structure that grows year-round) and a stem, which is the **bine** (an annual vine-like structure that dies off at the end of each season). The bine contains hairs that point down and help the stem attach to structures as it grows. Unlike a vine, which grabs onto surfaces, a bine wraps clockwise around an object as it grows toward light and touch (other objects such as plants, fences, and walls) (Fig. 5.3). The bine can extend 5.5–7.5 meters (m), which means it can grow as much as 0.3 m per day at the height of the growing season.

The hop plant is **dioecious**, which means it has both male and female versions. Flowers on the female plant, known as **strobili**, are much larger and contain much more hop oil than those on the male version (Fig. 5.4). If

FIG. 5.3. Hop yard in the Yakima Valley showing early spring growth. Note that the hop bines have just started wrapping around the strings hanging down from the poles in the field. (Courtesy Williamborg)

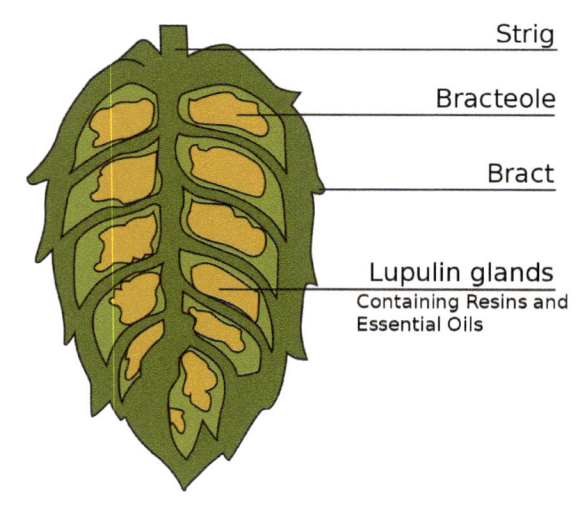

FIG. 5.4. A hop flower with parts of the hop cone noted. Note the lupulin glands inside the hop cone. (Courtesy Herr Schnapps)

pollination occurs, the plant produces seeds that have little brewing value, so separating the plants is very important (unless the grower is cross-pollinating to create a new variety). In other words, when we drive by a hop field, all the plants that we see are female. The strobili look like miniature green pine cones. Inside the cone along the **strig** (the stem of the cone) and beneath the leaves (**bracts** and **bracteoles**) are the **lupulin glands**, which excrete a yellow resin known as hop oil. Commercial hops have been selected to produce hop oils in as much as 20–30% of the mass of the flower. Hops are typically grown in temperate climates near the 40–50° latitudes. Places such as the Yakima Valley in Washington, the Hallertau region in Germany, and New Zealand are in this range of latitudes and well known for the hops they produce.

In the field, the growing season begins when the plant recovers from winter dormancy by sending up shoots. Then during the vegetative growth stage, which lasts from the end of May to early July in the Northern Hemisphere, the bine grows up to reach its maximum height. In 5–6 weeks, the bine will grow up to 7.5 m tall. That's more than a meter a week! By mid-July, most of the vegetative growth is lateral to the main stem. At the end of July, the plant begins to flower. The plant shifts its energy to producing these flowers, which in mature plants can result in up to 50% of the total mass of the plant above ground. By mid- to late August, the hop bines are cut down and taken back to the barn. There, the hop flowers are plucked from the bines and dried.

Hop cones are dried 8–12 hours to reduce the moisture level to 10%. This process is a lot like kilning. Warm air (55–65°C) is blown through a bed of hop cones; the humidity of the air circulating from the hops determines the moisture left in the hops. After the hops have cooled, they are baled and sent to cold storage until they are used. Before being used, the hops can be ground and pressed into pellets and extracted with liquid carbon dioxide; the oil is used in the brewery or as a stand-alone product.

Brewers want to use the lupulin glands of the hops, because they produce material that consists of two main categories of compounds: the essential oils and the alpha and beta acids. The **alpha acids** and **beta acids** have very similar chemical structures (Fig. 5.5). The compounds differ

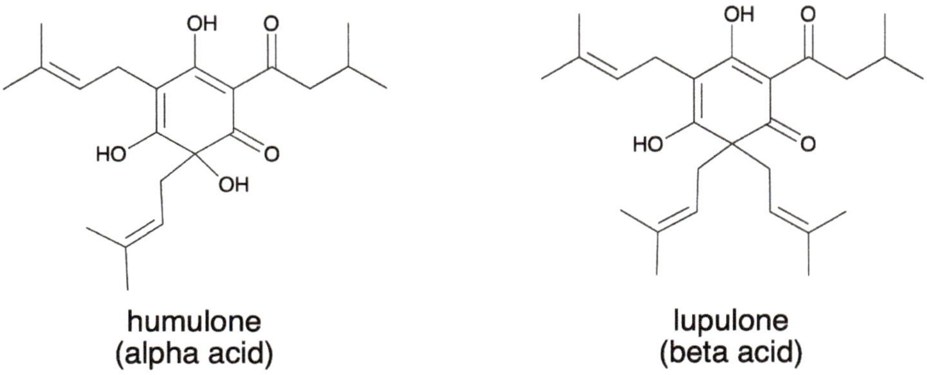

strig Stem of the hop cone.

bracts Leaves of the hop cone.

bracteoles Small leaves in the hop cone that protect the lupulin glands.

lupulin glands Areas in the hop cone along the strig from which oils are secreted.

alpha acids Compounds in hop oils that isomerize to iso-alpha acid during the boil process.

beta acids Compounds in hop oils that are not very water soluble but are prone to oxidation, which can result in a variety of flavors.

humulone
(alpha acid)

lupulone
(beta acid)

FIG. 5.5. Alpha and beta acids. Note the similarities and differences between these two compounds. Humulone can undergo a reaction to form iso-humulone (iso-alpha acid—the bitter compound) during the boil. Lupulone does not undergo that reaction. (Courtesy S. E. Johnson and M. D. Mosher—© ASBC)

in that an alpha acid is converted to iso-alpha acid during the boil process and contributes significantly to the bitterness of the finished product. A beta acid cannot undergo a similar conversion during the boiling process.

A multitude of compounds are found in the essential oil, including the terpene and terpenoid classes of organic molecules that provide beer with its flavor and aroma. Most notable are the molecules humulene, caryophyllene, farnesene, and myrcene. Figure 5.6 shows the structures of these compounds and indicates their associated flavors. Modified versions of these molecules are also present in hop oil. Those compounds (including geraniol, linalool, and nerol) produce a myriad of additional flavors and aromas (Fig. 5.7). Because each hop cultivar produces a different composition of hop oil and different amounts of alpha and beta acids, each hop oil has a unique flavor and aroma.

humulene
(balsamic, flowery, herbal, spicy, woody)

caryophyllene
(woody, spicy, floral, terpene)

farnesene
(woody, citrus, sweet)

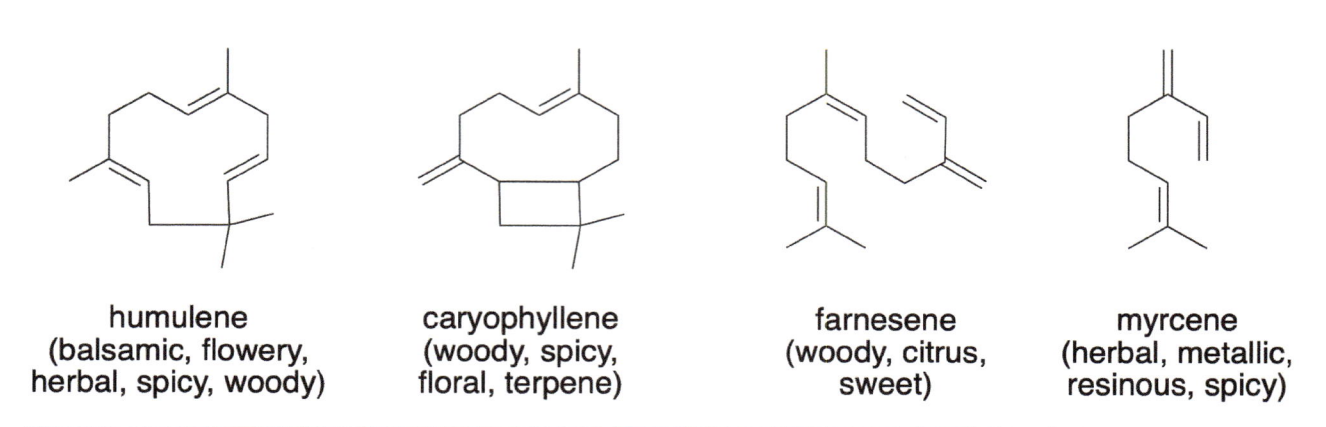

myrcene
(herbal, metallic, resinous, spicy)

FIG. 5.6. Terpenes in hop oil. Some of the common aroma and flavor descriptions are provided below each compound name. The lack of –OH groups in these molecules indicates that they are not very water soluble. (Courtesy S. E. Johnson and M. D. Mosher—© ASBC)

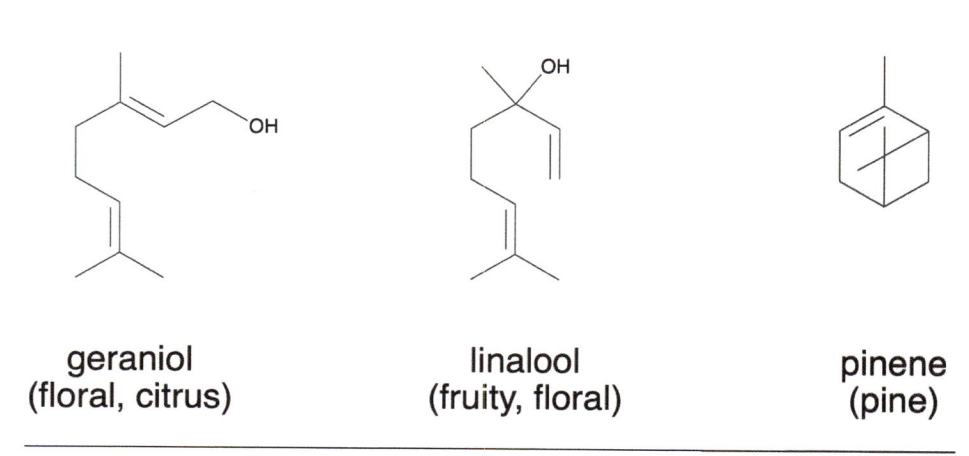

geraniol
(floral, citrus)

linalool
(fruity, floral)

pinene
(pine)

FIG. 5.7. Modifications of terpenes, such as myrcene, produce other hop oil compounds. These compounds contain the –OH group. Their water solubility is greater than that of the other terpenes. (Courtesy S. E. Johnson and M. D. Mosher—© ASBC)

Wort: The Perfect Growing Environment

Wort is the liquid that's obtained after the mash. Then during the boil, hops are added for aroma and flavor, and so are other components required by the final beer specifications. The makeup of that liquid is very important to the yeast during fermentation.

So, what's in the wort? What does the yeast require to grow and survive? Which compounds contribute to flavors, and which compounds aren't needed in the fermentation step? Knowing the answers to these questions is not only useful for ensuring good yeast health and producing clean beer, but it also helps in understanding the key components that may require monitoring during production.

Mashing involves steps to convert the starch in the malt into both fermentable and unfermentable sugars, also known as carbohydrates. Fermentable sugars from the mashing process include significant amounts of **glucose** and **maltose**. **Maltotriose** is also produced in some quantities. If any adjunct sugars are added to the wort during the boil, these compounds are also included in the wort. Thus, if any **sucrose** is added from fruits or cane sugar, it's available to the yeast for fermentation. These sugars are accompanied by **limit dextrins** and other larger pieces of the starch molecule that are not fermented by yeast (Fig. 5.8).

In addition to sugars, the wort contains degradation products of the barley seed, including pieces of the cell walls that held the starch granules, or **beta-glucans**. These materials can be relatively small subunits of the large beta-glucan molecule or even large pieces that were not filtered out earlier (Fig. 5.9).

Proteins and degradation products of those proteins arise from enzyme activity and from heating during the boil. The large proteins are usually classified into four groups based on size. The **albumins** and **globulins** are small proteins that tend to be fairly soluble during the mash and get extracted into the wort. The **glutelins** and **hordeins** are much larger and less soluble. The hordeins, in particular, tend to degrade more easily during mashing. The degradation products of proteins are smaller fragments of the protein known as peptides. The proteins can also be degraded to the individual amino acids that make up the proteins. The peptides and amino acids contribute to the amount of **free amino nitrogen (FAN)** in the wort.

Polyphenols (also known as tannins) are found in malt and other grains and remain in the wort during the mashing process (Fig. 5.10). They can also arise from adding hop materials during the boil. These organic compounds are part of the husk and vegetative matter. They are relatively soluble in water when the pH is neutral or alkaline, but small quantities are soluble even at the lower pH range of wort and beer. These compounds contribute a burned toast or tea-like flavor to beer and contribute to haze (both temporary and permanent).

Other components of the wort include ions (remember these from Chapter 4) that come from the malt and hops but primarily from the water used to prepare the mash. The typical ion makeup includes larger quantities of cations such as calcium and magnesium and smaller quantities of sodium and potassium. There tend to be trace quantities of metal ions such as iron,

wort The food- and nutrient-rich liquid that's obtained after the mash.

glucose A fermentable sugar that can't be broken down into smaller sugars.

maltose A fermentable sugar composed of 2 glucose units.

maltotriose A fermentable sugar composed of 3 glucose units.

sucrose A fermentable sugar composed of 1 glucose unit and 1 fructose unit.

limit dextrins Large, branching portions of the starch molecule that are not fermentable.

beta-glucans Pieces of plant cell walls that encompass starch granules.

albumins Smaller proteins that tend to be fairly soluble during the mash and get extracted into the wort.

globulins Smaller proteins that tend to be fairly soluble during the mash and get extracted into the wort.

glutelins Larger proteins that tend to be less soluble during the mash.

hordeins Larger proteins that tend to be less soluble during the mash.

free amino nitrogen (FAN) The number of amino acids that the yeast will ultimately have access to during fermentation; depends on the type of malt and the mashing process.

polyphenols Compounds that are part of the grain husk and vegetative matter; one of the two precursors to haze in beer.

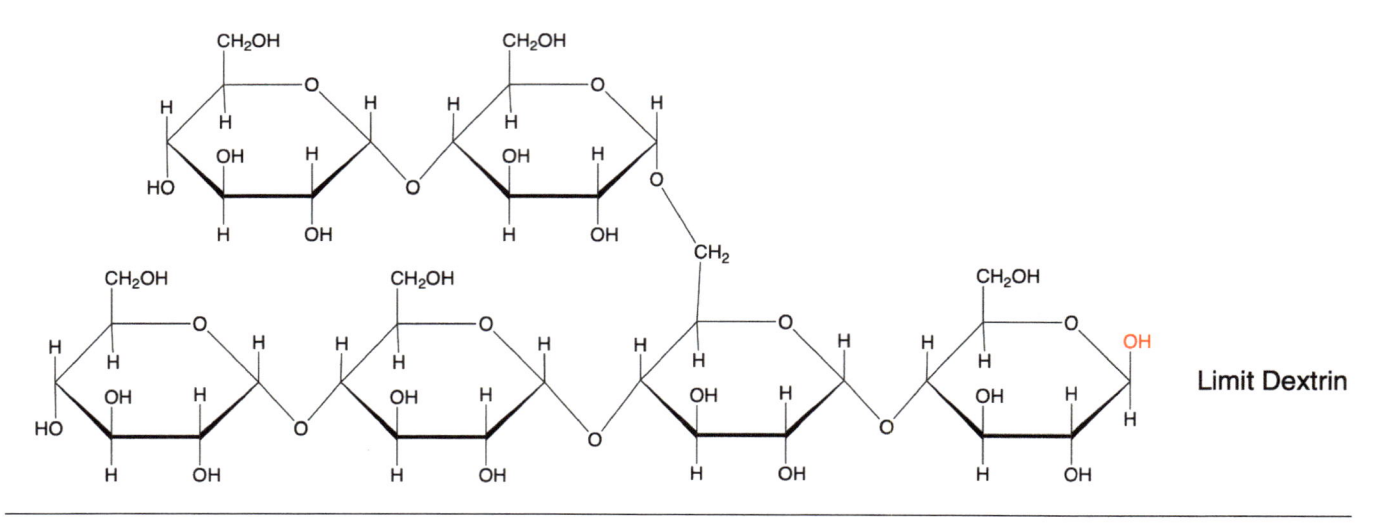

FIG. 5.8. Fermentable and unfermentable carbohydrates in wort. Glucose, maltose, and maltotriose are fermentable sugars. Larger pieces of starch, such as maltotetrose and limit dextrin, are unfermentable. These fermentable and unfermentable sugars are small pieces of the starch molecule that exists in the endosperm of the barley seed. The OH labeled in red can be oriented up or down for each sugar. (Courtesy S. E. Johnson and M. D. Mosher—© ASBC)

chromium, zinc, copper, and a host of other cations. Anions are also present in the wort, including sulfate, phosphate, carbonate, and chloride.

In short, there are a multitude of different compounds and ions in wort. Each comes from the malt, the hops, or the water used in the mash. What they contribute to the flavor of the beer is important, but their presence in the wort is essential for yeast health.

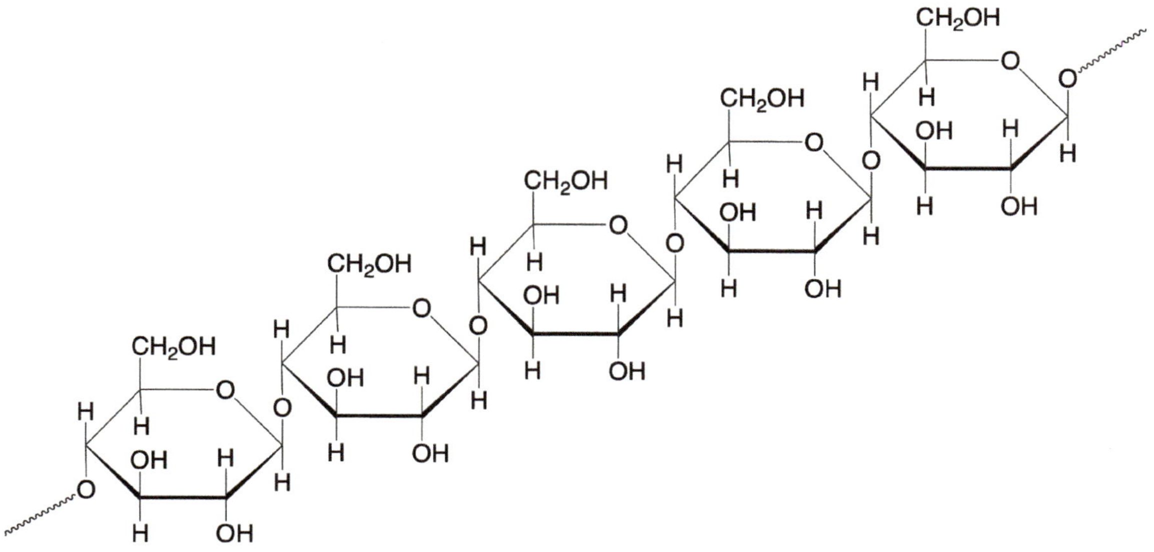

FIG. 5.9. Beta-glucan in wort. This is a small piece of the beta-1,4-glucan found in barley. Other pieces may be longer or shorter. Note the similarities and differences between glucan and starch. What sugar are both based on? (Both are polymers of glucose.) (Courtesy S. E. Johnson and M. D. Mosher—© ASBC)

quercetin

prodelphinidin B3

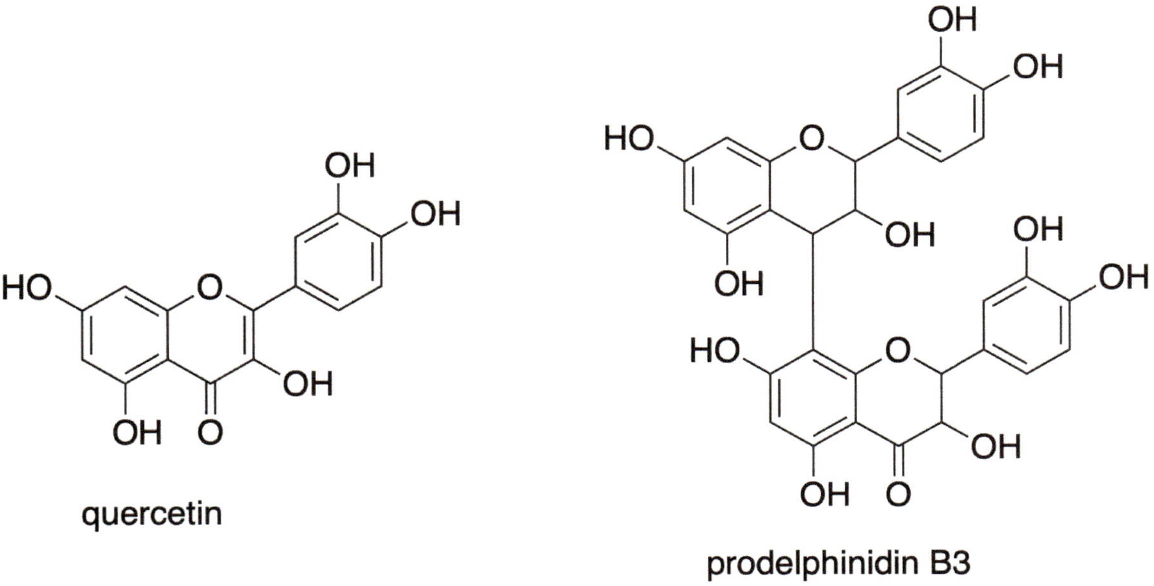

FIG. 5.10. Examples of polyphenols found in wort. The positions of the −OH groups can vary in the polyphenols, as can the number of molecules linked together. Note that the number of −OH groups makes these molecules fairly water soluble. (Courtesy S. E. Johnson and M. D. Mosher—© ASBC)

Types of Microorganisms

Many different microbes exist in the natural world. Some are pathogenic and cause diseases or produce toxic compounds that can harm humans. None of the known pathogens can survive in the common beers produced today because of the low pH, the presence of hop acids, and the presence of alcohol. However, this isn't completely accurate for low-alcohol beers; research has shown that pathogenic organisms can survive in these beers. And in fact, some microbes can survive and thrive in the relatively harsh environment of higher-alcohol beers. These microbes cause spoilage and often negatively impact the flavor of the beer.

Unfortunately, wort can contain human pathogens and almost any type of bacteria. The boiling process significantly reduces the number of microbes in the wort to a level at which there are no concerns about their surviving the boil. In other words, the brewer typically doesn't worry about microbes in boiled wort as long as the boil lasts for at least 20 minutes.

Bacteria are ubiquitous throughout the natural world. They infect beer and cause off-flavors, many of which we have already discussed. So, what are bacteria, and how can they proliferate so easily?

Bacteria are much smaller than the brewer's yeast used in fermentation. Although each brewery will have its own list of common spoilers (bacteria that can get into the beer), most of the bacteria in beer production belong to one of the following genera: *Pediococcus, Lactobacillus, Acetobacter, Megasphaera, Pectinatus,* and *Zymomonas*. The first two, *Pediococcus* and *Lactobacillus,* are the most common and likely will contribute to the majority of all bacterial infections of beer. Table 5.1 identifies the major genera of bacteria often found as contaminants in beer (although many of the genera have multiple spoilers) and the results of their infections. The table and Figure 5.11 also indicate the shapes of the major species.

Bacteria are prokaryotes: organisms that mostly contain a single circular genome. Members of the prokaryote class of organisms are typically 0.2–5.0 millimeters (mm) in size. They are single-celled organisms. Surrounding the cell is the plasma membrane, a fluidlike structure for which the main function is to keep material outside the cell (Fig. 5.12). The plasma membrane

TABLE 5.1. Common bacteria that can infect beer and wort and the results of their infections[a]

Genus	Shape	Off-flavor(s)	Other indicator(s)
Pediococcus	coccus	Butter; sour	Ropy;[b] hazy
Lactobacillus	bacillus	Sour	Turbid; hazy
Acetobacter	streptobacillus	Harsh sour	Vinegar odor
Pectinatus	bacillus	Sulfury; cheesy sour	Rotten egg odor
Megasphaera	streptococcus	Sulfury; oily	Rotten egg odor; turbid
Zymomonas	bacillus	Sour apple; sulfury	Rotten fruit odor

[a] Courtesy S. E. Johnson and M. D. Mosher—© ASBC.
[b] The term "ropy" refers to a slimy characteristic when the beer is poured.

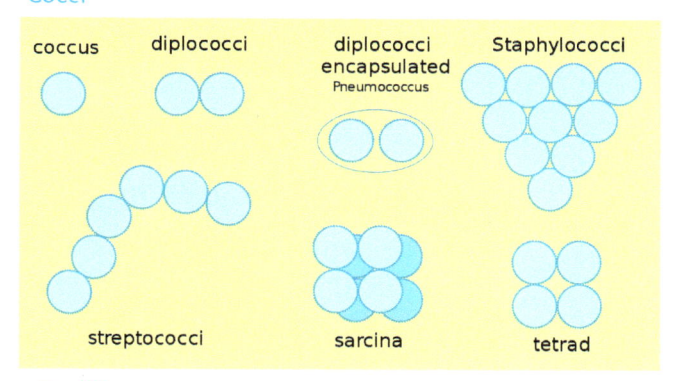

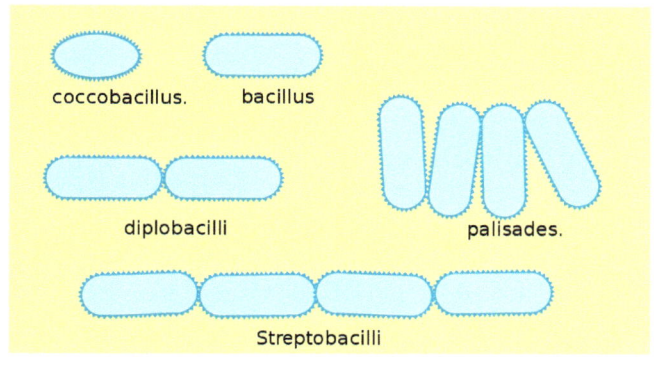

FIG. 5.11. Basic morphological differences among bacteria, showing the most often found forms and their associations. (Courtesy Mariana Ruiz Villarreal)

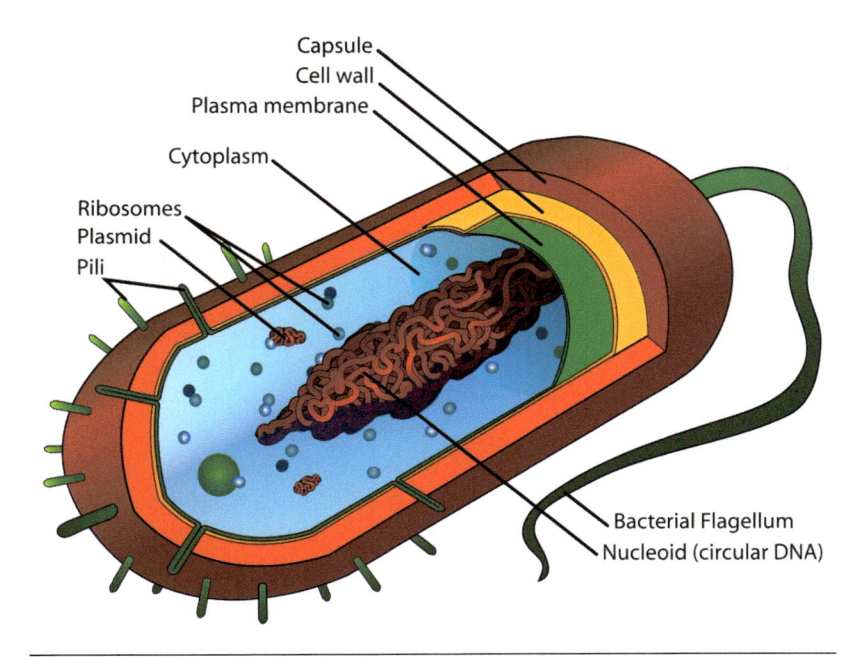

FIG. 5.12. Structure of a bacterial cell showing key elements. (Courtesy Mariana Ruiz Villarreal)

is surrounded by a cell wall, which is structural and much more rigid than the membrane. Some bacteria are further surrounded by a capsule, a viscous layer that aids in preventing cell desiccation and adhesion to surfaces. Projecting from the surfaces of some bacteria are structures that aid in movement and sensing. These include tiny fiber-like structures, known as pili, which can attach the cell to other cells and surfaces, as well as wriggling tails called flagella, which allow the cell to move through its medium. Inside the cell is a fluid known as cytoplasm. Floating in this liquid is a mass of DNA gathered into a bundle called a nucleoid, along with small packages of DNA outside the nucleoid known as plasmids.

As shown in Figure 5.11, bacteria come in many shapes. The two main shapes are spherical and rodlike. The spherical form is known as *coccus* (when there is only one) or *cocci* (when there is more than one). These bacteria can stay separated, gather in chains, or pair with other bacteria. The rod-shaped form is known as *bacillus* (singular) or *bacilli* (plural). The names of many genera of bacteria reflect whether they are spheres or rods. For example, the genus *Pediococcus* is dominated by spherical-shaped bacteria, while the genus *Lactobacillus* is so named because the microbes are rod shaped.

The Yeast Cell

Yeasts are members of the class of organisms known as fungi and as such are eukaryotes. Eukaryotes have a much more complex structure than prokaryotes, which include the bacteria. However, many of the structures that make up eukaryotic cells are quite similar to those that make up prokaryotic cells. The biggest differences are that eukaryotes generally have a nucleus, lack a cell wall, and lack a capsule.

Saccharomyces cerevisiae The scientific name for a fungus capable of converting wort into beer by fermenting sugars into ethanol and carbon dioxide.

There are many different yeasts, but we will focus on **Saccharomyces cerevisiae**, commonly known as brewer's yeast. Unlike most eukaryotes, *S. cerevisiae* has a cell wall that aids in protecting the cell while it's in a harsh environment, such as beer. Figure 5.13 illustrates the structure of the yeast cell. As shown in part A of the figure, yeast has a cell wall outside the membrane that's made up of protein; lipids (fats); and polysaccharides (polymers of small sugars) made up of mannan, chitin, and glucan.

chitin A hard, rigid, structural compound; makes up a bud scar on the yeast cell.

Yeast cells reproduce by budding (Fig. 5.13, part B). A daughter yeast cell forms from the existing cell wall and is filled with the appropriate organelles, nucleus, and other materials it needs to survive. When it has completely formed, it separates from the mother cell, leaving a bud scar made up of **chitin**. Chitin is rigid, structural, and very hard; it's the same polysaccharide that crab shells are made of.

invertase An enzyme on the outer surface of the yeast cell that converts sucrose into glucose and fructose.

Located on the outer surface of the yeast cell are proteins that can be used for flocculation and enzymes that can interact with the environment, including an enzyme known as **invertase**. Invertase converts sucrose (cane sugar) into two smaller molecules, glucose and fructose. Invertase is necessary because sucrose cannot cross the cell wall, but glucose and fructose can. Maltose and other fermentable sugars can cross the cell wall and the membrane. Using invertase, yeast cells can consume sucrose in addition to maltose.

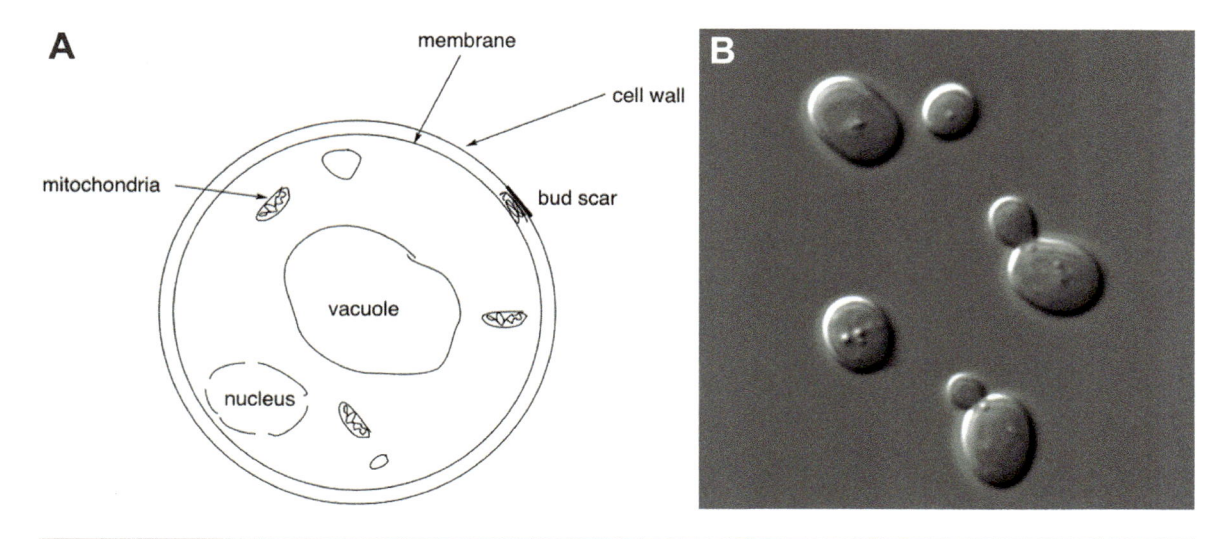

FIG. 5.13. The structure of brewer's yeast. **A,** The major parts of the cell. **B,** A budding yeast cell. (**A,** Courtesy S. E. Johnson and M. D. Mosher—© ASBC. **B,** Courtesy Wikimedia Commons.)

Inside the yeast cell is a nucleus, which contains the DNA for the organism (Fig. 5.13, part A). Inside the cell are also organelles, including storage granules in which fats and other compounds are placed. Mitochondria, the energy-producing organelles, are distributed throughout the cell. The most readily apparent structure in the cell is the vacuole; it's often the largest structure visible under a microscope. The vacuole is the storage location for fermentable sugars.

Basic Metabolism

Metabolism is the biochemical word that refers to the biological reactions in a living organism. An organism's metabolism either breaks down compounds (**catabolism**) or makes new compounds (**anabolism**). When compounds undergo catabolism, three major outcomes occur:

1. The food that has been taken in is broken down into small molecules.
2. Those molecules are excreted as waste or used to build new molecules.
3. Some molecules with highly energetic bonds are made during the catabolism. These molecules help drive other biological reactions in the organism.

In the first steps of metabolism in yeast, maltose enters the yeast cell and is hydrolyzed into 2 molecules of glucose. The glucose is passed to the mitochondria as energy, which is needed by the cell. The glucose then undergoes a series of biological reactions to harvest that energy. The first series of reactions is **glycolysis**.

In the presence of oxygen (**aerobic conditions**), the yeast cell uses the products of glycolysis to complete another series of reactions, the **Krebs cycle** (also called the citric acid cycle). Then the products of that cycle are passed

metabolism The biological reactions in a living organism.

catabolism The breaking down of compounds via metabolism.

anabolism The making of new compounds via metabolism.

glycolysis The metabolism of glucose into pyruvate for the cell to use as energy.

aerobic conditions An environment characterized by the presence of oxygen.

Krebs cycle A sequence of reactions by which most living organisms produce energy when presented with aerobic conditions.

oxidative phosphorylation Reactions that occur after the Krebs cycle to add phosphorous to the molecule; results in the production of ATP.

on to the last series of reactions, **oxidative phosphorylation**. Oxidative phosphorylation is particularly beneficial for organisms. Overall, three biological series of reactions produce a number of different outcomes:

1. **A tremendous amount of energy for the cell.** This energy can be used by the cell to grow and reproduce.
2. **Excess products from the Krebs cycle.** These products can be used to prepare other molecules using anabolic reactions, including fats, oils, polysaccharides, amino acids, proteins, and other molecules the cell needs to survive.
3. **Waste products that are excreted from the cell.** These include some of the excess products from the standard biological reactions and other side reactions.

The energy for the cell is stored in organic molecules. Two of the most important energy molecules are **adenosine triphosphate (ATP)** and **nicotinamide adenine dinucleotide (NADH)**. These molecules cycle between the low-energy adenosine diphosphate (ADP) and the high-energy ATP as they force other reactions to move forward. The NADH cycles between the low-energy NAD^+ and the high-energy NADH forms of the molecule.

adenosine triphosphate (ATP) An organic molecule used to store energy in living organisms; the low-energy form is ADP.

nicotinamide adenine dinucleotide (NADH) An organic molecule used to store energy in living organisms; the low-energy form is NAD^+.

anaerobic conditions An environment characterized by the absence of oxygen.

fermentation The metabolism pathway that occurs under anaerobic conditions to convert pyruvate to ethanol.

When oxygen is scarce or absent (**anaerobic conditions**), yeast cells continue to use metabolism to create energy and make the molecules needed to live. The yeast cells still consume maltose and other fermentable sugars. Those sugars are converted into glucose, and the glucose enters the first series of biological reactions (glycolysis). However, without oxygen, the Krebs cycle cannot be completed. Some of the reactions will occur, but the ultimate purpose of the cycle won't be achieved. Oxidative phosphorylation will not occur at all. To generate the energy needed to survive, the yeast cell follows a pathway known as **fermentation**. In this state, the cell doesn't produce enough energy and products from the series of biological reactions to grow and create new yeast cells. However, it can survive.

Glycolysis

Glucose begins the first series of reactions in metabolism (Fig. 5.14). The reactions add a phosphate group and then rearrange the molecule into a fructose-containing a phosphate. Another phosphate is added, and the resulting molecule is broken into two pieces. Each piece then undergoes more phosphate and rearrangement reactions, until it forms a molecule of pyruvate. The entire process produces 2 molecules of ATP and 1 molecule of NADH per glucose molecule. The pyruvate is then converted into a molecule known as acetyl coenzyme A (acetyl CoA) and produces 1 molecule of NADH.

The Krebs Cycle

Acetyl CoA starts the Krebs cycle. In this series of reactions, the acetyl group (a 2-carbon unit) of acetyl CoA is added to a molecule with 4 carbons. The 6-carbon molecule rearranges and eliminates 2 molecules of CO_2 and generates 2 high-energy NADH molecules. The resulting 4-carbon molecule rearranges back into the molecule, which can accept another acetyl CoA. Then the cycle runs again.

Along the way, some of the products made during the cycle can be redirected to make other compounds. Pyruvate, ketoglutarate, and oxaloacetate are key compounds that can be used to make other compounds. Acetyl CoA and some of the amino acids can be redirected to prepare fats.

Oxidative Phosphorylation

The NADH generated by the other processes of metabolism is utilized by a final process in the cell. The molecule is oxidized, and the result generates electrons that are donated to oxygen gas (O_2). The result of the process is the formation of water and ATP. Based on 1 glucose molecule, the outcome in oxidative phosphorylation is the production of 32 ATP molecules.

Overall, when oxygen is present, 1 glucose molecule results in the formation of 36 ATP molecules, 6 water molecules, and 6 molecules of CO_2. The overall process also consumes 6 molecules of molecular oxygen, which is a tremendous boost of energy per molecule of glucose. The result is enough

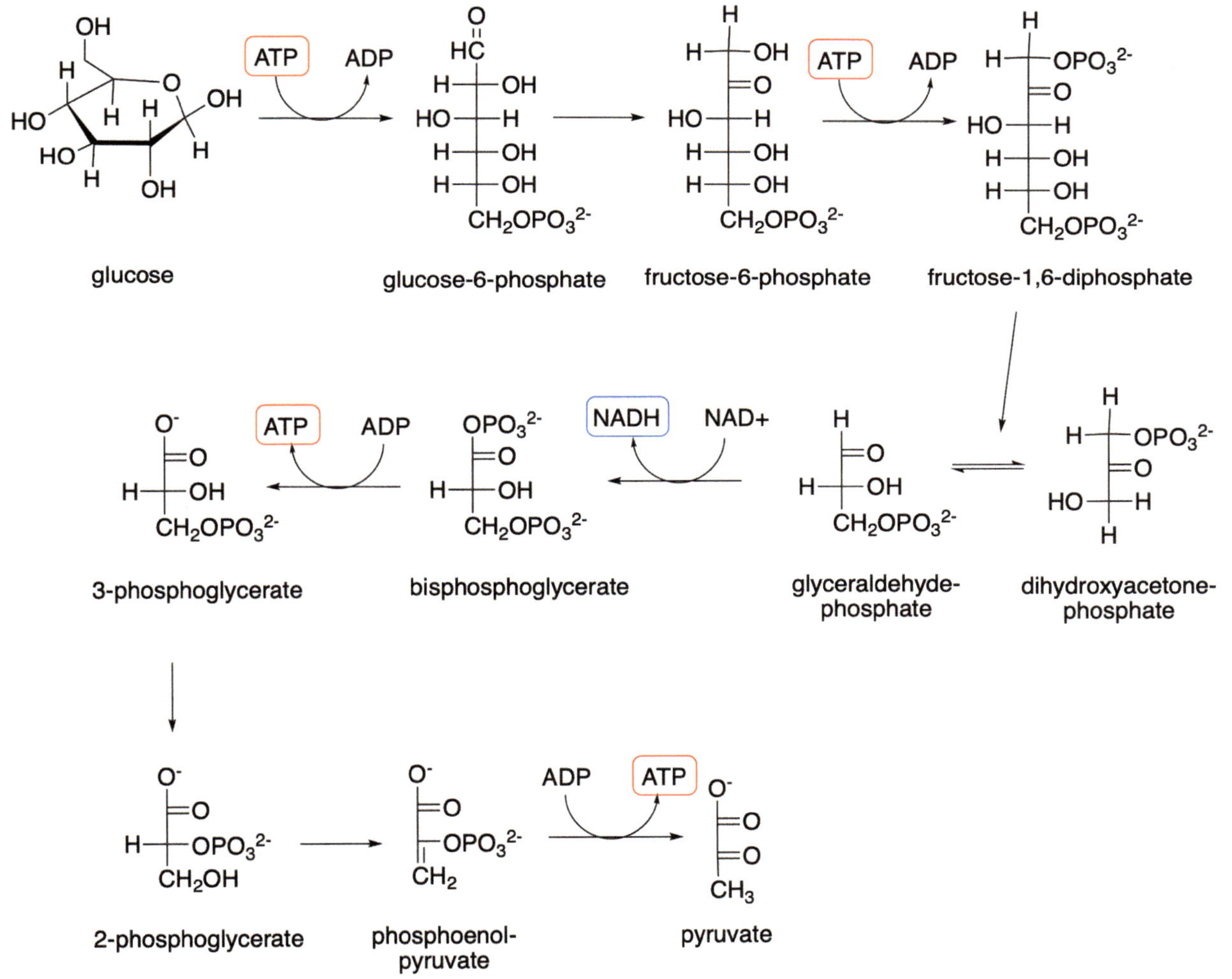

FIG. 5.14. Glycolysis. The process begins with glucose in the upper-left corner and proceeds across and down according to the arrows. Note that 2 ATP are consumed in the series, but because the fructose-1,6-diphosphate is broken into two pieces (and each piece continues the series), the net result is the production of 2 ATP and 2 NADH. (Courtesy S. E. Johnson and M. D. Mosher—© ASBC)

energy to grow; to make amino acids, proteins, and fats; and to thrive in an environment. In fact, so much energy is produced that the consumption of glucose doesn't have to be great to allow the yeast to reproduce:

$$\text{glucose} + 6\ O_2 \rightarrow 6\ H_2O + 6\ CO_2 + 36\ ATP$$

Fermentation

When the concentration of oxygen in the wort is depleted, yeast cells cannot complete the Krebs cycle or oxidative phosphorylation. In addition, without oxygen, the cells cannot prepare fats and many of the other products required to grow. The yeast cells enter a survival mode and consume glucose so they can continue to live. The glycolysis pathway will provide the basic energy that the cells need and allow the production of pyruvate, which can still be used to perform other reactions.

The amount of NAD^+ increases during fermentation. This low-energy state of the yeast cell signals a need to create more energy. In this case, the pyruvate produced from glycolysis undergoes two reactions that provide the needed energy. The first reaction produces 1 molecule of CO_2 and 1 molecule of acetaldehyde (CH_3CHO). The acetaldehyde is then reduced to ethanol as a new molecule of NADH is formed. The ethanol is a waste product and excreted from the cell. Likewise, the CO_2 is not used by the yeast and escapes the cell.

The overall result is the conversion of 1 molecule of glucose into 2 molecules of ethanol, 2 molecules of CO_2, 2 molecules of water, and 2 molecules of ATP (shown in the formula below), which produces considerably less energy than when oxygen is present. Note as well that the overall process doesn't make any NADH, as is to be expected, because the oxidative phosphorylation reactions are not available to use this molecule:

$$\text{glucose} + 2\ ADP \rightarrow 2\ H_2O + 2\ CO_2 + 2\ \text{ethanol} + 2\ ATP$$

As the amount of fermentable sugars decreases, the use of other materials in the wort becomes important. The yeast cells take in molecules such as diacetyl and reduce them to generate more NAD^+. By generating more NAD^+, the cell is able to metabolize more glucose. The overall result is the formation of more energy.

Requirements for Metabolism

Yeast cells require more than just maltose in the wort to survive. As they grow, they need additional amino acids, which they use to make proteins. They also require phosphate anions, which are used in glycolysis and in the conversion of ADP to ATP. Metal cations are also required. For example, **zinc**, iron, and copper are necessary micronutrients that bind to certain enzymes to make them active. Without these metal cations, the enzymes can't perform their specific biological reactions. The cell's metabolism and other processes slow down and even stop working altogether when the concentrations of these metals aren't large enough. Unfortunately, too much of a good thing is bad. Too much zinc, for instance, and the yeast die.

zinc A micronutrient needed for healthy yeast growth.

CHAPTER 6

Working in the Lab

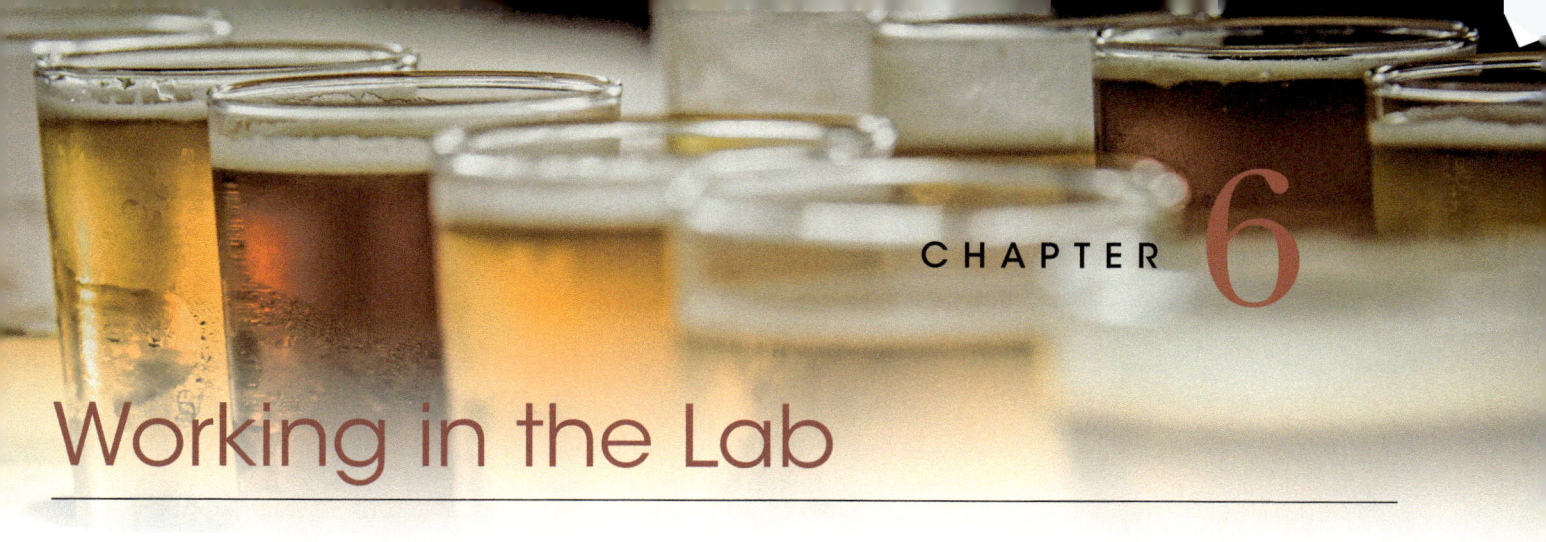

A lab can be a daunting place to work, and a brewery laboratory is no exception. However, for someone who is familiar with common items in the laboratory, who understands the chemical glassware commonly used, who knows the purpose of each instrument, and who knows the operation of the lab, working in the lab is actually fun. Various kinds of equipment, such as graduated cylinders and volumetric flasks, may be recognizable from the science classes we took in high school. Regardless of what's available in the lab, it's important to know the different types of glassware and what they are used for, as well as how they should and should not be used. The same is true for the other features of the lab. This chapter will cover these topics in detail.

This chapter will also explore a common fear that many people have when they first begin to work in a chemical laboratory: breaking something. Glassware—no matter how high the quality or the type of material from which it's made—is still glass. Even worse, because new laboratory staff are usually not familiar with the glassware they are using, they may not be aware of how much it costs. Yes, glassware can be expensive, but lab workers should not be deterred from using it. With training, they will gain the needed confidence and pay careful attention to the tasks they perform.

Let's consider the glassware used to measure out 100 milliliters (mL) of water. A 100 mL beaker may cost less than $1, but a 100 mL volumetric flask may cost as much as $50. Both flasks will hold 100 mL of water or any other liquid, so the beaker seems like a better purchase. In this chapter, we will consider that difference from both the cost perspective and the utility perspective. Let's start our exploration of the lab by diving into the world of laboratory glassware.

Laboratory Glassware

To use the appropriate piece of glassware during an analysis, the brewery laboratory staff needs to understand the difference between accuracy and precision. We discussed these concepts earlier in this text, but a brief summary will be helpful here. In terms of the glassware used in measuring a liquid's volume, accurate glassware is that which obtains the same volume repeatedly. Precise glassware is that which can distinguish between two very

borosilicate glass Relatively fragile glass that can't withstand large, rapid temperature changes.

Pyrex A brand of glass that can withstand large rapid temperature changes.

similar volumes. As those volumes get more and more similar, the glassware needs to be more and more precise to distinguish between them.

Two main types of glass are used in the lab: **borosilicate glass** and **Pyrex**. Both are made from a relatively similar mixture of chemical compounds, including silica (SiO_2), boric acid (B_2O_3), soda ash (Na_2CO_3), and aluminum oxide (Al_2O_3). Unfortunately, on its own, borosilicate glass is relatively brittle and unable to handle large changes in temperature without shattering. This problem has been fixed in Pyrex, to which small amounts of calcium, magnesium, and iron oxides have been added. Introducing these metal cations into the crystal structure of the glass makes Pyrex less likely to break when dropped, hit, or subjected to thermal stress. However, Pyrex is still glass and will break under enough force. Plastic "glassware" could be used to perform the measurements needed, but plastic items are much less able to handle the temperatures of boiling water. They are also pliable, which even at room temperature directly affects their ability to be precise and accurate.

The increased resistance to temperature and shock makes Pyrex a better choice for use in the lab, but it comes at a cost. Pyrex items can cost two to four times as much as borosilicate items, which are readily available and inexpensive. For single-use test tubes and other glassware that doesn't need to be resistant to temperature changes, using borosilicate items is often the best way to go. For everyday use and for uses that involve significant temperature changes, such as heating to boiling and then cooling in ice, Pyrex items must be used. They cost more, but the cost is necessary to avoid breakage.

Types of Glassware

beaker Glass shaped into a cylinder that looks like a cup.

test tube Glass shaped into a tall, slender cylinder; requires a rack to stand upright.

test tube rack A device made to hold multiple test tubes.

powder funnel A funnel with a large opening; typically used in transferring solids.

liquid funnel A funnel with a narrow opening; typically used in transferring liquids.

funnel support A device made to support a funnel above the piece of glassware into which a chemical is transferred.

ring clamp A metal circle that can be attached to a vertical pole.

ring stand A vertical pole attached to a heavy base so that it resists tipping.

What major types of glassware are used in the lab? There are many types—more than can be covered in this discussion. Figure 6.1 illustrates the most commonly used pieces of glassware.

Beakers are useful for holding used liquids and for measuring out liquids when accuracy and precision are not of concern. Many beakers have volume markings on the outside, but these markings can vary as much as plus or minus 10–15% in terms of accuracy. That means when a beaker is used to measure a 100 mL sample of water, it could contain anywhere from 85 mL to 115 mL.

Test tubes tend not to have markings denoting volume. However, they are useful for performing reactions on small volumes and for performing multiple analyses using the same reagents over and over. Test tubes can be used to create a color scale for use in colorimetry, and they are quite useful for storing yeast. The biggest issue for test tubes is that they require the use of a **test tube rack** to stand upright. Unless they are in some kind of a holder (even a beaker), their contents will spill.

Funnels come in two styles. A **powder funnel** has a large opening at the bottom, which makes it ideal for transferring solids from one container to another. It also works well for transferring nonfiltered liquids. A **liquid funnel** has a small opening, which makes it useful for filtering a solution. (The method for accomplishing that task is discussed later in this chapter.) A **funnel support** is used to hold a funnel above the piece of glassware into which the material will be transferred. A gap between the funnel and the glassware is required to avoid creating a vapor block, which will stop the flow of liquid. Alternatively, a **ring clamp** and **ring stand** can be used for the same purpose. When liquids

Working in the Lab **95**

are being filtered by vacuum, a **Büchner funnel** and **vacuum filter flask** are used. A Büchner funnel is ceramic or plastic and has a flat, perforated bottom; filter paper is required that exactly covers the bottom. This type of funnel attaches to the vacuum filter flask, which has a one-holed stopper. The flask has a nipple on the side, where a vacuum can be applied.

An **Erlenmeyer flask** and a **Florence flask** have similar uses in the lab. Each has a small opening, which allows the user to swirl liquid inside it without spilling or splashing. The shape of the Florence flask makes it particularly useful for swirling. An Erlenmeyer flask is also very useful for heating liquids, which tend to evaporate more slowly when heated in this type of flask than when heated in a beaker.

Büchner funnel A ceramic or plastic funnel with a flat, perforated bottom.

vacuum filter flask A heavy-walled flask that supports a Büchner funnel.

Erlenmeyer flask A conical-shaped beaker.

Florence flask A spherical-shaped beaker with a flat bottom.

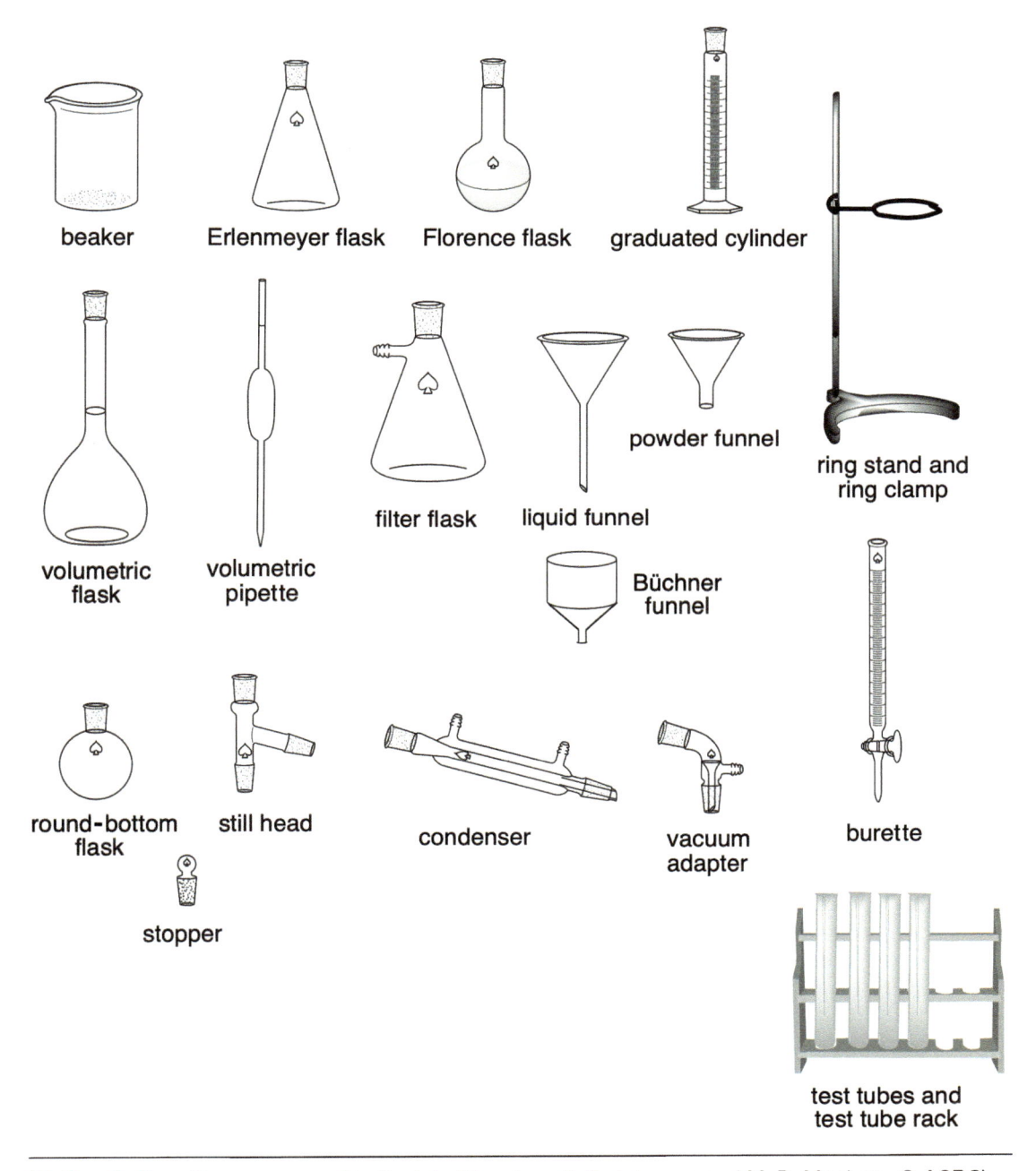

FIG. 6.1. Glassware used in the lab. (Courtesy S. E. Johnson and M. D. Mosher—© ASBC)

volumetric flask A flask that is similar in appearance to a Florence flask but carefully machined and marked with a specific volume.

volumetric pipette A tube that typically has a wide portion in the middle and is carefully machined and marked with a specific volume.

graduated cylinder A cylinder marked along the side with volumes.

graduated pipette A tube marked along the side with very specific volumes.

burette A tube that's similar in appearance to a graduated pipette but contains a stopcock at the bottom.

round-bottom flask A spherical flask used for heating liquids in the laboratory.

condenser A tube surrounded by a jacket that can be filled with cold water or another fluid.

still head An h-shaped tube that allows connecting three glassware items.

vacuum adapter A bent tube fitted with a nipple, where a vacuum can be applied.

Volumetric flasks are denoted by their very long necks, which have only single volume markings. When a volumetric flask is used correctly, the volume printed on it is both accurate and precise. Depending on the flask, the level of precision can be to one, two, three, or even four decimal places beyond the unit measurement. If used incorrectly, however, a volumetric flask has very little precision. Also, volumetric flasks can be quite expensive.

Similar to volumetric flasks are **volumetric pipettes**. This type of pipette requires using a bulb to withdraw liquid from a cache of liquid (perhaps a beaker). It has only one marking near the top, which denotes the specific volume with great precision. When used correctly (as illustrated later in this chapter), volumetric pipettes are very precise. Volumetric pipettes can be marked "to contain" (TC) or "to deliver" (TD) on the glass. Glassware of any kind that's denoted TC will contain the exact amount reported when used properly. But when the content is poured out into another vessel, some of it will remain inside the container or adhere to the sides. No matter how much you shake or tap, that additional liquid won't come out. As a result, the volume in the new container isn't the same volume that was in the TC measuring device. TD glassware delivers the volume you are measuring. If you measure exactly 100 mL in a volumetric pipette that's TD and then put that volume in another container, you will transfer exactly 100 mL. Unfortunately, the requirement for delivering the exact volume is that the liquid must be water (or a very dilute solution in water). Measuring 100 mL of hexane with a TD pipette won't result in exactly 100 mL of hexane in the other container. It will be close but not exact, so precision disappears in this case. Like volumetric flasks, volumetric pipettes can be expensive.

Graduated cylinders are less precise than volumetric glassware, but they still maintain some level of precision. They are also faster to use, and when the precision doesn't need to be especially high, they are the best choice for measuring liquids. The marks on the side of a graduated cylinder denote the volume from the bottom up. Unfortunately, most graduated cylinders are TC, so the volume poured from the cylinder isn't exactly the volume indicated on the side.

Graduated pipettes are basically graduated cylinders that have a scale on the side. A graduated pipette is used by filling the pipette using a bulb, recording the volume in the pipette, and then releasing the desired volume. The bottom of the graduated pipette usually doesn't have a scale, so the user must be careful not to go below the bottom mark. Just like a graduated cylinder, a graduated pipette is less precise than a volumetric version, and it costs less, as well.

A **burette** can also be useful in the lab. A burette is basically a graduated pipette with a stopcock at the bottom. The scale printed on the side is upside down, so the "0" volume mark is at the top. A burette is filled by pouring in liquid from the top. To use it, the volume is recorded and then a certain amount is delivered by opening the stopcock. After the final volume is read, the amount of liquid that was delivered is determined by subtraction.

Specialized glassware can also be found in the lab, including **round-bottom flasks**, **condensers**, **still heads**, and **vacuum adapters**. Each of these pieces usually has a cloudy surface, where it can be connected to another piece of glass that has the same cloudy surface. In the past, making that connection was accomplished by using corks and grease or by finding two pieces of glass that fit each other. Currently, all the joints have the same slope, so they fit together snugly. This is known as the **standard taper (⚶)**, which is noted on the glass with an "S" overprinted on a "T." The numbers beside this symbol represent the diameter of the opening in millimeters and the length of the joint in millimeters. In the brewery laboratory, an ⚶ of 19/22 or 24/40 is a common size. Simply making sure that all the glassware has the same standard taper will ensure that individual pieces can be assembled to make the equipment needed for the experiment.

Other items typically found in the lab include a **spatula** (or scoop) used to handle solids; a **thermometer; stoppers** (glass, rubber, or cork); a **test tube holder** (a device that allows the user to handle test tubes when they are hot); a pair of **tongs** (used to handle other hot items); and a **Bunsen burner**. Each of these items is used to assist in performing the jobs conducted by the lab.

The Meniscus

Precise glassware is used primarily to measure liquids and/or prepare liquid solutions. It's important to note, however, that even though many beakers and Erlenmeyer flasks have volume markings, they provide gross approximations and should not be used for making accurate and precise liquid measurements. Solids are easily measured out using a balance or scale, but for liquids, the task is more difficult. Obtaining the mass of a liquid on a balance is possible, but the liquid tends to adhere or wet the sides of the glassware in which it's placed. Thus, an accurate amount of liquid delivered into a container isn't determined. Because of this, graduated or volumetric glassware is used to measure liquids. Each measurement is made by reading the meniscus of the liquid when it's in the glassware.

The **meniscus** is the curve in the upper surface of a liquid in a tube at the mark where a measurement is being taken. A concave meniscus forms when the liquid interacts strongly with the glassware, as in the case of water. The water then rises along the edges of the glass, giving a concave appearance to the upper surface of the liquid. A convex meniscus forms when the liquid doesn't interact with the glassware and contracts down below the surface of the liquid. Most liquids form a concave meniscus; mercury and a few other liquids form a convex meniscus.

The appropriate way to record a measurement is to read and report the marking at either the bottom of a concave meniscus or the top of a convex meniscus. Figure 6.2 shows how to read a concave meniscus. The most important part of reading the meniscus is to make sure that your eyes are parallel to the markings on the glass. If your eyes are not parallel, you will not read or report the volume accurately.

Using Graduated Glassware

It's appropriate to report any measurement in a graduated cylinder, graduated pipette, or burette to one decimal place further than that provided by the markings on the glass. Doing so requires the analyst to estimate that last number. In fact, it's rare to have the last number in a measurement be a 0. For example, if the meniscus of the liquid rests exactly halfway between 24.5 mL and 24.6 mL, the measurement should be recorded as 24.55 mL.

standard taper (⚶) Standardized dimensions of a glass joint that include the angle of the joint.

spatula A stick made of glass, metal, or wood that is wide at one or both ends; used to scoop solids.

thermometer A device used to measure the temperature of a substance.

stopper A plug inserted into a glassware item to close it; made of glass, rubber, or cork.

test tube holder A twisted wire device that can be used to hold a test tube.

tongs A scissor-like utensil used to hold hot objects.

Bunsen burner A gas-powered device that produces a single open flame.

meniscus The curve in the upper surface of a liquid in a tube.

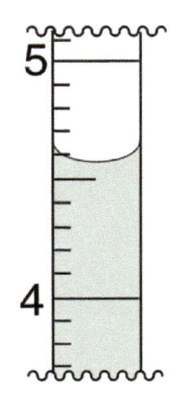

FIG. 6.2. Reading a meniscus. Note that the reading is accurate only when your eyes are parallel to the markings on the glassware. Here, the bottom of the concave meniscus appears to be at 4.57 mL. (Courtesy S. E. Johnson and M. D. Mosher—© ASBC)

Failing to report this last number is a common error in using graduated glassware. Initially, this doesn't seem to be much of an issue, but it can result in final values for an analysis that are off by 10% or more (Fig. 6.3). That's a big difference, and it can result from failing to estimate that last number. For example, suppose that a graduated pipette is used and the initial volume measurement is 1.51 mL and the final volume measurement is 2.59 mL (so 1.08 mL volume delivered). If the analyst doesn't write down the last decimal in each number, the error in the volume will be 8%. If that pipette is used multiple times to perform an experiment, the overall error in the experiment will continue to increase to the point where the analysis becomes meaningless.

Using Volumetric Glassware

When volumetric glassware is used, the analyst is concerned only with one mark on the glass. Filling the glassware so that the meniscus sits perfectly on top of that mark will result in an exact volume measurement. So, how do we make sure our measurements are exact?

Let's assume we need to make 1 liter (L) of a 1.000 molar (M) solution of sodium chloride in water. Remember from the chemistry chapter (Chapter 4) that the molarity of the solution is the number of moles of the solute (in this case, NaCl) per liter of solution. We calculate that 1 mole of NaCl is 58.44 grams (g) of NaCl, and so we obtain that mass of salt and add it to a 1 L volumetric flask (Fig. 6.4). We then add some distilled water to the flask and swirl it to try to dissolve the salt. We add only 300–400 mL of water (no need to measure it exactly), because all we need to do is dissolve the salt. When it has dissolved, we add additional distilled water until we are about 2 centimeters (cm) below the mark on the neck of the flask. Then we fold over a piece of blotter paper or paper towel until it can fit into the neck. Very carefully, we lower the paper into the flask and rub it against the sides of the neck above the

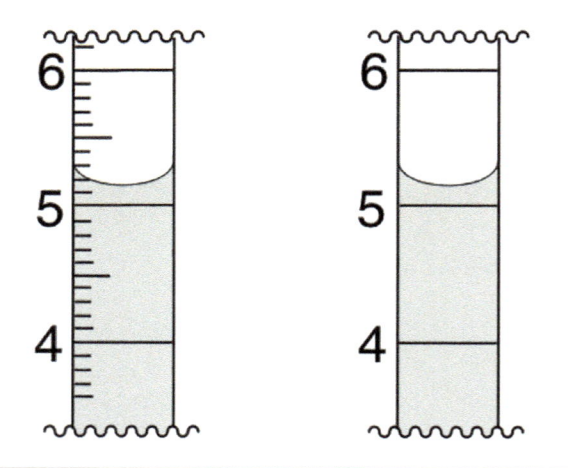

FIG. 6.3. Precision in glassware. The scale on the right suggests that we read the level of the liquid to one decimal and get 5.2 mL as the volume. The scale on the left suggests that we can see an additional decimal, and we would get 5.25 mL as the volume. (Courtesy S. E. Johnson and M. D. Mosher—© ASBC)

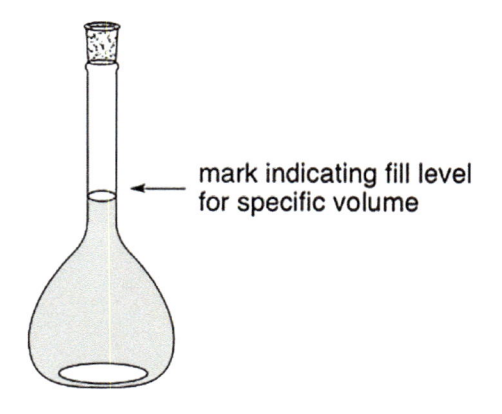

FIG. 6.4. Volumetric flask. The mark indicating the fill level is valid only at a given temperature (typically printed on the flask). (Courtesy S. E. Johnson and M. D. Mosher—© ASBC)

mark. The point of doing this is to remove water that may have adhered to the sides of the neck. Once the neck above the mark has been completely wiped down, we use a disposable pipette and add water to the flask by dropping it directly into the solution. We avoid making any contact with the neck while doing this. If we accidentally get some water on the neck above the mark, we remove it with the folded paper before we continue. We also do everything we can to avoid splashing while adding the drops of water. Finally, we add drops and pay very close attention to the meniscus until it sits exactly on top of the mark. But the solution isn't ready yet. We still need to mix it thoroughly. We do this by placing a stopper in the flask and inverting the flask multiple times. After inverting the flask, the volume will appear to be lower than the mark. This is okay. It's the result of some of the liquid being stuck to the lid and sides of the neck. No additional solvent should be added at this point.

To be accurate, we need to make sure that the water or other solvent we are using is at the appropriate temperature. The volume of a liquid changes as the temperature changes. Thus, to measure the volume as accurately as possible, the temperature of the liquid must match the temperature recorded on the glassware we use. Typically, this is 22°C (72°F), but it may be different depending on the glass manufacturer.

Note that when you make the solution, the liquid level might be lower than the mark. This is because some of the liquid is now adhering to the neck and the stopper. This doesn't matter, because the solution is mixed completely. Don't "top off" the volumetric flask, or the concentration won't be what was calculated. We just have to accept that instead of 1,000 mL (1 L), we might have only 999 mL available to use.

To obtain a specific volume of a solution, we can use a volumetric pipette. This, of course, assumes that we have access to a pipette that's calibrated to the volume we need. Common volumes for commercially available pipettes are 1 mL, 2 mL, 5 mL, 10 mL, 20 mL, 25 mL, 50 mL, and so on.

To use the pipette (Fig. 6.5), some of the solution from the stock bottle is poured into a beaker. It's important to have a little more in the beaker than is needed. The right hand should be placed near the top of the pipette so the thumb can be placed on the end. The left hand should squeeze a bulb and place it on the end of the pipette. Rather than push the bulb all the way onto the end of the pipette, it's better to push gently to maintain a tight seal. The end of the pipette is placed into the beaker just above the bottom but not all the way down. The pipette should not be forced toward the bottom of the beaker. It's very fragile, and contact such as this may damage or break the pipette. The bulb should be slowly released, so that it sucks the liquid up into the pipette.

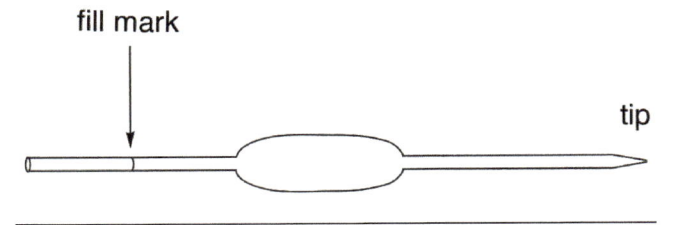

FIG. 6.5. Volumetric pipette. (Courtesy S. E. Johnson and M. D. Mosher—© ASBC)

The level of liquid is allowed to rise above the mark on the pipette. Then, in one quick motion, the bulb is pulled away from the pipette with the left hand while the thumb of the right hand is placed on the opening of the pipette. It takes some practice, but the process of removing the bulb and sealing the end of the pipette with the thumb should not change the volume of liquid in the pipette by very much. Finally, the thumb is slowly tilted back and forth until the seal is slightly broken. This causes the level of liquid in the pipette to move down slowly toward the mark on the pipette. When the meniscus lies exactly on the mark, the thumb seal should be tightened so the liquid stops moving.

The pipette is then withdrawn from the beaker and the tip is touched against the side to remove any excess liquid. Then the pipette is gently moved to the flask where the liquid is needed and the thumb is removed. Holding the pipette still causes all the liquid in it to drain into the flask. Next, the pipette is gently moved to the top of the flask and the tip is touched against the glass to remove any last remaining liquid. There should be a last drop or two inside the pipette. That liquid is supposed to remain, because it's factored into the location of the mark on the glass.

Two mistakes are commonly made when using a pipette. The first is shaking, jostling, or quickly moving the pipette when it's full of liquid. These kinds of motion can cause a small amount of liquid to shoot from the tip of the pipette. Evidence that this has happened can sometimes be indicated by the presence of bubbles near the tip of the pipette. If this does happen, the pipette should be emptied and refilled. The second common mistake is to use the bulb to squeeze out the content of the pipette. Doing so can decrease the time it takes to deliver the liquid, but if the process is continued beyond the large portion of the pipette, the amount of liquid remaining in the pipette may not be accurate. Thus, the measurement of the exact volume will be off. In this case, the liquid should be returned to the beaker (if possible) and the pipette refilled.

Filtering Liquids

There are times that a liquid must be separated from a solid in the lab. The liquid could be decanted in much the same way that a bottle of homebrew might be poured so the yeast cake doesn't get transferred. But this method doesn't get all the liquid out of the mixture. It also doesn't keep all the solids away from the liquid just transferred. In cases in which a clear liquid free from any solids is needed, we use the technique of filtering.

Gravity filtering is a method that isolates the liquid without concern for the quality of the solid that's isolated (Fig. 6.6). To accomplish the task, a piece of filter paper is folded in half, and a small corner of the fold is torn off and discarded. Then the paper is folded in half again. The resulting paper looks like one-quarter of a pie. One leaf of the paper that isn't torn is pulled up while the sides of the piece of pie are gently pushed toward each other. This makes a small cone that's almost the exact dimensions as the glass liquid filter. An alternative to this method is to use fluted filter paper, which looks similar to a coffee filter. Fluted paper can be purchased prefolded or made by folding the paper in half and then folding it in an accordion-like method. (Many online videos show how to make fluted filter paper from nonfluted filter paper.) Using fluted filter paper can decrease the filtration

gravity filtering Draining a liquid using a funnel and filter paper.

FIG. 6.6. Gravity filtration setup. (Courtesy S. E. Johnson and M. D. Mosher—© ASBC)

time by allowing air to pass through the funnel along the outside of the filter and help avoid pressure differences. The additional surface area of the folded filter paper also increases the flow rate of the liquid.

The filter paper is then placed inside the glass filter. At this point, the filter paper is wetted with a small amount of liquid. The liquid that is used must be the same as in the solution to be filtered. For most applications, this means using water. However, if pure water is used to wet the paper, the concentration of the solution will change. Alternatively, a small amount of the solution to be filtered can be used to wet the filter paper. Doing so keeps the concentration of the solution the same. Whichever method is used, wetting is necessary to allow the filter paper to adhere to the glass and to improve the filter's performance. To use the filter, just pour the solution into the filter paper. It works best if the filter is about three-quarters full of liquid.

Vacuum filtering is another method used to separate liquids and solids (Fig. 6.7). While this method is used primarily to isolate a solid at the expense of the liquid, it can be modified to maximize the amount of liquid collected. In this method, a Büchner funnel is fitted with a piece of filter paper that just covers the holes in the funnel. It is attached to a vacuum filter flask and the vacuum hose is attached. The filter paper is wetted by pouring a small amount of the solution to be filtered into the center of the paper. Then the vacuum is turned on and the solution is poured into the funnel. The funnel should be filled at least halfway to ensure that the solution passes through the filter paper instead of around it. The level in the funnel is topped off periodically until all the solution has been added. When all the liquid has passed through the funnel, the hose is removed. Then the vacuum is turned off. It's important to always remove the hose before turning off the vacuum.

vacuum filtering Draining a liquid using a funnel, filter paper, and vacuum; requires using a Büchner funnel.

Distillation

Some experiments require purifying a liquid by distillation, which involves boiling a liquid in one flask, condensing the vapors in a tube, and then transferring that condensate into another flask. This process is used to prepare alcoholic spirits such as vodka and whiskey. In the lab, distillation is

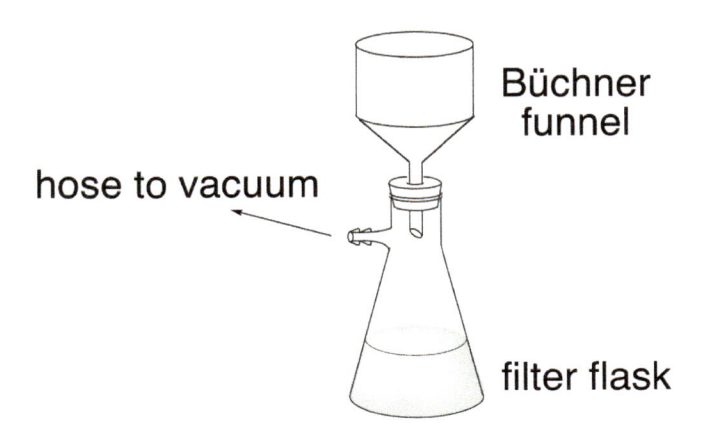

FIG. 6.7. Vacuum filtration setup. (Courtesy S. E. Johnson and M. D. Mosher—© ASBC)

performed on beer samples to separate analytes of interest with low boiling points from the bulk of the material in the beer. For example, one method for determining alcohol by volume involves distilling the beer before measuring the concentration of ethanol. (Details on the process used to distill a liquid are provided later in this chapter.)

Handling Chemicals

Handling chemicals can be incredibly dangerous if the analyst isn't trained appropriately. As mentioned before, it's essential for anyone handling a chemical to be familiar with the safety data sheet (SDS) information for that chemical. Knowing the safety information for each chemical is necessary to limit or prevent injuries and health risks.

That being said, most chemicals in the brewery laboratory are fairly straightforward to handle. Even so, adhering to some specific practices can make working with chemicals even less hazardous. In addition, knowing proper handling techniques can also increase the precision of the particular analyst and thus any analyses he or she performs.

Obtaining the Mass of a Solid or Liquid

Arguably the most important pieces of equipment in the laboratory are the balance and the scale (Fig. 6.8). These two instruments are not the same, although here on Earth, they provide the same result.

A **balance** consists of a pan on which something is placed and then counter-weighted with items of a known mass. When the pan becomes perfectly balanced, the **mass** of the item on the pan is determined. This can be done by reading a number on a digital display, by adding up the locations of sliding bars, or by counting the weights added to a second pan. The most basic

balance A piece of equipment used to measure the mass of an item.

mass A measure of the amount of matter.

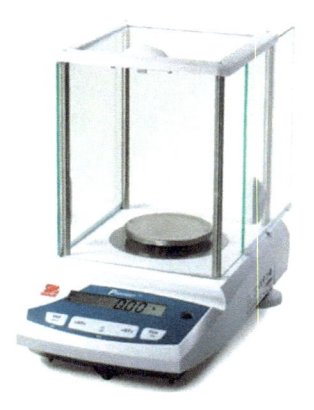

FIG. 6.8. Balances and scales are the most important instruments in the lab. Left to right, Triple beam balance, analytical balance, and scale. (Left and middle, Courtesy OHAUS—Reproduced by permission. Right, Courtesy Polder—Reproduced by permission.)

balance involves two pans and a pile of weights. Current technology allows the use of a cell that gives off an electrical signal based on how much the cell is displaced by the items on the pan.

A **scale** measures the **weight** of an item based on how much a spring moves when the item is placed on the pan. The spring is often attached to a scale that can be used to determine its displacement and hence, the weight of the item on the pan. For example, a hand-held scale can be used to determine the weight of a fish. The fish is placed on the scale, and the scale is then lifted up to eye level. The weight is determined by the location of the needle on the scale. Behind the scenes, the needle is attached to a spring that stretches with the weight of the fish.

Because of how a scale works, it determines only the weight of the item, not the mass. A balance, on the other hand, determines the mass of the item, not the weight. Here on Earth, the difference between weight and mass is only conceptual, but if we leave Earth, the conceptual difference becomes evident. The mass of an object is the amount of material it contains. The mass will remain the same no matter where the object is located. The weight of an object is the effect of gravity on a mass, which means it will change with a change in gravity. For example, the *mass* of a 25 kilogram (kg) bag of malt on Earth is 25 kg. It also *weighs* 25 kg here on Earth. But for the brewery of the future on Mars, where gravity is only 37.5% that on Earth, the distinction becomes obvious. The bag of malt will still have a mass of 25 kg but will weigh only 9.4 kg.

Because of the intricacies of balances and the precision to which they measure the mass of an object, they are much more expensive than scales. Moreover, scales are often much less precise and accurate than balances. So, the two have different uses in the brewery. Scales are used to determine the weight of malt, adjunct, and other ingredients in the preparation of a batch of beer. In these cases, the precision (and accuracy, within reason) in determining the mass of an object is less significant. Balances have uses that focus on obtaining the masses of small amounts of materials, for which precision and accuracy are vital, such as those used to make solutions to analyze beer in the lab.

To obtain the mass of a solid sample for use in the laboratory, a number of key steps must be followed. First, the pan and surrounding stage of the balance should be clean and free of debris. Dirt and debris interfere with the balance's ability to measure the mass of the item being weighed, cause corrosion of the balance, contaminate the samples, and might poison the user if accidentally touched. Cleaning immediately after use can be accomplished with a small, dry paint brush for minor spills of dry chemicals or a wet sponge for more messy spills.

Next, the pan should be lined with a disposable piece of **weigh paper** or a small beaker should be placed on the pan. Commercial weigh paper can be purchased, or in a pinch, wax paper can be cut into small squares. The weigh paper is folded and scored gently to create a pouring spout. Then the paper is placed on the pan and any doors on the balance are closed. The balance can be **tared** to "zero out" the mass of the paper or beaker and allow only the mass of the chemical to be reported.

Next, using a clean spatula or scoop and holding the bottle of chemical close to the pan, the chemical is added to the paper or beaker. Alternatively, the bottle can be held over the paper and tilted while rotating it slowly to

scale A piece of equipment used to measure the weight of an item.

weight A measure of the effect of gravity on an object.

weigh paper A piece of paper used to protect the balance or scale during a measurement.

tare To zero the balance or scale.

dispense the chemical. Care should be taken to avoid spilling any of the chemical on the pan during the process. If some is spilled, the paint brush should be used to sweep it from the pan. When the appropriate amount is obtained, any doors are closed to avoid drafts on the pan. The value of the mass is read from the balance and recorded to as many decimal places as the balance provides. The weigh paper can then be picked up with the hands and the chemical poured from the paper into the appropriate flask. As always, the lid is replaced on the bottle of chemical and the balance is then cleaned.

To obtain the mass of a liquid on a balance, the process is the same, but a small beaker is chosen to be the receptacle on the pan. Instead of pouring the liquid from the bottle, the liquid should be transferred using a disposable pipette. Any excess liquid removed from the bottle can be returned if it touched only the clean disposable pipette. So, best practice is not to overshoot the mass of the liquid in the beaker. Excess liquid in the beaker should be disposed of, rather than returned to the original bottle.

Obtaining the Volume of a Liquid

Often, the easiest way to measure the mass of a liquid is to obtain a specific volume of the liquid using a graduated cylinder, graduated pipette, or volumetric pipette. The specific glassware chosen depends on the level of precision needed in the measurement. A graduated cylinder has the least precision, and a volumetric pipette has the most. (The processes for using these items were outlined in the previous section.)

Once you know the volume, you can calculate the mass of the liquid if you also know the density of the liquid (Fig. 6.9). Multiplying the volume by the density will determine the mass of the liquid, assuming that the temperature of the liquid when its volume is measured is the same as the temperature of the density measurement. Because the density changes as the temperature changes, slight differences in temperature result in slight differences in the final mass that's calculated. In other words, if the mass of the liquid is determined by measuring the volume, the precision of the final mass will be less precise than if it were obtained on the balance.

$$density = \frac{mass\ (g)}{volume\ (mL)} = \frac{mass\ (kg)}{volume\ (L)}$$

$$1\frac{g}{mL} = 1000\frac{kg}{m^3}$$

$$mass = density \times volume$$

FIG. 6.9. Density units and calculations. Density is the mass of an object divided by its volume. In the lab, typical units are g/mL, which is 1/1,000 the value when the units are kg/m^3. The mass of a liquid can be determined from its volume using the density, assuming the units of the density contain the same unit as the volume. (Courtesy S. E. Johnson and M. D. Mosher—© ASBC)

If only the volume is needed, it can be obtained simply using a graduated cylinder, graduated pipette, or volumetric pipette. The level of precision in the measurement should determine the piece of glassware used for the operation. As noted before, it's important to remember that using TD and TC versions of glassware will influence how the sample is obtained and the accuracy of the volume.

Preparing a Solution from a Solid

A solution can be easily prepared from a solid. If you know the concentration and obtain the correct mass of the solid, you can prepare the solution using a volumetric flask. This process was outlined earlier in this chapter and is shown in Figure 6.10.

Preparing a Solution by Dilution

Often, a stock solution is prepared at a concentration that's much larger than needed in the method of analysis. The stock solution can then be kept for long periods of time and used when needed for an analysis. By diluting the stock solution, a solution of smaller concentration can be made for direct use in an analysis. The dilute solution can be prepared weekly or even daily as needed.

Calculating the amount of the stock solution required to create a less-concentrated solution is easily accomplished using a proportionality equation. That equation is known as the dilution equation,

$$C_1 \times V_1 = C_2 \times V_2$$

where C_1 is the concentration of the stock solution, V_1 is the volume of the stock solution, C_2 is the concentration of the desired solution, and V_2 is the volume of the desired solution. The concentration and volume can be in any units, such as molarity or parts per million (ppm). The key, however, is that both C_1 and C_2 must be the same units (and V_1 and V_2 must be the same units, as well). Then by plugging values we know into the equation, we can calculate the value that we don't know.

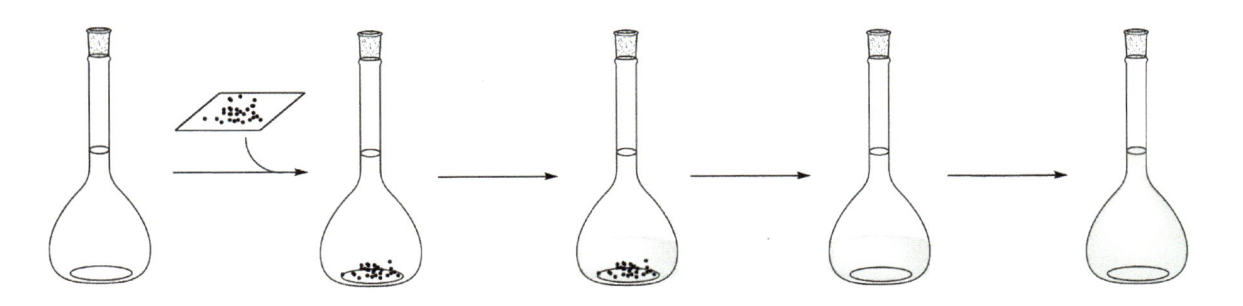

FIG. 6.10. Preparing a solution. A solid is added to a volumetric flask. Some solvent (in this case, water) is added, and the mix is swirled to dissolve the solid. Once the solid has dissolved, the solvent is added to the mark on the volumetric flask. (Courtesy S. E. Johnson and M. D. Mosher—© ASBC)

Let's assume we want to make 100 mL of a 0.25 M solution of glucose, and we have a stock solution that's 1.0 M glucose. We plug the numbers into the equation and solve for the value that's missing:

$$(1.0 \text{ M}) \times V_1 = (0.25 \text{ M}) \times (100 \text{ mL})$$

$$V_1 = \frac{(0.25 \text{ M}) \times (100 \text{ mL})}{(1.0 \text{ M})}$$

$$V_1 = 25 \text{ mL}$$

Based on our calculation, we need to obtain 25 mL of the stock solution (Fig. 6.11). We use a volumetric pipette to withdraw exactly 25 mL and then add it to a 100 mL volumetric flask, because we want to make 100 mL of the new solution. Finally, we add water up to the mark.

This dilution equation works with any concentration unit and any volume unit. Just as a reminder, the units we use for C_1 must be the same that we use for C_2 (similarly for V_1 and V_2). We might have to convert units to make sure we hold to this rule. For example, if we know the volume of a solution in liters and want to obtain a certain amount in milliliters after dilution, we will need to convert these units to be the same.

Distilling Liquids

As noted before, distillation is the process used to purify a liquid or to separate two liquids that boil at different temperatures. In the brewery laboratory, most of the distillations involve heating a water-based solution to remove or collect a low-boiling compound. In these cases, using a thermometer is essentially unnecessary. We know the boiling point of water, which is all the thermometer will show us during the process.

To perform a distillation, the specialized glassware is assembled as illustrated in Figure 6.12. A clamp is first attached to the round-bottomed flask

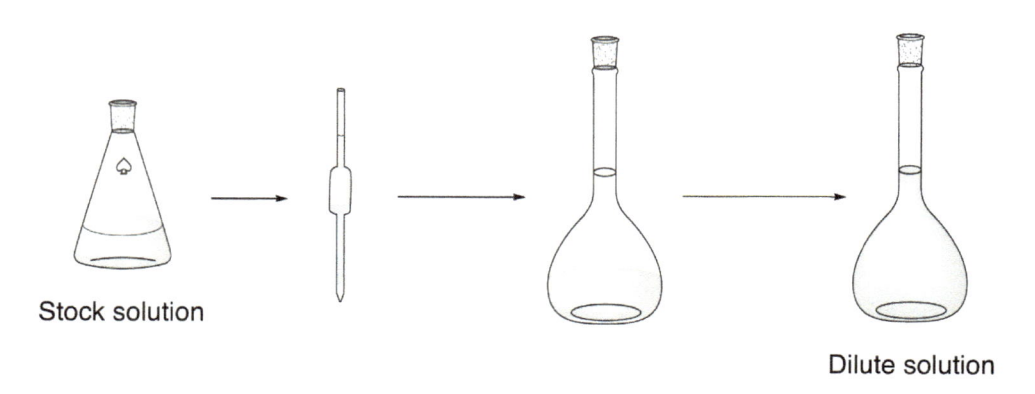

FIG. 6.11. Preparing a diluted sample. A specific amount is removed from the stock solution and placed in a volumetric flask. Then solvent is added to the mark on the volumetric flask to make the dilute solution. (Courtesy S. E. Johnson and M. D. Mosher—© ASBC)

and then to the ring stand. A heating mantle is placed under the flask and held in place with the ring clamp. An electrical cord connects the heating mantle to a variable transformer. The transformer is then connected to the wall socket.

The liquid to be distilled is added to the round-bottomed flask along with five or six boiling stones. Then the still head is placed on top of the round-bottomed flask and a stopper is placed on the top of the still head. A thermometer can be placed there, if desired, but since most of the material that will be distilled is a solution in water, the temperature will be known. Water boils at 100°C (212°F) at sea level and slightly less at higher elevations.

Hoses are attached to the nipples on the condenser, and then the condenser is placed on the still head. It can be held in place either by using another clamp and ring stand or by tying it to the still head. (Rubber bands often work quite well for this task.) Once firmly attached, the hoses are connected to the faucet and the sink. The hose that's attached at the lowest point on the condenser should also be attached to the faucet. The other hose is placed in the sink and the water turned on to a trickle. The vacuum adapter is attached to the end of the condenser in a fashion similar to that done for the condenser. Finally, a flask or graduated cylinder is placed under the end of the vacuum adapter. The entire apparatus is now set up to perform a distillation.

The distillation is initiated by turning the power on at the transformer. Power is applied to the heating mantle, which then heats the round-bottomed flask. The rate of heating can be adjusted using the dial on the transformer. Care should be taken not to heat the solution too rapidly, such that the boiling can't be controlled. If the boiling occurs too rapidly, the heating mantle can be removed by simply lowering the ring clamp on which it sits.

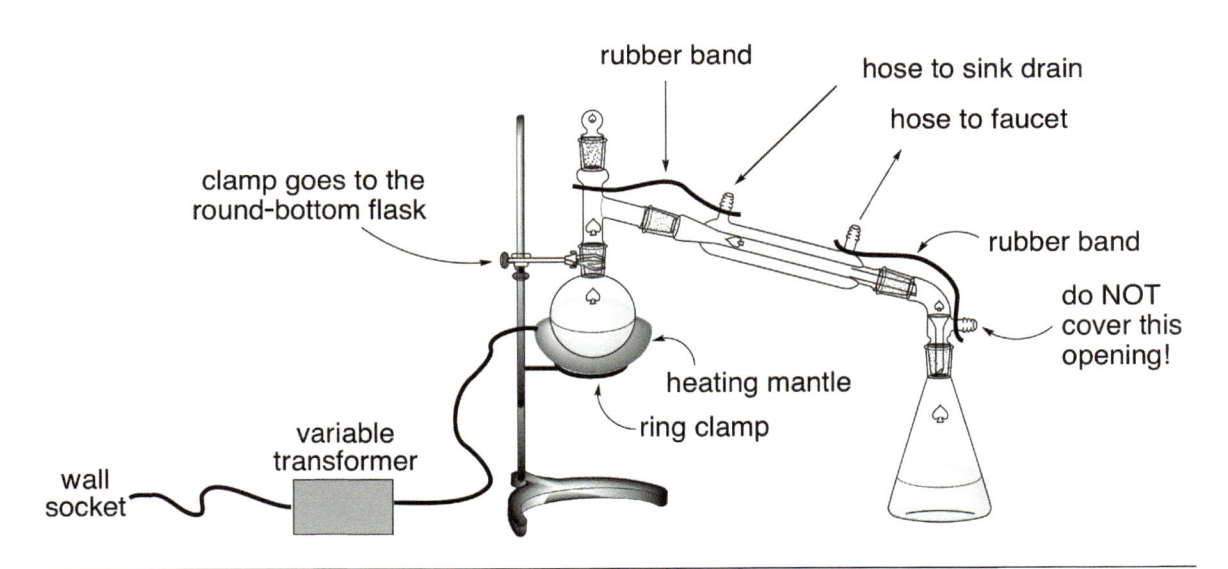

FIG. 6.12. Simple distillation setup. Note that the glassware is all standard taper and fits together snugly, so any vapors will not escape. Note, as well, that the heating mantle is plugged into a variable transformer instead of directly into the wall socket. (Courtesy S. E. Johnson and M. D. Mosher—© ASBC)

The vapors of the liquid will rise up in the round-bottom flask and then enter the still head. They will be visible initially as they condense on the sides of the glass and form a film. Eventually, they will rise high enough to enter the condenser, where they will be completely converted back into liquid. The liquid will run down the length of the condenser and through the vacuum adapter and then drip into the collection flask.

The initial stages of the distillation are often the most important. Losing any of the vapors during these stages results in a significant change in the analysis results. It's important, therefore, to ensure that each joint is firmly seated and that the rate of heating is fast enough to complete the distillation in a short amount of time but slow enough so it doesn't boil over and send foam through the entire distillation setup. If foam does contaminate any part of the apparatus other than the round-bottomed flask, the entire setup should be turned off and cleaned and the process started again with a new sample.

Once the appropriate amount of distillate has been collected, the heating mantle is lowered and turned off. When cool to the touch, the apparatus can be disassembled and then cleaned. It's good practice to clean the apparatus as soon as possible. Failure to do so often results in baked-on residues that become difficult to clean.

Handling Microbes

One of the important tasks often undertaken by the brewery laboratory is to handle, evaluate, and grow microbes. The microbes under investigation could be beer spoilage organisms or brewer's yeast needed for the propagation tanks. Whatever the microbe may be, handling it requires using specialized glassware and dedicated laboratory spaces that are free from contaminating microbes. In this section, we will explore those pieces of equipment and illustrate their use.

Sterile Versus Sanitized Versus Clean

clean Free of soil, grime, dirt, and residue.

sanitary Characterized by a significantly reduced number of microbes; a quantity less than found in the normal environment.

sterile Free of microbes.

Before using the equipment, the instruments, and the space for the handling of microbes, a number of steps must be completed, including cleaning and sanitizing with the goal of sterilizing. **Clean** means free of soil, grime, dirt, and residue. **Sanitary** means that the quantity of microbes present is less than in the normal environment. While the goal is to eliminate all microbes and create a **sterile** environment, that's extremely difficult to do. Instead, we focus on creating a clean and sanitized environment in which the quantity of microbes is only a small fraction of what would be found in the normal environment. Creating a sterile environment is almost impossible. Moreover, working in a sterile environment without recontaminating the area or the equipment requires significant changes in how things are done. Instead, the brewery space should be dedicated to microbe work, and protocols should be in place to keep it as sanitary as possible.

Four characteristics of cleaners impact their ability to perform their job:

1. **Temperature.** The temperature of the cleaner has a large influence on its ability to clean. As the temperature increases, the ability of the cleaner improves dramatically. This makes sense: Hot water cleans much better than cold water.
2. **Time.** The time a surface has been exposed to the cleaner also has an impact. As the time devoted to cleaning increases, the cleaning improves.
3. **Concentration.** Diluted or weak cleaners don't clean well. In fact, there is an optimum concentration that works best for many cleaners, and doubling the concentration doesn't always double the cleaning ability. Also, depending on the chemical, a high concentration can damage the equipment.
4. **Physical action.** Physical action has a positive effect on the ability to clean. Scrubbing, for example, does a much better job of removing dirt than simply soaking.

To reduce the quantity of microbes to an acceptable level, it's necessary to clean all equipment, instruments, and glassware, plus the area itself. "Clean" means that the area has been washed with a cleaner (for the appropriate amount of time, temperature, and concentration and with physical action). Cleaning removes dirt and grime from items and exposes nooks and crannies where microbes can hide.

The ingredients in cleaners, as shown in Table 6.1, attack the protein residue, hop resin grime, carbohydrate material, calcium, and other salt buildup on equipment. Caustic cleaners are best for most items, and the best ones contain **detergents** that help to solubilize materials that aren't very water soluble. These cleaners also contain **wetting agents** that help reduce the surface tension of water and make it more likely that the cleaner will get into nooks and crannies. Cleaners also contain chemicals that act as **chelating agents**, which bind metal cations and drag them into solution. If the surface being cleaned has hard-water buildup or is caked or covered with salts, then an acidic cleaner is the best choice. While this kind of cleaner is less common, it's good to have available in case the surfaces of the items being cleaned need it. Knowing the type of material that makes up the soil allows selecting the best cleaner.

detergent A chemical that solubilizes oils and fats, particularly during cleaning.

wetting agent A chemical that decreases the surface tension of water.

chelating agent A chemical that binds to metal ions.

TABLE 6.1. Characteristics of cleaners and sanitizers used in the brewery[a]

Caustic cleaners	Acidic cleaners	Sanitizer	Oxidizer
Sodium hydroxide	Phosphoric acid/ Nitric acid (1.2:1)	Alcohol >70% or ozonated water	Bleach or saturated steam
Detergent	Wetting agent		
Wetting agent	Chelating agent		
Chelating agent			

[a] Courtesy S. E. Johnson and M. D. Mosher—© ASBC.

Once a surface has been cleaned and all the dirt removed, it should be rinsed with water to help remove the cleaner and then sanitized to reduce the microbial level to an acceptable quantity. While perfect sanitation results in a sterile surface, obtaining sanitary levels that reduce the microbial count by 99.999% can be acceptable. This can be accomplished by either mechanical or chemical means.

Mechanical means include treatment with steam; in this case, steam is directed onto the item. The temperature of the steam and the time of exposure dictate the ability of the steam to sanitize the surface. This method is potentially useful for small items that can be placed in a chamber into which the steam can be added. This sanitizing chamber, known as an **autoclave**, typically operates at 120°C (248°F) at 140 kilopascals (kPa) with injected steam. After 20 minutes of exposure to these conditions, the items are virtually sterile.

Using this method isn't possible for many items and surfaces in the lab, so chemical means must be used to sanitize the items. Chemical sanitizers tend to be acidic. Some must be rinsed off before surfaces can be used; others can be used without rinsing. These sanitizers work by attacking and killing the microbes.

> **autoclave** A device (or the use of the device) that sterilizes objects by applying heat, moisture, and pressure.

Glassware Used in the Micro Lab

Glassware used in the analysis and handling of microbes includes the standard equipment used in the analytical lab. It also includes specialized equipment, such as an autopipette, petri dish, and microscope.

An **autopipette** is a useful tool in the micro lab. It's essentially identical to the graduated pipette used to measure out solutions in the analytical lab. However, an autopipette is made to transfer extremely small volumes with some degree of precision, if used correctly. Common autopipettes can deliver volumes from microliters (0.001 mL) to milliliters (1.000 mL). They require the use of disposable pipette tips.

To use an autopipette, the dial on the autopipette is turned until the appropriate volume is selected. Then the solution vial and the target vial are opened and set up close to each other. The autopipette is grasped with the right hand and placed inside a pipette tip. Gentle tapping will seat the pipette tip onto the autopipette. Next, the plunger is depressed to the first stop and the tip is placed in the solution. The plunger is then slowly released to withdraw the sample. In some cases, it may be necessary to repeat this step two or three times to make sure the sample is withdrawn correctly. Once the sample is withdrawn, the pipette tip is pulled out of the solution and touched to the side of the vial above the liquid. Doing this removes any drops of liquid that have adhered to the side of the tip. The tip is directed to the target vial and the plunger is depressed to the first stop and then pushed further to the full stop. This last push expels air from the tip and ejects the entire liquid sample from the tip. Again, the tip is withdrawn from the liquid and touched to the inside of the vial to ensure that all the liquid has been transferred. Then the autopipette is held over a beaker or other receptacle and the button on the side of the autopipette is depressed. This ejects the tip from the autopipette.

> **autopipette** A mechanical pipette operated to deliver a specific volume of liquid.

Taking proper care of the autopipette will ensure that it remains precise over time. It should always be stored in an upright position to stop liquid from getting inside the autopipette and ruining the mechanism. In addition, the correct pipette tip should always be selected to avoid this liquid ingress.

Proper Use of the Microscope

A **microscope** is one of the most important pieces of equipment that every brewery laboratory should have, regardless of whether the lab is handling microbes. A microscope allows the brewery to count the number of yeast in a sample. This gives information that can be used to determine if the brewer is pitching the correct number of yeast; underpitching and overpitching cause issues with beer production and result in off-flavors in the final product. Chapter 15 includes the methods used to count yeast and provides information on how to perform that task.

Taking care of the microscope is very important. The eyepieces and lenses and all the surfaces should be wiped periodically with lint-free lens paper and lens cleaner that's designed specifically for these sensitive glass pieces. Doing this helps reduce the buildup of grease and dirt that can obscure the view of anything using the microscope. The light should also be maintained and the bulb replaced as it becomes dim with time.

A sample is prepared for viewing on a microscope by placing a **microscope slide** on a clean surface. The best way to do this is to place the slide on a lint-free paper towel, which ensures having a clean surface and also facilitates cleanup. For a liquid sample, a drop is added to the slide and a small **cover slip** is added on top of it. This thin square of glass is very important. It reduces the thickness of the sample, which makes focusing easier, and it keeps the microscope lens from getting wetted by the sample. To avoid air bubbles under the cover slip and to limit finger grease that can obscure images, only the sides of the cover slip should be handled. One side of the slip is placed on the slide, and then the slip is laid down on the sample. Like the cover slip, the slide should be handled only on the edges.

For a dry sample, the cover slip is placed directly on the slide. A cover slip is used to help keep a dry sample in place and to protect the microscope lens from accidentally touching the sample. Hair is a good example of a sample that can be prepared via dry sampling. Simply put the hair on the slide, cover it with the cover slip, and observe.

A microscope is operated by placing a slide containing the sample on what's called the stage. The slide is held in place on the stage with clips. To begin, the lowest magnification is chosen. The sample is viewed by looking through the eyepiece and focused using the focusing knobs. The stage is moved with the appropriate knobs until the sample is centered in the eyepiece. Then the next-higher magnification is chosen and the process repeated until the sample is centered and focused. These steps are repeated until the magnification is 400×. At this magnification, yeast cells are clearly visible. By adjusting the focus and the light source, it may even be possible to see the organelles inside the yeast samples.

Because most beer-spoilage bacteria are much smaller than yeast, they appear only as dots in the field at 400×. It's necessary to increase the

microscope An instrument used to view microbes and yeast.

microscope slide A glass plate on which a sample is placed for viewing under a microscope.

cover slip A thin square of glass that's placed on top of a sample on a microscope slide.

magnification to at least 600×–1,000× to see them and their structure. At this level of magnification, the yeast cells are very large and fill the eye piece. The use of a special oil may be required to prevent the magnification lens from getting scratched by the slide or cover slip. Once the viewing is complete, if the 600× magnification (or higher) was chosen, it's important to wipe the excess oil from the lens. Doing this helps to reduce the buildup of oil on the microscope.

Petri Dishes and Agar

Growing samples of microbes takes some practice, but when done correctly, it can provide information on microbial infection and allow the brewery to collect, save, and grow yeast. We will cover this topic again in Chapter 14; for now, we will cover the basics of the process.

The tools needed to grow microbes properly include **petri dishes**, test tubes and a rack, a Bunsen burner, and a **loop**. The petri dishes, test tubes, and other items of glassware that hold liquid that will come in contact with the sample must be sterilized. (The loop doesn't need to be sterilized now; it will be sterilized immediately before use.) Remember that sterilization is one step beyond sanitization. For glassware, this means placing it in an autoclave or, in the absence of an autoclave, placing it in a pressure cooker and boiling it for 30 minutes. The additional pressure is important to get the temperature high enough to ensure complete sterilization. Whichever method is used, the glassware must be placed inside a sterilized bag after being sterilized. If it isn't, microbes in the air will contaminate the surfaces.

Specific recipes of **agar** enhance or inhibit the growth of certain types of microbes, such as Universal Beer Agar. Many agars are available commercially, or agar can be made by adding together the appropriate amounts of chemicals. First, the powdered agar mixture is massed on the balance and added to an Erlenmeyer flask that's at least twice the size of the amount of agar to be prepared. Then the appropriate amount of water is measured in a graduated cylinder and added. The solution is mixed until everything is dissolved and then autoclaved to sterilize the mixture.

Alternatively, if an autoclave is not available, a magnetic stir bar is added to the flask. Aluminum foil is placed on the top of the flask, and the flask is placed on a stirring hot plate. Heat is applied until the liquid begins to boil and then adjusted to maintain a simmer for 30 minutes.

The agar mixture is removed to the bench and allowed to cool until it's about 50°C (120°F). Then it's poured into sterile petri dish bottoms or into sterile test tubes until they are one-third to one-half full. Lids are placed on the petri dishes, or caps are placed on the test tubes. The petri dishes are stored with their lids on until they come to room temperature and then inverted and stored upside down. The test tubes are placed in a rack in such a way that they are held at a 45–70° angle until the agar is set. These so-called **slants** are useful for storing and handling yeast. Once the test tubes have set, they can be stored vertically. The agar plates and slants should be used as soon as possible. After 3 days (if the agar is boiled) or 1 week (if an autoclave is used), the sterility of the agar comes into question.

To use a petri dish containing agar, the Bunsen burner is lit and the plates and samples are brought out. The end of the loop is put into the

petri dish A low, flat dish with a lid.

loop A hand-held stick made of wire that has been bent into a circle at the end.

agar A gelatin-like substance that's used to support the growth of microbes.

slant A gelatin-like substance that's poured into a test tube and solidified at an angle.

burner flame and held there until it glows red. This ensures sterilization of the loop. The sample container is opened and the loop is inserted into the liquid sample. A sample is collected with a little swirling. The loop is then removed with the right hand, while the left hand opens the agar plate (or test tube). Figure 6.13 shows how to perform the plating. The loop is moved back and forth over a one-quarter piece of an imaginary pie. The petri dish is turned one-quarter of the way. The loop is then placed back in the flame and sterilized again. Next, the loop is streaked once across the area of the first piece of the pie and then moved back and forth over the next one-quarter piece of the pie without any further overlap. This is repeated two more times until the entire pie has been covered. The result is that a large number of microbes are placed into the first one-quarter of the pie, and with each subsequent piece, the number of microbes becomes significantly less. Immediately after the last swipe is made, the lid is returned to the dish. The petri dish is then stored upside down.

To use a test tube containing agar, the Bunsen burner is lit and the tubes and samples are brought out. The loop is sterilized in the flame as before. Then the sample container is opened and the loop is inserted into the liquid sample. The loop is then withdrawn and rubbed across the surface of the agar in the test tube. Immediately after the loop has been swiped, the lid is returned to the test tube.

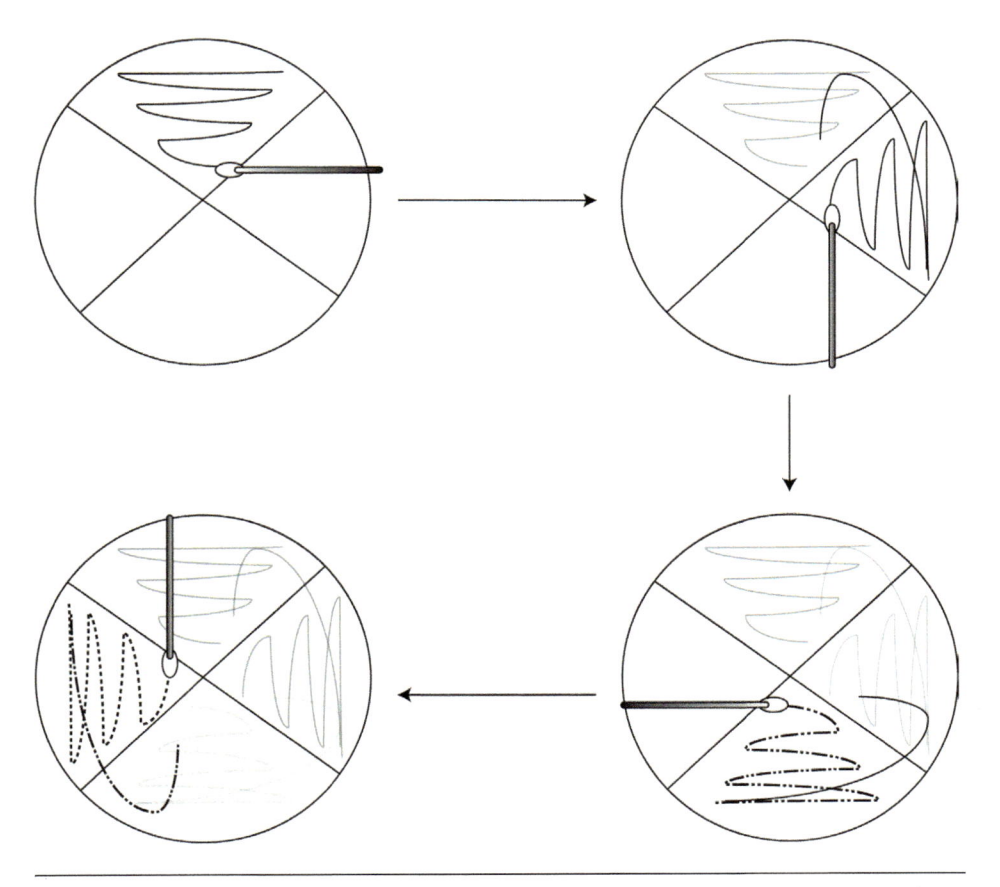

FIG. 6.13. How to properly apply sample to an agar-containing petri dish. (Courtesy S. E. Johnson and M. D. Mosher—© ASBC)

Broths and Growing Yeast

The micro lab is often charged with growing fresh yeast samples for use in the propagation tanks. Fresh yeast is required about every 7–10 fermentations, which means the lab will likely need to grow fresh yeast about every week and perhaps much more often.

The lab starts by taking a slant of yeast from the refrigerator. As described earlier, these slants were made from a sample of yeast and stored in the refrigerator for this purpose. Slants need to be remade every 3–4 weeks to ensure that the yeast maintain viability. A Bunsen burner and loop are also needed, and so is a broth in which to grow the yeast.

The broth is wort with a gravity in the range of 1,040–1,048 kg per cubic meter (kg/m^3) (1.040–1.048 g/mL or 10–12°P) that's lightly hopped. Adding hops provides a small antibacterial effect to the wort. The wort can either be made in the brewery or prepared from dried malt extract in the lab. The wort from the brewery is placed in a small vessel known as a **Carlsberg flask**, which can be heated to sterilize everything.

A 100 mL sample of this wort is placed into a sterilized Erlenmeyer flask. The wort can be sterilized in the Erlenmeyer flask using an autoclave, or it can be boiled gently for 20 minutes. Once the wort has been sterilized, a foam stopper is placed in the Erlenmeyer flask and it's allowed to cool to room temperature. The loop is then sterilized by heating it in the Bunsen burner. The loop is placed in the slant, and a small amount of the yeast is scraped off the agar. The foam stopper is removed and the edges of the Erlenmeyer flask are passed through the Bunsen burner to sterilize them. Then the loop is placed into the wort and swirled to release the yeast. The foam stopper is returned to the flask. The flask is then placed in a shaker table (or a magnetic stir bar is added and the flask is placed on a stir plate). During the 12–24 hours of vigorous stirring or shaking, the yeast grow and multiply. This creates a large number of yeast, which can be used to inoculate a larger sample of wort.

The general rule of thumb is to increase the volume by 10 for each successive growth stage. Thus, the next sample of wort would be 1 L in size. The 100 mL sample of wort and yeast is retrieved from the shaker or stir plate, the edges of the flask are passed through the Bunsen burner, and the mixture is poured into a 1 L sample of wort that has also been flamed at the opening. This large flask is shaken or stirred again for 12–24 hours. This sample can then be added to a 10 L sample and eventually to a 1 hectoliter (hL) vessel in the propagation tank farm. Modifications to this rule of thumb can be made but are based solely on the characteristics of the yeast. Keeping the 10-fold increase in volume maintains the most optimized conditions for growth without stress.

Carlsberg flask A sample vessel used to hold quantities of wort that can be sterilized.

Instruments and the Laboratory

CHAPTER 7

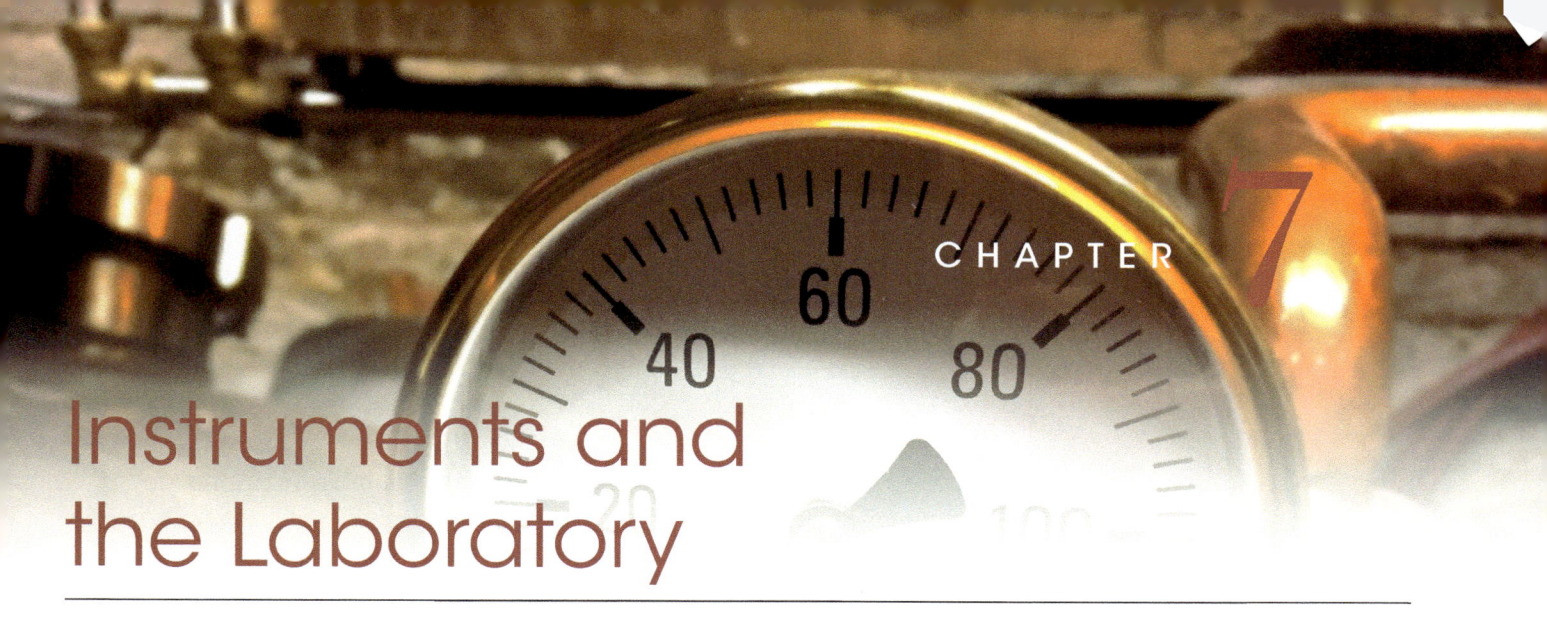

A well-equipped laboratory has not only glassware and chemicals but also instruments capable of measuring properties of solutions that would be impossible to observe otherwise. Instruments can obtain precise and accurate data reliably and repeatedly, and they can report from one result to thousands of results each day, depending on the analyses being performed. Many analyses are simply not possible without the aid of instruments in the laboratory. Lab personnel should be familiar with every piece of equipment: how it works, how to perform standard maintenance on it, how to tell if it's working properly, and how to troubleshoot if it doesn't seem to be working properly. This chapter will address many of these topics.

Instruments in the Laboratory

Instruments can be extremely sensitive, are sometimes delicate, and can be expensive, depending on their complexity. Thus, the operator of any instrument should be adequately trained on its proper use, operation, and maintenance. Instruments that don't work can't be fixed by whacking them or just running more samples, and instruments that don't respond quickly won't speed up if a button is pushed harder. An untrained analyst may not be aware of what it costs to repair the damage that might result from mistreating or mishandling an instrument.

instruments Sensitive machines with precisely machined parts that measure the concentration or identity of an analyte; often include on-board computers that communicate with laptop and desktop computers dedicated to given instruments.

Instruments Versus Equipment

A piece of **equipment** is considerably different from an instrument. Equipment can be sensitive and delicate, just as instruments can, but equipment doesn't typically measure the **analyte** in a sample. Equipment includes stirring plates, hot plates, balances, and other items that are quite necessary in the lab. Like instruments, equipment should be treated with care and cleaned and maintained according to manufacturers' requirements. Performing such maintenance will ensure that equipment continues to operate properly over its lifetime.

equipment Simple laboratory tools, such as spatulas, balances, hot plates, mixers, and so on.

analyte A compound being analyzed within a solution.

Many different instruments are useful in the brewery laboratory. They can be grouped into categories based on the principles that govern operation of the instrument. Table 7.1 outlines those categories and gives examples of instruments. The four categories are as follows:

1. **Instruments that perform an analysis based on the color intensity of a solution.** The instrument produces a larger response as the intensity of the color increases. This category of instruments gives very accurate results if the color is specific for a single component of the mixture. Unfortunately, this category of instruments can also be very nonspecific and respond to a wide range of compounds within the matrix. In addition, analytes of interest may be obscured by other chemical compounds in the solution matrix.
2. **Instruments that physically separate the individual components within a sample and then measure the quantity or relative amounts of each component.** Methods that use instruments in this category are highly specific for a particular compound within the matrix. Some compounds can obscure the analyte the analyst wants to study. Instrument methods have been created that will eliminate or significantly reduce those interferences.
3. **Instruments that perform an analysis based on the physical properties of the entire sample.** Methods that utilize these instruments are not at all specific for a particular component within a matrix.
4. **Instruments that perform dual methods.** These instruments combine the features of two categories to provide highly specific analyses.

The instruments in each category are fairly similar in terms of design and layout. The principles of operation are also fairly uniform within a category. This uniformity is useful because there are many different manufacturers of instruments. For example, there are multiple manufacturers of spectrophotometers. Although the specific instructions on how to use the software for each version of the spectrophotometer are usually quite different, the general principles underlying how the instrument works and what steps to follow to perform an analysis are usually quite similar.

TABLE 7.1. Selected categories of instruments in the brewery lab[a]

Category	Basis of operation	Examples of instruments
1	Analyze color intensity of a solution	Ultraviolet- (UV-) visible spectrophotometer, fluorimeter, turbidimeter, colorimeter
2	Separate and measure compounds	Gas chromatography (GC), high-performance liquid chromatography (HPLC)
3	Analyze physical properties	Hydrometer, refractometer, viscometer
4	Perform dual methods	Gas chromatography-mass spectrometry (GC-MS)

[a] Courtesy S. E. Johnson and M. D. Mosher—© ASBC.

We will discuss the operation of each category of instruments in a future chapter. The colorimeter is discussed in Chapter 10, the spectrophotometer in Chapter 11, chromatographic methods in Chapter 12, and instruments used in measuring physical properties in Chapter 13. Two general principles that govern all of these methods are based on the precision and accuracy required for an analysis. The precision and accuracy of the instrumental method is only as good as the laboratory personnel's attention to preparation of the sample. Proper use of the most precise glassware will provide the best samples for instrumental analysis. Proper techniques in transferring liquids and solids will result in the most precision for an analysis.

Notice that we have neglected to mention the accuracy of a particular analysis. That's because that accuracy is usually out of the hands of the laboratory worker. That accuracy is based on the method used. Poor methods of analysis are not accurate, whereas good methods of analysis give accurate results. That's why some of the methods that have been discovered, explored, and proven to provide accurate results are approved by organizations such as the American Society of Brewing Chemists (ASBC) and the European Brewery Convention (EBC). Lab workers are responsible for following proper operating procedures to be as precise in their techniques as possible. And if they are, the result will be an analysis that's as accurate as possible. In Chapter 8, we will explore in more detail basic computer programs that can be used for recordkeeping and discuss ways in which a lab can confirm that it's providing results that are both precise and accurate.

Pros of Using Instruments for Analysis

Using instruments in the lab greatly improves the precision and accuracy of an analysis. As we will see in Chapter 10 when we examine colorimetry as an analytical technique, using an instrument results in a noticeable improvement in the quality of the results. The precision of the technique improves, and thus the accuracy of the results improves. This is true for almost every method of analysis that utilizes instrumentation.

In addition, some methods that utilize instrumentation focus on reporting the results of a specific analyte, whereas the noninstrument-based analysis might provide combined data for a group of similar compounds. For example, the percentage of alcohol in a beer can be determined by measuring the density of a beer compared with the density of the wort from which it was made. The difference in specific gravity is then related to the percentage of alcohol. Unfortunately, the change in specific gravity isn't specific only for ethanol. Other compounds dissolved in the beer—such as flavoring oils, hop oils, and aroma compounds—will change the density of the beer. And depending on the quantities of these oils that are added to the beer, the difference between the measured percentage of alcohol and the actual percentage of alcohol can be significant. The error in using specific gravity for alcohol by volume (ABV) calculations is even more dramatic in higher-alcohol beers. This makes the method of analysis of percentage of alcohol via specific gravity a nonspecific method for analysis.

An instrument-specific method to determine the percentage of alcohol in beer would isolate the ethanol from the sample and report only that information. Thus, other compounds in the sample matrix would not affect the results. An example of this type of analysis is gas chromatography with a flame ionization detector (FID) or mass spectroscopic (MS) detector. (Detectors will be discussed further in Chapter 12.) In this instrument-specific method, a sample is injected into the instrument, which separates all the components in the sample. The ethanol quantity is measured using a detector. (This method is very specific.) Remember that in this method, each compound in the matrix is isolated before the analyte quantity is measured. Not even the addition of a large quantity of flavoring oil to the beer will change the quantity of ethanol that's measured by the instrument. In other words, the instrument will report 5.0% ethanol in a beer containing 5.0% ethanol.

Another advantage of using instrument-based methods can be the speed of the analysis. For example, the amount of calcium in brewing water can be determined by a variety of methods. One noninstrument version (known as Eriochrome Black T) involves adding small quantities of a chemical that binds to the calcium in the solution until a color change occurs. It can take 10–15 minutes to get an adequate value for the amount of calcium, not counting the sample and solution preparation. However, an instrumental version inductively coupled plasma–optical emission spectrometer (ICP–OES) can measure the concentration of calcium in about 1 minute. In this particular case, the instrument version is faster and very specific but also extremely more costly. Ultimately, adding the appropriate instrumentation to the brewery laboratory helps to ensure that the specific measures of the quality of the product are precisely and accurately measured, with the goal of reproducibly preparing each brand of beer in the brewery's portfolio.

Cons of Using Instruments for Analysis

Using instruments in the brewery laboratory can be significantly more expensive than using noninstrumental techniques. In addition to the cost of the instrument, costs associated with the use of the instrument also must be taken into account. Those include the costs of reagents, compressed gases, electricity, water, and other supplies required for every analysis. Additionally, many instruments should be used by personnel trained in their maintenance and operation. While noninstrumental methods also have costs, they tend to be less expensive. As the number of instrument-analyzed samples increases, so does the cost of the supplies. There is a slight economy of use for the cost of supplies: If only a few samples are analyzed each day, the cost of supplies is usually much more than when many samples are analyzed each day. For example, if 1 liter (L) of a solution is made for a specific analysis but that analysis needs only 250 milliliters (mL) of that solution, then the other 750 mL will be wasted. However, if four samples can be evaluated by the method, then all the prepared solution can be used. The cost of the solution will be the same but less solution will be wasted, making the analysis more cost effective.

In addition to the costs of supplies, there are costs associated with the upkeep of the instrument. For example, sample holders must be replaced due to breakage and wear. In some cases, disposing of sample holders may be more economical than cleaning reusable ones. Some parts have finite lifetimes; for example, chromatography columns lose their ability to separate components.

Finally, there are costs associated with maintenance. These costs include weekly preventive maintenance, periodic software updates, and occasional repair of computer hardware and instrument parts. Replacement instrument parts can be expensive.

Brewers should consider purchasing demonstration ("demo") and refurbished instruments to decrease the initial costs. According to conventional wisdom, a new car loses 10–20% of its value when it's driven off the lot. The same principle holds true for demo and used equipment. In some cases, an instrument still in stock with the vendor can be purchased at a savings if the next model has just been released. Purchasing an older model or used instrument will significantly reduce the initial expense. Typically, a demo or refurbished instrument has a warranty, which will help alleviate concerns about whether the instrument will work.

It's also important to consider that each instrument requires a physical space in the brewery laboratory (Fig. 7.1). In addition, time is required to set up the instrument and to attach any gas and water lines and sometimes, a dedicated power line. While there are portable versions of some instruments, they typically are not as accurate and precise as desktop versions. However, in some cases, the portable version can produce results that are sufficient for brewery needs. Some of these instruments, such as the ICP–OES, require relatively significant floor space and once set up, should not or cannot be moved. Even small instruments, such as the ultraviolet-visible (UV-vis) spectrophotometer, are about the same size as a hard-sided briefcase and require a dedicated space on the bench. It's also important to consider the computer

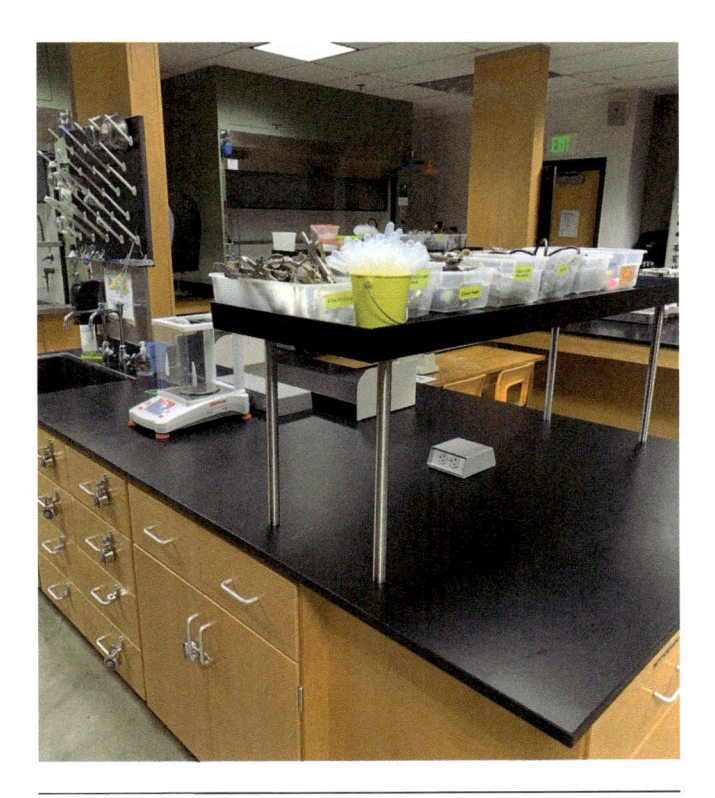

FIG. 7.1. Good use of laboratory space, in which everything has its own bench space. It's important to avoid cluttering the bench with supplies. (Courtesy S. E. Johnson and M. D. Mosher—© ASBC)

TABLE 7.2. Pros and cons of instrumental methods[a]

Pros	Cons
Increase precision and accuracy	Are expensive to purchase
Can be specific for a particular analyte	Are expensive to maintain and use
Can be faster than noninstrumental methods	Require dedicated lab space

[a]Courtesy S. E. Johnson and M. D. Mosher—© ASBC.

that operates the instrument. Some instruments can interface with a laptop or work over Wi-Fi. However, they still require a space on the bench, which then can't be used for chemical preparation. It's best to plan or allocate space for instrument purchases, especially when space is an issue.

The pros and cons of instrumental methods are listed in Table 7.2. In sum, the advantages of using instruments in the brewery far outweigh the disadvantages.

Laboratory Conditions for Optimal Instrument Operation

Even though much of the work in a laboratory may still involve the use of noninstrumental methods, a laboratory should be set up to accept instrumentation. That means designating areas in the lab where glassware will be set up for titrations, distillations, and extractions. Similarly, space will be needed to make solutions and to prepare samples. That space should be maintained in the lab even if instruments are added. Instrument performance (both accuracy and precision) is greatly improved when care is taken to provide adequate conditions for an instrument (Fig. 7.2).

The laboratory also should be a place where a $10K spectrometer will be safe. In this context, "safe" means that chemicals won't be spilled on the instrument, that water won't flood the bench and soak the electrical connections, and that staff can work around the instrument without hindrances.

An instrument room that's separated from the wet chemical lab is ideal but often impossible to accomplish. So when instruments need to reside in the same space where other lab work occurs, careful attention must be paid to the placement of the instruments and arrangement of the lab. First, the instruments should be located at one end of the lab and away from the area where chemical preparation and manipulation will take place. Also, each instrument should have its own dedicated space. This means space not only for the instrument but also for the computer controlling the instrument and for sample preparation or manipulation near the instrument. For example, as discussed in Chapter 3, a relatively small UV-vis spectrophotometer (the size of a briefcase) requires about 5–6 linear feet of bench space when a computer, a workspace, and a buffer space are added.

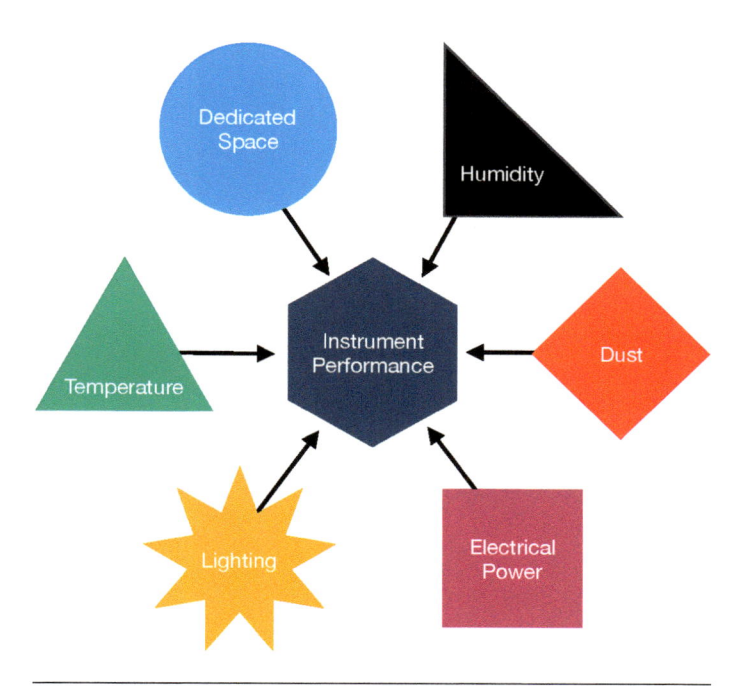

FIG. 7.2. Factors that affect instrument accuracy and precision. (Courtesy S. E. Johnson and M. D. Mosher—© ASBC)

Temperature

The temperature of the laboratory is very important when instruments are used. Although many instruments operate best when the temperature is 65–68°F (18–20°C), the most important issue is to ensure that the lab maintains a constant temperature. Instruments determine analytes by using sensitive detection devices. We will explore the different types of instruments and how they work in detail in Chapters 10 through 13, but for now, some generalizations can be made. For example, the analyte is often present only in a small quantity. Thus, the result of detecting it causes a minute change in the instrument. That change is converted into an electrical signal that's sent to an amplifier. The amplifier then increases the electrical signal many times, resulting in a reading that the user can observe.

Because the instrument's original response to a small concentration of analyte can be minute, any change in how the instrument operates can result in a minute change to the response. Fluctuating temperatures in the laboratory can affect how the instrument responds to the detection of an analyte over time. Any change in the response will then be amplified along with the original signal that the user sees. In other words, the quantity of the analyte will appear to fluctuate as the temperature fluctuates in the lab. The result might be an instrument that reports a concentration of analyte that's high or low compared with the actual value.

Given this sensitivity to temperature, airflow from heating, ventilation, and air conditioning (HVAC) can affect the performance of instruments. Improper placement of an instrument can cause it to respond differently when the HVAC is blowing cold air, warm air, or no air.

Electrical Power

Like fluctuations in the temperature of the lab, fluctuations in the electrical current can cause variability in the instrument's response. Because the instrument relies on an electrical signal to report the detection of an analyte, any oscillation in that electrical signal could result in reduced precision. The normal oscillation of 60 hertz (Hz) for alternating current is accounted for by the instrument. Here, however, fluctuations in the current coming from the line are problematic. For example, acceptable voltage variations in the United States are from 92% to 105% of expected voltage. This means that the voltage may fluctuate over the course of using the instrument each day. While some buffering from these changes is provided in the instrument, not all changes can be addressed. It's useful to consult with the manufacturer about necessary power requirements before buying an instrument.

Sensitive equipment can't handle most surges and spikes. In addition to causing immediate damage to the computer boards and other equipment inside the instrument, spikes and surges result in variations to how the instrument responds. If a measurement using the instrument is underway during a surge or spike, the data produced will not be precise. Even the instrument itself or another device plugged into the same line can cause a spike or surge when it turns on a fan. Spikes last only a brief time (on the order of nanoseconds), but they come with huge increases in voltage. In some cases, the voltage can jump 5,000–10,000 volts. Surges, on the other hand, produce much smaller increases in voltage (10–35% increase) but can last up to seconds.

Simple surge protectors, such as those purchased in electronics and office supply stores, can help to reduce the impact of some line surges and spikes. But many of these devices offer a false sense of protection, because they can't switch fast enough to prevent all spikes or surges from reaching the instrument. The protection they offer saves only the integrity of the computer and some of the less-sensitive computer boards. They don't offer protection from large spikes and power variations. For these types of protection, a voltage regulator or line conditioner with surge protection is needed. These devices often come with a battery backup, so if the power fails or becomes unstable over a period of time, lab personnel will have time to power down the system.

Humidity

Static electricity buildup on electrical equipment is a constant issue and might result in an instrument discharging this buildup through itself. The transfer of static electricity from lab personnel can also discharge into an instrument. To some degree, grounding the instrument can help reduce the damage from these discharges, but the biggest reduction in static electricity comes from reducing the moisture in the air.

Humidity is a measure of the amount of water vapor in the air. **Relative humidity** is the term typically used in weather reports for the value reported as the percentage of water in the air versus the maximum percentage of water vapor at the given temperature. When the relative humidity is 100%, no more water vapor can exist in the air. At this level of relative humidity, it's probably raining. Whatever the humidity, water vapor in the air condenses on everything in the lab (Fig. 7.3), forming a thin film of water on

---

relative humidity A value reported as the percentage of water vapor in the air versus the maximum percentage of water vapor at the given temperature.

instruments, computers, and all the parts inside them. That film of water naturally resists the buildup of static charge.

As the film of water gets thinner, the amount of resistance decreases. Some sources recommend at least 40% relative humidity to avoid static electricity, while others suggest that protection from static discharge doesn't occur until the relative humidity reaches 60%. Because lab instruments should be in a temperature-controlled room, the range of 40–60% humidity provides good protection from static. Taking additional protective measures—such as eliminating carpet, providing good airflow within the lab, and ensuring adequate grounding of all equipment—can virtually eliminate problems with static electricity discharge.

Dust

A dirty environment is no place to perform a sensitive analysis. Dirt, grime, and dust are significant pollutants that should be eliminated from the workspace. They are particularly detrimental to the operation of instruments (Fig. 7.4).

Dust buildup on an instrument can cause significant impacts. Dust can clog vents in the instrument case and reduce the ability of the instrument to regulate its internal temperature. Dust can also get inside the instrument and build up on fans, computer boards, and other internal parts. Dust obscures

FIG. 7.3. In extreme humidity, water can condense onto surfaces, similarly to how dew drops form. Condensation can damage instrumentation. (Courtesy 41330 from Pixabay)

FIG. 7.4. Dust buildup on a keyboard. If the dust is here, it's everywhere else, too, including inside the computer or the instrument. (Courtesy S. E. Johnson and M. D. Mosher—© ASBC)

the ability of water vapor to coat the sensitive components and can increase the risk of static discharge. Dust can also act as a conductor between adjacent locations on a computer board, resulting in arcing and massive failures. In addition, dust can coat optical windows inside the instrument and, like dirt on a window, reduce the intensity of light that passes into the instrument. All of these conditions reduce the ability of the device to provide accurate and precise results.

To combat dust buildup, each instrument should be cleaned according to the manufacturer's recommendations and directions. Optical windows should be cleaned periodically using the manufacturer's recommendations; an optical window should never be cleaned with water unless specifically directed by the manufacturer.

Lighting

It doesn't seem like the lighting in a lab should affect what happens inside an instrument, but it can. Some instruments are not affected by lighting, but spectrophotometers can be affected enough to reduce the precision of their measurements. This class of instrument measures the amount of light that interacts with a sample, so stray ambient light entering the instrument may influence the measurement.

Instruments should be placed in the lab so they are not in the path of direct lighting or sunlight. However, the lighting of the lab must be adequate. All the lights in the lab should be regularly maintained. Flickering lights can cause an issue of changing light intensity.

Rarely is a lab located in a room with exterior windows, but in some breweries, that's the only option. The best option is to use a room with windows for an office or taproom. Having windows in the lab significantly impacts the precision of the instrumentation and other analyses because of light from the outdoors. An instrument placed in front of a window is subject to direct light during part of the day and to indirect light for most of the rest of the day; then overnight, there is no light from the window. In addition, heat buildup from direct sunlight can significantly damage the electronics within an instrument, in addition to causing issues associated with light penetration. Moreover, the changing temperature of the instrument as it cycles from hot to cold over the course of a day causes wear, reducing the instrument's lifespan and increasing the need for regular maintenance.

Even an instrument that doesn't use light to conduct measurements can be negatively affected by being placed in front of a window. In short, an instrument should not be placed anywhere near a window. Even if the window is completely covered, it's best to consider placing the instrument in another location.

Instrument Upkeep and Maintenance

Instruments essentially measure the concentrations of analytes. Good instruments can reproducibly provide the same value for the measurement of a single sample. In other words, instruments help provide precision in an analysis.

However, instruments are only able to perform the task given to them via the method they are running. Any accuracy that arises from the instrument is due to the method of the analysis, not solely from the instrument. A **method** is the list of steps and processes that the instrument follows as it performs an analysis. In addition, the lab worker must prepare the sample for the analysis. Often, this relies on conversion of the analyte into something that can be measured by the instrument. If that conversion is done poorly (perhaps the analyte isn't converted entirely or something is made that isn't easily measurable), then the method won't have a high degree of accuracy. In short, the instrument is responsible for the precision, and the method is responsible for the accuracy. Unfortunately, the lab worker can impact both if he or she has poor technique (Fig. 7.5).

method The steps and processes given to an instrument to successfully report results that are relevant.

To keep an instrument in good working order, it should be maintained daily as recommended by the manufacturer. Maintenance might include exterior cleaning, flushing out liquid lines, and emptying waste containers. Reagent bottles associated with the instrument are best stored sealed. The manufacturer's recommendations should be followed.

It's also important to follow the instrument manufacturer's schedule for periodic **preventive maintenance**. Preventive procedures can range from simple checks to exhaustive regimes. In some cases, a trained technician may be best suited to conduct these procedures; a maintenance contract is typically available from the manufacturer or another qualified agency. After any major maintenance has been completed, standard samples should be analyzed and compared with previous results and with the factory specifications for that standard sample. If the data from the instrument doesn't match the specifications for the instrument, some fine-tuning of the performance may be needed.

preventive maintenance The steps required by the manufacturer to ensure that an instrument continues to operate optimally.

The preventive maintenance schedule is different for each instrument. Typically, that schedule is provided in the manual for the instrument. If the schedule is in doubt, it's best to contact the manufacturer. If this still leaves some doubt, it's often best to choose the shorter times between preventive maintenance services. So while it might seem useful to provide a preventive maintenance schedule in this text, it would not be practical to write one for

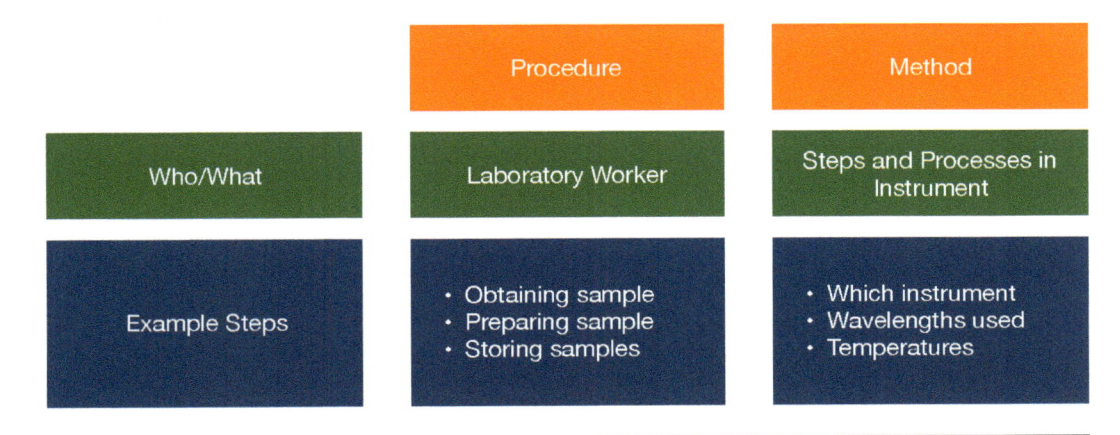

	Procedure	Method
Who/What	Laboratory Worker	Steps and Processes in Instrument
Example Steps	• Obtaining sample • Preparing sample • Storing samples	• Which instrument • Wavelengths used • Temperatures

FIG. 7.5. Differentiation between the procedure and the method for analysis. (Courtesy S. E. Johnson and M. D. Mosher—© ASBC)

every instrument that could be used in a brewery lab. Rather, the brewer should consult the owner/user manual for a given instrument and create a schedule of specific activities to perform at specific times. That schedule should then be posted near the instrument and recorded in a computer- or paper-based maintenance log. Doing so will help the user track exactly how many more hours or measurements can be taken before the next preventive maintenance must be performed.

CHAPTER 8
Reporting Your Results

Scientists explore the world using the **scientific method**, which is essentially a series of steps. First, they pose a question about something; this provides the direction or topic for the research. Then, being as unbiased as possible, they answer the question based on their prior knowledge, information they find in the library, and/or understanding of other scientists' work. The answer that they propose is known as a **hypothesis**.

The bulk of a scientist's work, then, involves performing experiments to confirm or refute a hypothesis. The scientist performs an experiment, collects the data, and evaluates the data from the experiment. After comparing the evaluation of the experiment with the hypothesis, the scientist modifies the hypothesis to account for the new data, creates a new hypothesis that verifies the new data, or confirms the existing hypothesis. He or she repeats this process until every possible experiment has been performed and every experimental result confirms the hypothesis. At this point, the hypothesis becomes a tested hypothesis, but the scientific method is not complete.

Next, the scientist must report the results and the proposed hypothesis. Doing this allows additional experiments to be performed on the original question and the hypothesis, gives other scientists a chance to comment, and potentially confirms the experimental results that have already been performed. When confirmed, the tested hypothesis becomes either a theory or a law. A **law** is a tested hypothesis that describes the answer but doesn't explain why. For example, Darcy's law describes only what happens during sparging. It relates the pressure drop across the grain bed given a specific flow rate, bed permeability, wort viscosity, and so on. It does not, however, indicate why the pressure changes during sparging. A **theory** is a tested hypothesis that describes why a specific answer is correct. For example, the formation of skunk flavor in light-struck beer is a theory. It explains why the skunk flavor arises in beer.

Scientists who fail to report their results for others to read and explore don't allow others to comment on their hypotheses. Thus, their hypotheses are often incomplete, not fully tested, and unable to contribute to the greater good of the field. In other words, their hypotheses never become theories or laws. In the worst cases, bias creeps into their hypotheses, leading to claims that could not be confirmed by others even if the information were shared. Moreover, any meaningful results that might help explain the world around us remain unknown.

scientific method A series of steps that scientists follow to determine the answer to a question.

hypothesis A proposed answer to a question that is based on experimental evidence.

law A tested and confirmed hypothesis that predicts a principle but does not explain why.

theory A tested and confirmed hypothesis that explains an answer to a question.

Although this discussion appears to be localized to the nonbrewing world, it isn't. The brewing laboratory is deeply involved in performing the scientific method on a daily basis. Lab personnel work with questions that focus on the brewery product and brewing process: What is the level of diacetyl in that fermenter? How many yeast cells are there in the propagation tank? What is the concentration of fermentable sugars in the wort? The hypotheses produced by lab personnel result from these questions. The level of diacetyl is still too high to release the beer from that fermenter. The number of yeast cells in the propagation tank is adequate for the next batch of beer. These and many other questions and hypotheses form the basis of the work that the brewery lab performs.

The brewery lab must report its results to perform its job completely, adequately, and thoroughly. Every analysis must be recorded in a location from which it can be easily retrieved. Those analyses must be considered when the sample is evaluated. In addition, every evaluation must be shared with others, studied by others, and confirmed by others before being accepted as correct. Because of this, everyone in the lab should be familiar with the process by which results are reported to ensure that each analyst is reporting the information uniformly.

Following the scientific method also means that brewery lab personnel should operate with a complete lack of bias in performing every analysis. Often when lab personnel obtain a sample from the brewery floor, they overhear others talking about a particular product that isn't performing well. They may hear brewers discussing their perception about the most recent mash-in. Whatever they hear or observe, lab personnel must not consider that information as they perform their jobs.

This chapter will explore how brewing laboratories report their data and evaluations of the results. It will uncover some basic charting and graphing processes and provide a primer on the use of spreadsheets to convey information usefully.

Graphing with Microsoft Excel or Another Spreadsheet Program

spreadsheet A computer program that allows the entry, tabulation, and storage of data and calculations.

Being able to use a **spreadsheet** program is an important skill for brewing laboratory personnel. The authors would argue that this skill is vital to the success of the lab. Spreadsheets are used to relay, calculate, and evaluate the results of the analyses performed in the lab. While a lab can operate without any computer evaluation of the results, this isn't recommended. Reports can be generated by hand, and graphs and calculations can be prepared using old-school methods. But the time devoted to doing these tasks properly will be significant, the risk of making errors will be great, the relay of information to others will be slow, and the ability of the lab to respond to changes quickly will be hindered.

Computer-based spreadsheets are readily available for every computer system. If you use a Mac, a PC, or a computer with another operating system, there is a spreadsheet program for you. Some spreadsheets come preloaded on new computers, others are freely available for download from the internet, and others can be purchased at a local computer or office supply store or purchased online and then downloaded. Deciding which software program to acquire should be based on an evaluation of the spreadsheet's capabilities:

1. **Compatibility.** The program should be compatible with the computers used in the lab and in the brewery as a whole. This compatibility means that data can be shared to any computer through an intranet or by other means. (Just like we learned in kindergarten, sharing is an important and useful skill.)
2. **Tables and calculations.** The software program chosen for the brewery should be capable of creating tables and performing calculations on different cells, including addition, subtraction, multiplication, and division. At the minimum, the spreadsheet should provide these four mathematical functions. Other mathematical functions are quite useful, such as calculating an average and determining the standard deviation. Being able to perform more advanced calculations, such as IF statements and text lookups, increases the utility of the program.
3. **Charts and graphs.** The software program should be able to create charts and graphs of the data. Although creating charts and graphs using a software program isn't required (and neither is creating them by hand), the visual representation of the data is often more useful in an analysis than the specific numbers. Important types of charts and graphs include the scatter plot and the bar graph. The **scatter plot** is a visual way to relate the data in terms of two variables along *x*- and *y*-axes (such as the value of an analysis versus time). The **bar graph** allows the user to see the value of an analysis versus the batch number. These two types of graphs seem similar; however, the scatter plot uses a linear scale to plot the time and the bar graph does not.

scatter plot An *x-y* plot of data points.

bar graph A plot of data with specific values on the *y*-axis that are broken into different categories on the *x*-axis.

Many spreadsheet programs offer essentially the same types of data manipulation (often using very similar commands), but the most commonly used spreadsheet is Microsoft Excel. A free software package known as OpenOffice Calc provides features similar to Excel. Yes, there are software packages designed exclusively for use in a brewery, but many of them lack the ability to input laboratory data beyond simple notetaking. There are also programs known as laboratory information management systems (LIMS), which are incredibly helpful to a laboratory because they facilitate the managing and ordering of chemicals and other supplies and provide an in-house database or template that helps with tracking data. These tasks can be performed in a spreadsheet program but are often made easier by using a system designed specifically for laboratory management.

For the purposes of this text, we will focus on the use of spreadsheets. Their utility for any brewery lab, ease of use, and availability on any budget are three reasons for selecting this focus. In fact, Excel and OpenOffice Calc can perform all the tasks needed by brewery lab personnel. These programs are also very versatile in that they can be modified extensively and customized with templates and macros that can perform almost any task. Some translation of the following discussion may be needed for those using other software packages.

Entering Results

The results of analyses should be recorded immediately after they are produced. In the past, results of analyses were often written down in a notebook, where each page listed the specific analysis being performed, the date of each result obtained, a systematic name for each sample that indicated where and when it was obtained, and the result of each analysis on that sample. In today's computer-based world, the paper notebook has been replaced with the computerized notebook. There are many ways to enter results, but we will focus on just one of them.

To begin, a separate file should be created for each batch of beer. Doing so allows tracking of the batch for other purposes. The name of the file should indicate the name of the product and include the date the batch was started. If multiple batches of the same product were started on the same day, letters should be used at the ends of the file names to distinguish the individual batches. For example, if a brewer starts two batches of Crazy Pilsner on the same day, the file names could be crazypilsner-010118A and crazypilsner-010118B. Using systematic names for the files makes archiving and finding things much easier.

The master file for a batch of beer should include all the information about it. Different sheets within the file should be denoted Mash, Boil, Cellar, Filter, and Package or written to include any delineation the brewer needs to accurately track and report data for that batch. Then within each sheet, cells should be denoted for the specific values that are recorded. For example, the Boil sheet should have cells indicating the time of the start of the boil, the time of the end of the boil, the time of the first hop addition, a comment field for the type of hop added, a comment field for the lot number on the hops used, and so on. The sheet should also include a cell for each analysis performed. For example, the Cellar sheet might include cells for temperature, gravity at a specific time, diacetyl concentration, cell count per milliliter on the pitch yeast, and so forth. Everyone in the brewery should have access to these master sheets, and the person who performs a specific measurement should update each field as the data are collected. The file is a living document throughout the entire brewing process.

The lab should also have a set of files focused on the analyses that it performs. Those files should hold all the results of a specific analysis for a particular week, a specific month, a quarter (3 months), or an entire year based on a number of factors. If the instrument or method used to perform the analysis changes, a new spreadsheet for that analysis should be created. If the sheer number of results performed by the lab exceeds the ability of the spreadsheet to handle calculations easily, a separate spreadsheet should be created. If the instrument is on a quarterly preventive maintenance schedule, a new file should be created each quarter. Columns within the file should include the date of the analysis, sample identifier, analysis number, and result, plus a comment field or two.

Using a spreadsheet allows the digital transfer of data between the master spreadsheet and an analysis spreadsheet. In Excel, this transfer is done by selecting the cell that's the target of the data. An equals sign (=) is typed in that cell, the cell in the other spreadsheet is selected, and the Return key is pressed. When both spreadsheets are open on the same computer, the cell in the target spreadsheet automatically populates with the value entered in the working spreadsheet. The authors recommend denoting the master spreadsheet for the specific batch as the target spreadsheet and the working spreadsheet as the analysis spreadsheet. Thus, when lab personnel enter the values for the analysis into the analysis spreadsheet and save the spreadsheet, the master spreadsheet will be updated. If the files are not co-located on the same computer but are linked via an intranet or the Internet, the master file can be updated by simply updating the file. (Using options on the "Data" tab, the links can be set to automatically update.)

Using a master template for each batch, the links to the analyses spreadsheets can be set up at the start of the batch. The result is that anyone with access to the master spreadsheet for the batch, on any computer that's linked via an intranet or the Internet to the lab computer, will get real-time updates on lab analyses as the batch progresses through the system. This allows the lab to communicate rapidly with the brewers, cellar personnel, and any other employees of the brewery who need that information.

Standard Curves

Analysis spreadsheets are great tools that can easily perform further evaluations of laboratory analysis methods. One evaluation commonly used during a laboratory analysis is the creation of a standard curve. For example, a series of standards can be analyzed routinely using the method. Those values are used to correlate the concentration of the analyte with the response that's reported by the instrumental method. The correlation is known as a **standard curve**.

Interestingly, a standard curve prepared using samples with known concentrations of analyte measured on the ultraviolet-visible (UV-vis) spectrophotometer is known as a **Beer-Lambert law plot**. The UV-vis spectrophotometer measures the intensity of specific wavelengths of light absorbed by an analyte. August Beer and Johann Heinrich Lambert explored the relationship between the intensity of absorbed light with the concentration of the analyte in the sample. They found that at low concentrations, the relationship is directly proportional and linear. An equation known as the **Beer-Lambert law**, also known simply as Beer's law, was developed to express this relationship:

$$A = \varepsilon \times b \times c$$

where A is the absorbance of the specific wavelength of light; ε is the proportionality constant (also known as the molar absorptivity) in units of per centimeter per molar; b is the path length of the sample in centimeters; and c is the concentration of the analyte in the sample in units of molarity.

Let's consider each variable in the Beer-Lambert law in more detail:

1. **Absorbance** is a scientific term that relates the amount of light that passes through a sample to the amount of light that enters the sample. (Note that this is not *absorbency*, which refers to the ability to soak up a liquid—for example, by diapers.) Mathematically, absorbance is the base 10 logarithm of the intensity of light that enters the sample (the

standard curve An *x-y* plot of data points for a series of known concentrations.

Beer-Lambert law plot A standard curve prepared using samples with known concentrations of analyte that have been experimentally measured.

Beer-Lambert law Provides the direct relationship of molar absorptivity, path length, and concentration to the absorbance of a solution.

absorbance A measurement of the amount of light that is absorbed by the sample.

incident intensity The concentration of light that enters the sample.

transmitted intensity The concentration of light that passes out of the sample.

incident intensity, $I_{incident}$) divided by the intensity of the light that leaves the sample (the **transmitted intensity**, $I_{transmitted}$). In other words, the absorbance, A, is represented by this expression:

$$A = \log_{10}\left(\frac{I_{incident}}{I_{transmitted}}\right)$$

If all the light passes through the sample without being absorbed, then the incident intensity and transmitted intensity are equivalent. So, $I_{incident}/I_{transmitted} = 1$, and the logarithm of 1 is 0. In other words, the absorbance in this case is 0. If the intensity of light that passes out of the sample is half the intensity of light that enters the sample, then the absorbance is 0.301 ($A = \log(2)$). As the intensity of light that passes out of the sample drops, the absorbance increases. (It should be noted that there are no units for the value of absorbance.)

As we will discover Chapter 11, the absorbance value most accurately reflects the amount of light absorbed by a sample when it's between 0 and 1. For example, assume the ratio of incident to transmitted light for a concentrated sample is 100:1 (and then $A = 2.00$):

$$A = \log_{10}\left(\frac{I_{incident}}{I_{transmitted}}\right) = \log_{10}\left(\frac{100}{1}\right) = \log_{10}(100) = 2.00$$

The instrument will be forced to discern the difference between ratios at this level as the concentration changes. Unfortunately, a 10% change in the intensity of light that's absorbed by the sample at this concentration is barely discernible by most instruments. For example, a 10% decrease in the amount of light that's transmitted through the sample will give a 110:1 ratio, and the absorbance value will be 2.04 ($A = \log(110)$). This isn't a very large change in the absorbance for a very large change in the intensity. A better example is to use a much more dilute sample, for which even a 1% change in intensity can be easily identified. Let's assume, for example, that in a dilute sample, the ratio of incident to transmitted light is 1.10:1 ($A = 0.041$). A 10% change in that ratio at this concentration will result in $I_{incident} : I_{transmitted}$ 1.21:1 ($A = 0.083$). This is a much larger change (more than double the original absorbance) for the same percentage change in the intensity of light.

molar absorptivity A proportionality constant that relates the ability of a compound to absorb light.

2. The **molar absorptivity** value in the Beer-Lambert law is basically a proportionality constant. It has units of per centimeter per molar (1/cm•M). It correlates the concentration of the analyte in the sample to the intensity of light that it absorbs. In other words, it's a measure of the analyte's ability to absorb light. Some analytes absorb light easily and have very large molar absorptivity values. Others do not absorb light well and have small molar absorptivity values. While the molar absorptivity value can be estimated for a particular analyte, the only way to know the exact value is to make a solution with a known concentration, measure the absorbance of that solution, and then solve the Beer-Lambert law for molar absorptivity. However, as we will see later, we don't really need the molar absorptivity constant to create a standard curve.

3. The **path length**, b, in the Beer-Lambert law reflects the thickness or depth of the sample. A narrow sample tube won't absorb as much light through its sides as a wide sample tube. Fortunately, the instruments typically admit only a sample tube of one size, commonly 1 centimeter (cm) thick. Thus, the path length that the light must travel through the sample is the same for every sample that's run. And when $b = 1$ cm, the value doesn't change the relationship between the concentration and the absorbance.

4. The **concentration** in the Beer-Lambert law is in units of molarity. We discussed this concentration unit and how to calculate the value for a solution in Chapter 4. And when we use the Beer-Lambert law to determine the molar absorptivity of a particular analyte, we must use the units of molarity. However, when a standard curve is created and we are less interested in the specific value of the molar absorptivity, any units can be used for the concentration. As long as the units for c are the same for every standard sample, then the standard curve will be valid. The relationship between A and c in the Beer-Lambert law will still be valid with other concentrations, but the proportionality constant won't be the molar absorptivity. When the standard curve is used to determine the concentration of an unknown sample, the concentration will be reported in the same units used in the standard curve.

path length The thickness of a sample where the light traverses it.

concentration The amount of a particular analyte per a given volume.

To prepare a standard curve, a series of samples are prepared that span the range of all the concentrations of interest. Then, following the manufacturer's instructions for using the instrument, the absorbance of each sample is measured and a plot created from the data. The concentration of the sample is recorded on the x-axis on the plot, and the measured absorbance is logged on the y-axis. Using a spreadsheet simplifies this process. The numbers are entered such that one column of the spreadsheet represents the x-axis (concentration) and the next column in the spreadsheet represents the y-axis (absorbance) (Fig. 8.1).

In Excel, creating the plot involves highlighting both columns of data and then using the computer mouse to select "Insert…Chart…XY(Scatter)" from the menu. (The exact command used to do this and other tasks will differ based on the version of Excel and whether the program is running on a PC or a Mac.) A plot then appears on the spreadsheet (Fig. 8.2).

The equation for the **linear regression** of the data is created by clicking on the plot and adding a chart element. Selecting "Trendline…Linear"

linear regression A mathematical treatment of a set of data in which a trendline is drawn so it's as close to each specific data point as possible.

	A	B	C
1	Sample #	Concentration (ppm)	Absorbance
2	Standard 1	1.00	0.155
3	Standard 2	2.00	0.280
4	Standard 3	3.00	0.405
5	Standard 4	5.00	0.655

FIG. 8.1. Entering data for a standard curve.
(Courtesy S. E. Johnson and M. D. Mosher—© ASBC)

produces a line on the plot. Selecting "More Trendline Options" provides the choices "Display equation on chart" and "Display R-squared value on chart." When those choices are selected, the equation of the line in the form $y = mx + b$ will appear on the plot (Fig. 8.3).

The equation of the line illustrates the relationship between the absorbance value (the y value) and the concentration (the x value) of the analyte. The proportionality constant is the **slope of the line** (the m value; in this case, $m = 0.1243$). The value of b (in this case, $b = 0.0339$) in the equation represents where the line crosses the y-axis and is indicative of a background absorbance resulting from the matrix or the test itself. The equation of the line can then be used to determine the concentration of any unknown sample, if the absorbance of the unknown sample is known. Alternatively, the plot can be used graphically to determine the concentration by looking up the absorbance of the sample on the y-axis, translating over to the line, and then translating down to the x-axis. The concentration of the sample is determined by where that line intersects the x-axis (Fig. 8.4).

slope of the line The rise over the run; the change in the y-axis versus the corresponding change in the x-axis along a straight line.

Sample #	Concentration (ppm)	Absorbance
Standard 1	1.00	0.155
Standard 2	2.00	0.288
Standard 3	3.00	0.405
Standard 4	5.00	0.655

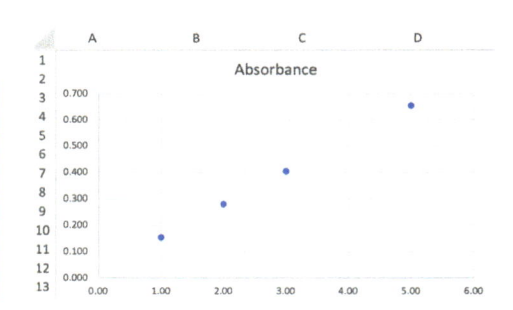

FIG. 8.2. Creating a standard curve in Microsoft Excel. Selecting "Insert x-y scatter plot" results in creating a plot of the data. (Courtesy S. E. Johnson and M. D. Mosher—© ASBC)

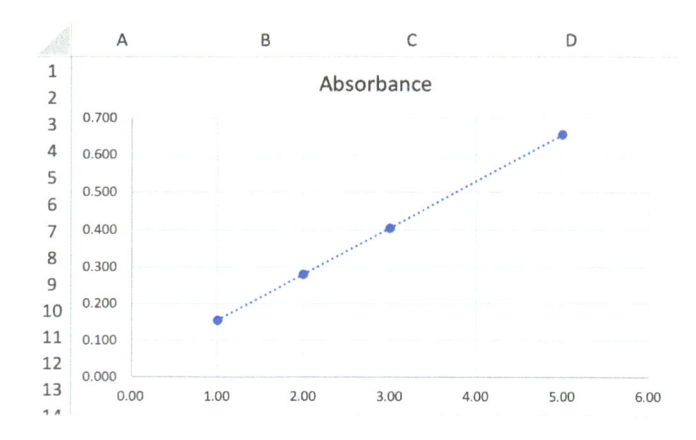

FIG. 8.3. Linear regression analysis of the data. After a linear trendline is inserted, the options to display the equation and R-squared value on the chart are checked. The equation of the best-fit line appears with the correlation coefficient. (Courtesy S. E. Johnson and M. D. Mosher—© ASBC)

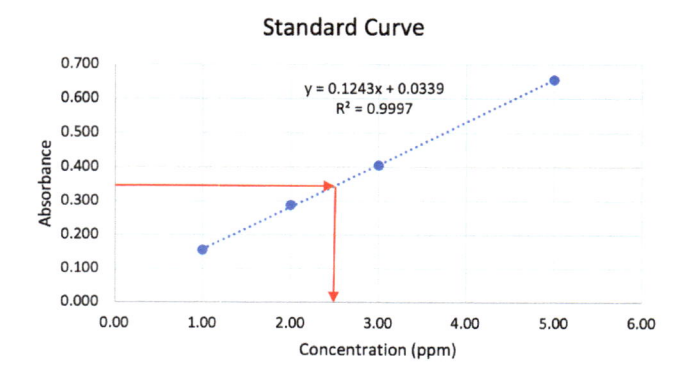

FIG. 8.4. Graphical use of the standard curve to determine the concentration of an unknown. As shown by the red lines, if the absorbance of an unknown sample is 0.35, the resulting concentration of analyte will be 2.5 parts per million (ppm). (Courtesy S. E. Johnson and M. D. Mosher—© ASBC)

Some precautions should be followed when using standard curves. The Beer-Lambert law works only as long as the absorbance and the concentration values form a linear relationship. This means that the data must be very close to the line that's calculated. The degree of deviation from the line is represented by the R-squared value. A perfect line shows an R-squared value of 1.000; if the data are 100% random and don't correlate at all, the R-squared value will be 0.000. As the data scatter away from the line, the R-squared value drops. There really isn't a rule of thumb for when the line isn't linear based on the R-squared value. But remember that the closer the R-squared value is to 1.000, the better the correlation of that standard curve. A brewing lab should decide the R-squared cutoff limit based on the history of the experiment.

In addition to using the computer to report the degree of linear correlation, it's necessary to perform a visual inspection of the line. The data points may vary in such a way that the relationship really isn't linear (Fig. 8.5). If it isn't, the standard curve shouldn't be used, because it doesn't accurately reflect the relationship between the concentration and the absorbance. Deviations from linearity can be addressed, but the level of error in the measurements increases dramatically as the linearity drops.

One method used to utilize data that follow a nonlinear correlation is to fit other line shapes to the data. In such a case, the method may still be used to determine the value of an analysis, but the response of the analysis to changes in the value becomes nonlinear. For example, in Figure 8.6, the data have been correlated using a polynomial expression. As shown in the figure, the analysis is better represented regarding small concentrations than large concentrations. This may be acceptable to the brewery laboratory, but the large concentrations that are less responsive in the method should be of concern.

To ensure that the standard curve fits the Beer-Lambert law as closely as possible, some steps can be taken. First, the concentrations of the standards and the samples should be relatively dilute. If the maximum absorbance is less than 0.500, the chances are good that the solutions are dilute enough to provide a linear relationship between the concentration and the absorbance. Second, the concentrations of the standards and the samples shouldn't span more than about a factor of 10. In other words, the Beer-Lambert law might be linear from 0.10 ppm to 1.0 ppm, but it might not be linear from 1 ppm to 100 ppm.

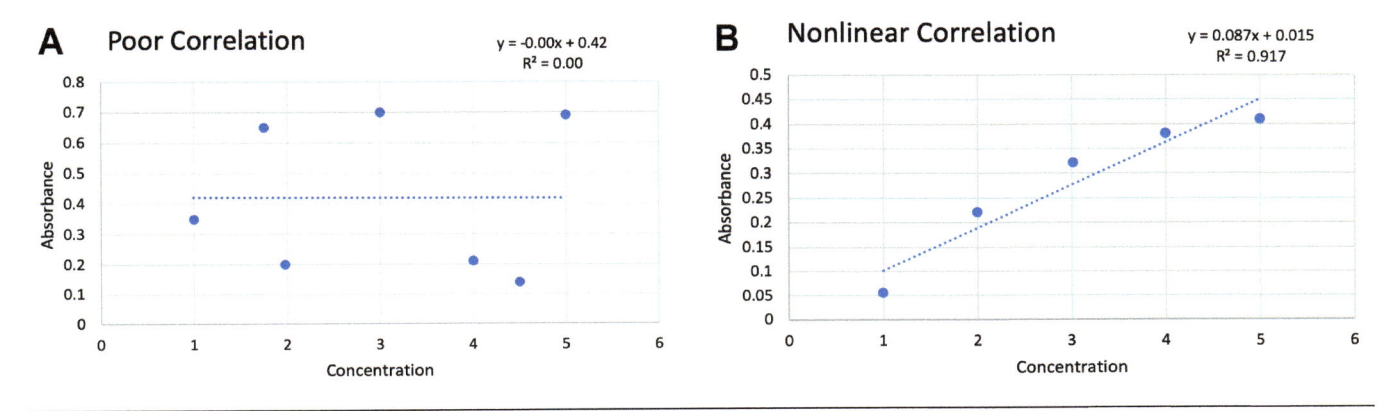

FIG. 8.5. A, A nonlinear relationship indicates a poor correlation of the data or poor agreement with the Beer-Lambert law. B, Note that the R-squared value for the nonlinear correlation still indicates a strong overall correlation, although the data are visually not linear. (Courtesy S. E. Johnson and M. D. Mosher—© ASBC)

The standard curve can be used to determine the concentrations of unknown samples. For example, let's assume that the standards (expressed in ppm) have all been determined and the resulting equation of the line is $y = 0.125x + 0.030$. The concentration of an unknown sample can be determined simply by obtaining the absorbance of that sample. Let's assume that an unknown sample has an absorbance of 0.330. The concentration of that sample in ppm can be determined by hand by solving for the value of x in the equation:

$$y = 0.1243x + 0.0339$$

$$0.330 = 0.1243x + 0.0339$$

$$0.2961 = 0.1243x$$

$$2.38 \text{ ppm} = x$$

Alternatively, the concentration can be calculated directly by the spreadsheet. When the absorbance is entered into the spreadsheet, the adjacent cell does the calculation immediately. To set this up, we solve the equation of the line algebraically for x. Doing this mimics the process we just completed as we solved for the value of x given the value of y:

$$y = 0.1243x + 0.0339$$

$$y - 0.0339 = 0.1243x$$

$$\frac{(y - 0.0339)}{0.1243} = x$$

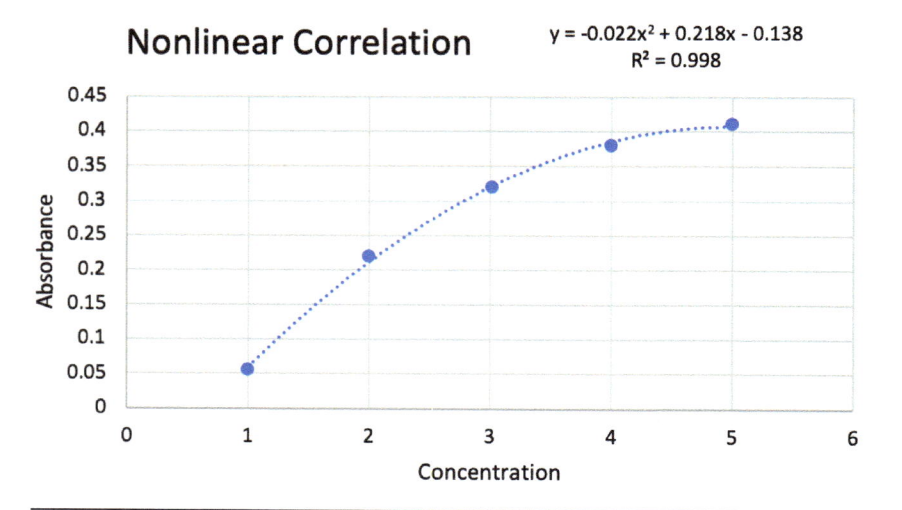

FIG. 8.6. A nonlinear relationship. Note that small absorbance values provide a good correlation to the concentration. For example, absorbance values of 0.10 and 0.15 span a range of concentrations from 1.2 to 1.5. At high absorbance values, the correlation to the concentration is less accurate. For example, absorbance values of 0.35 and 0.40 span a concentration range from 3.2 to 4.0. (Courtesy S. E. Johnson and M. D. Mosher—© ASBC)

Then we click in the cell where the concentration should be reported. If we type = followed by the equation that was just solved, the answer will be calculated directly in that cell. For our example, we will type $=(y-0.039)/0.1243$. However, instead of typing the letter y, we click on the cell that contains the absorbance for that sample (Fig. 8.7). When the Return key is pressed, the value of the concentration is then calculated (Fig. 8.8).

The utility of the spreadsheet is that the equation just entered can easily be copied into other cells to perform the calculation in them. For example, in Figure 8.9, the absorbance values for "Unknown 2" through "Unknown 10"

Sample #	Concentration (ppm)	Absorbance
Standard 1	1.00	0.155
Standard 2	2.00	0.280
Standard 3	3.00	0.405
Standard 4	5.00	0.655
Unknown 1	=(C6-0.0339)/0.1243	0.330

FIG. 8.7. Entering the equation into the spreadsheet. The absorbance value for "Unknown 1" is entered into the "Absorbance" column. The equation is then typed into the "Concentration" column. Note that the value C6 reflects the cell that contains the absorbance value for that sample. (Courtesy S. E. Johnson and M. D. Mosher—© ASBC)

Sample #	Concentration (ppm)	Absorbance
Standard 1	1.00	0.155
Standard 2	2.00	0.280
Standard 3	3.00	0.405
Standard 4	5.00	0.655
Unknown 1	2.38	0.330

FIG. 8.8. The result of the equation in the spreadsheet. After the equation is entered and the Return key is pressed, the spreadsheet completes the calculation. (Courtesy S. E. Johnson and M. D. Mosher—© ASBC)

A

	A	B	C
1	Sample #	Concentration (ppm)	Absorbance
2	Standard 1	1.00	0.155
3	Standard 2	2.00	0.280
4	Standard 3	3.00	0.405
5	Standard 4	5.00	0.655
6	Unknown 1	2.38	0.330
7	Unknown 2		0.252
8	Unknown 3		0.381
9	Unknown 4		0.299
10	Unknown 5		0.241
11	Unknown 6		0.507
12	Unknown 7		0.637
13	Unknown 8		0.308
14	Unknown 9		0.274
15	Unknown 10		0.223

B

	A	B	C
1	Sample #	Concentration (ppm)	Absorbance
2	Standard 1	1.00	0.155
3	Standard 2	2.00	0.280
4	Standard 3	3.00	0.405
5	Standard 4	5.00	0.655
6	Unknown 1	2.38	0.330
7	Unknown 2	1.75	0.252
8	Unknown 3	2.79	0.381
9	Unknown 4	2.13	0.299
10	Unknown 5	1.67	0.241
11	Unknown 6	3.81	0.507
12	Unknown 7	4.85	0.637
13	Unknown 8	2.21	0.308
14	Unknown 9	1.93	0.274
15	Unknown 10	1.52	0.223

FIG. 8.9. Copying and pasting the formula in the spreadsheet. Dragging down **(A)** fills the equation into the spreadsheet and automatically performs the calculations **(B)**. (Courtesy S. E. Johnson and M. D. Mosher—© ASBC)

are entered into the table. Next, the cell containing the formula is selected, and the mouse pointer is held over the edge of the cell until a bold plus sign (+) replaces the pointer. By clicking and dragging down to "Unknown 10," the formula is copied into those cells.

Plotting Results, Cumulative Sum Charts, and Trending

The results of the analyses can be distributed to everyone using the spreadsheet. However, a visual representation of the data is often much more important and useful. Using the spreadsheet, the brewery laboratory can provide several types of graphics, including charting the results to show trends in the data.

The first type of chart is a **simple plot** of the data as a function of time. This chart typically has very specific values that aid in representing the data. A standard chart is shown in Figure 8.10. We already covered this type of chart, but let's revisit it here. The **target value** of the analysis is recorded on the y-axis in the center of the chart. That target value depends on the specifications of the batch of beer. Two other lines are drawn on the chart, as well. The first is the **warning boundary**, which typically lays 2 sigma values above and below the target value. The second is the **action boundary**, which typically lays 3 sigma values above and below the target value. A **sigma value** is the standard deviation of the method of analysis, and it's determined by measuring a standard sample with a known value repeatedly. Often, five measurements of the sample are enough to estimate the standard deviation, although the best standard deviations are determined with at least 10 measurements of the standard sample. The values from each measurement can be entered

simple plot An *x-y* plot.

target value The desired value of a specification.

warning boundary A value for a specification that indicates the process needs to be scrutinized.

action boundary A value for a specification that indicates the process needs to be adjusted.

sigma value The standard deviation of the method of analysis.

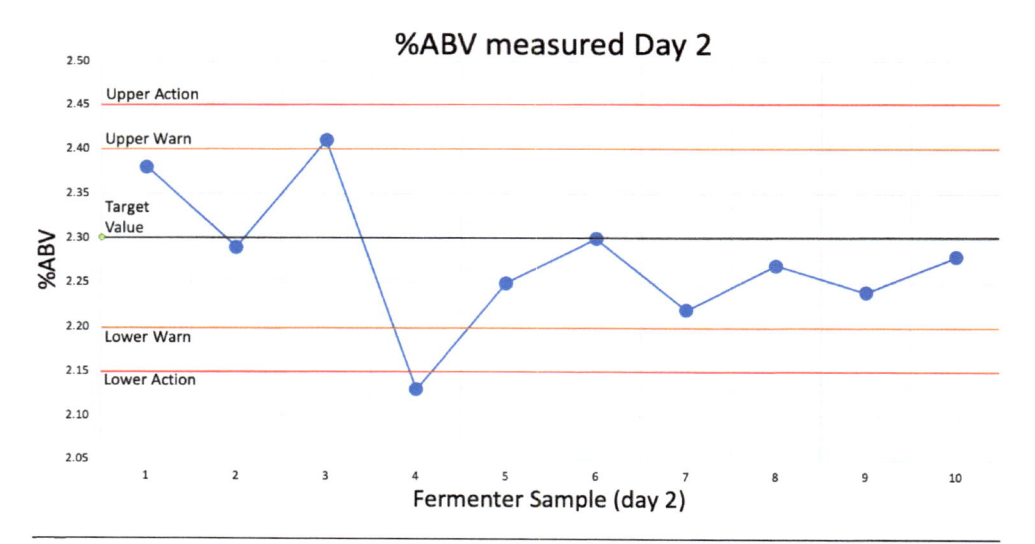

FIG. 8.10. A standard reporting chart. For brewery personnel viewing this chart, the third sample caused a warning, and discussion resulted in an adjustment being made. As shown in the chart, that adjustment resulted in a significant change, which caused the fourth sample to fall below the lower action level. Subsequent measurements were within the boundaries. (Courtesy S. E. Johnson and M. D. Mosher—© ASBC)

into the spreadsheet in a column. At the bottom of the data, the calculation of the average (the target value) and the standard deviation (1 sigma) can be determined. By typing =STDEVP(…) and replacing the … with the 10 measurements, the spreadsheet will give the sigma value of the method.

The warning boundary can be used to visually notify everyone when a value for the analysis is close to indicating a possible issue. If a value is above or below the boundary, then that particular sample should be scrutinized. The action boundary indicates that the sample is out of specification. This indication should trigger a full analysis of the sample, attention to the manufacturing process where the sample was collected, and potential adjustment to the process based on the requirements for the overall process.

Unfortunately, a standard reporting chart doesn't do much else than report the values of the analysis. A more useful chart is the cumulative sum chart, or **CUSUM chart**, which reports the cumulative sum of the deviations from the target. The result is a graphic representation of the trends in the analysis. A CUSUM chart can be created by hand. The difference between the sample's current value and the specification target value is determined and added to the difference from the previous sample's analysis (Fig. 8.11). This is determined for each sample. If the difference between the value of the sample's analysis and the specification target is positive over a period, the graph will show that trend. Similarly, if the difference is always negative, that trend will also appear. Thus, the brewer can react quickly to the trending and correct the process before the sample analysis reaches the warning or action boundary.

A CUSUM chart can be created easily using the spreadsheet. The data are entered into the first two columns of the spreadsheet, and the first row is reserved for the target value. The third column is the cumulated sum of the measurement difference from the target value. Specifically, given the table

CUSUM chart A chart that plots the cumulated sum of differences between measurements and the target value; shows trending in a measurement.

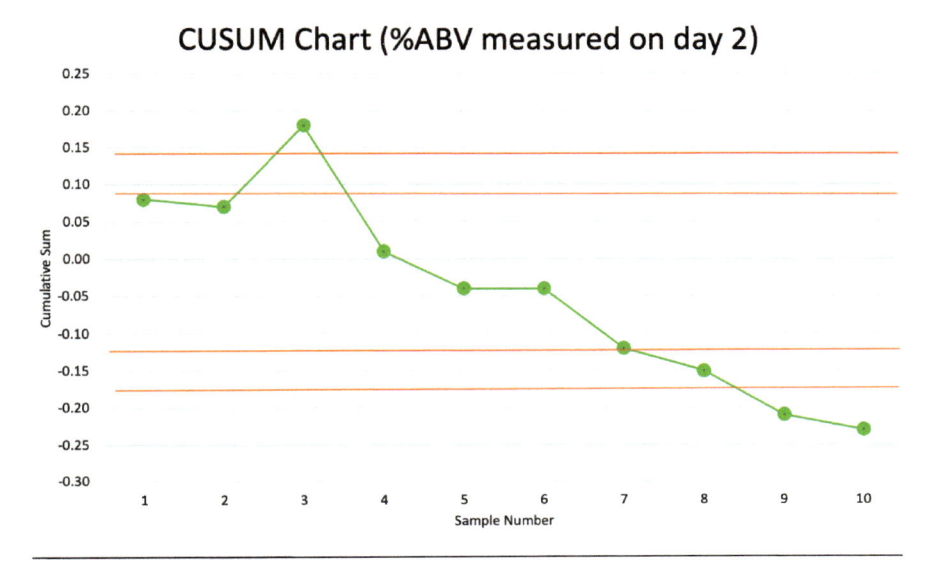

FIG. 8.11. A CUSUM chart. The data presented are the same as in Figure 8.10. The y-axis is the deviation from the target value. Note that the trend in the data indicates an inherent issue with the process at this point in the measurement. (Courtesy S. E. Johnson and M. D. Mosher—© ASBC)

	A	B	C
1	Sample	Absorbance	CUSUM
2	TARGET	2.3	
3	1	2.38	=(B3-B2)+C2
4	2	2.29	0.07
5	3	2.41	0.18
6	4	2.13	0.01
7	5	2.25	-0.04
8	6	2.30	-0.04
9	7	2.22	-0.12
10	8	2.27	-0.15
11	9	2.24	-0.21
12	10	2.28	-0.23

FIG. 8.12. A CUSUM table. The formula in cell C3 is selected and dragged down the column to fill in the values of the cumulative sum for the rest of the table. (Courtesy S. E. Johnson and M. D. Mosher—© ASBC)

shown in Figure 8.12, the formula to enter is =(B3-B2)+C2. This formula takes the difference of the specific measurement and the target value and then adds it to the running sum of the differences. Dollar signs ($) are added in front of the B and the 2 to fix that cell (B2) within the calculation. By clicking in the cell and then dragging down to fill the column, the values can be easily calculated. Clicking on cell C12 reveals the formula to be =(B12-B2)+C11, indicating that the individual cells were advanced as the formula was dragged down but that the reference cell containing the comparison value didn't change. A scatter plot of column A versus column C or a line-based bar chart of column C provides the CUSUM chart shown in Figure 8.11.

Tables and Record Keeping

Record keeping in the brewery lab is very important. As shown in the previous section, the specific data for each analysis can be entered into a spreadsheet, providing easy access to information for the entire brewery and useful data for further analysis. In addition to creating tables of data, the spreadsheet can aid in visually reporting issues with data.

Using Functions

Having a table of values is useful, but staring at a table and trying to identify trends can be difficult. Sure, creating charts can help give some visual clues as to the nature of the data, but using a more visual method is especially helpful.

For example, suppose a value that's entered into the spreadsheet is outside specifications. This value would be easy to identify if it were formatted using color. The spreadsheet can automatically do this. In Excel, select the column of the analysis data and then choose "Format…**Conditional Formatting**". Select "Classic" style and then choose "Format only cells that contain". Next, choose "cell value" and then select "greater than". The value of the warning line or the action line is entered and the color chosen to reflect either a warning or an action. This process can be repeated by adding new conditional formatting to the same column. When a number is entered into the column, it will change color based on the conditions set in the formatting.

In addition, the spreadsheet can handle almost any manipulation or calculation that a brewery laboratory wants to perform. For example, the **average** of a series of data can be determined by entering =AVERAGE(…) and then indicating the cells that contain the data to be averaged in the parentheses (replacing the …). Similarly, the standard deviation, sum, median, and a host of other mathematical functions can be simply entered into the spreadsheet to perform calculations on the data that are collected.

A spreadsheet can also perform **text-based reporting** of values. Suppose that the operator of the program wants to have text appear if a particular value is too high. The spreadsheet can do this, too. For example, if the gravity of a sample is more than 1.0450, the text warning "HIGH" should be reported. This can be done by entering =IF(xxx>1.0450,"HIGH"," "). The xxx is the cell being checked to see if the value is high. The formula will make the comparison, and if the value is greater than 1.0450, "HIGH" will appear in the cell. If the value isn't greater than 1.0450, a space will be entered in the cell. While the formula seems complex, the basic structure is relatively simple. We can adjust the formula to compare the cell with a different cell that might change. For example, we could perform the calculation to find out if the entered value is 5% greater than a standard value. The formula would be =IF(xxx>(yyy*1.05),"HIGH"," "), where yyy is the location of the cell with the standard value.

Many more calculations can be performed by the spreadsheet. It can count the number of times that a value has been high over the last week, it can determine the minimum or maximum value of a series of data, and it can determine the different quartiles in a set of data. The full capability of the software can be determined by using the Help feature within the program, reading a printed manual, and visiting various websites that support the software.

Conditional Formatting A set of rules in a computer program that adjust how a cell is displayed in a spreadsheet.

average The mean of a set of measurements; determined by adding all the measurements and dividing by the number of measurements.

text-based reporting Using tables, words, and phrases to report the results of an analysis.

Macros and Advanced Excel Functions

A **macro** is a set of commands that a user creates within the software. When the macro is initiated, or run, it performs that set of commands in the specified order. Yes, we can perform these commands by hand or type them one after another in the software, but if the sequence of steps is sufficiently long and/or repetitive, using a macro might be the best option. Using macros can greatly speed up working within a spreadsheet.

macro A set of operations that the user creates and the software performs.

This section won't cover everything you can do with macros in Excel but will instead introduce the topic and provide enough information to allow you to experiment with using macros. For those of us who use Excel daily, macros are the way to go. They increase the utility of the program, reduce time on the computer, improve productivity, and allow spreadsheets to be fully customized to laboratory processes. Since it's impossible to predict exactly what a spreadsheet needs to do for a given brewery lab, only the basics are included here.

The macro environment is easily accessed in an Excel spreadsheet, and it can be used by beginners without any programming knowledge. It's quite forgiving, allowing you to play with it to see what it can and can't do. And if you have used the programming language Visual Basic, macros can be programmed to perform just about any task.

Programming 101

The easiest method for setting up a macro in an Excel spreadsheet is to simply record the steps followed to perform a task. In other words, we tell Excel to start recording what we want to do; perform those steps using the standard commands in the program; and then tell Excel to stop recording. Those recorded steps become the macro, which we can access and run whenever we want.

Let's run through an example to see how this works. Suppose you want to create a macro to determine the standard deviation for a series of wort preboil gravities. Because brewery personnel routinely measure the preboil gravity, the spreadsheet is updated with this information for each batch. Creating and running a macro will allow us to verify that our instrumental method is accurate, precise, and/or simply within specifications. So, the brewery lab workers prepare a standard wort sample with a known gravity—say, 1.0400 grams per milliliter (g/mL). The workers will use the same method that the brewery personnel use to determine the experimental gravity of that wort sample. Assume the table in Figure 8.13 represents the 10 measurements obtained for the sample and the known gravity of the sample.

To record a macro to calculate the standard deviation, we must tell Excel that we want to do this. By default, the ability to do this is hidden, but the preferences for Excel can be adjusted to make sure that the "Developer" tab is visible. Once visible, the tab can be selected to set up the macro.

From the menu bar, we select the "Developer" tab. To start the macro commands, we click the Record Macro button. In the window that appears, we enter a name for the macro. The name can be anything but must not include spaces and must start with a letter. A name that represents the function of the macro is the best option. A shortcut keystroke can also be selected, but it's important to be careful in doing so. If the shortcut that's chosen already exists, the old shortcut will be overwritten. If the shortcut is reserved by the computer's operating system, the new shortcut will be rejected. In addition, the shortcut should be recorded somewhere conspicuous so that it isn't easily forgotten and can be retrieved easily if it is forgotten. Finally, providing a description of the macro will make it easier to verify the function of the macro later.

We set up the macro by clicking the Record Macro button and entering the information, as indicated previously. We click in the cell where we want to place the standard deviation of the gravities (in Figure 8.13, cell B14). Then we enter =STDEVP(B2:B11) and select all the cells that we want to use for the standard deviation of the measurement. Then we press the Return key, which enters the formula into that cell and performs the calculation. Finally, we click the Stop Recording button. Our macro is saved.

To run the macro, we make sure that we first click in a cell that's three cells below our set of data and that the cells containing the data have no more than 10 values. These limits are placed by the location in which we set up the macro (B14) and by the set of data we wanted to use for the calculation (cells B2–B12). In other words, if we are performing the standard deviation on a list of only five values, we make sure to click on a cell that's no higher than row 14 in the spreadsheet and at least three rows below the data. For example, in Figure 8.14, we want to calculate the standard deviation of the five values in the table. We click in cell G16 before typing the shortcut keystroke command for the macro. This cell is three cells lower than the data, and the cell is greater than row 14. Pressing the shortcut keystroke here will return the standard deviation of the five values. One small warning: If we are not at row 14 or greater when we enter the shortcut, a blue dot and a 0 will be entered in the cell. This result indicates that the macro was not positioned correctly.

	A	B
1	Trial Number	Gravity (in g/mL)
2	1	1.0402
3	2	1.0400
4	3	1.0399
5	4	1.0405
6	5	1.0401
7	6	1.0402
8	7	1.0398
9	8	1.0403
10	9	1.0400
11	10	1.0398
12		
13	Average	
14	Stand Dev	

FIG. 8.13. Experimental determination of the standard deviation of a method. A macro can be recorded to enter the standard deviation of the data, as well as other tasks. (Courtesy S. E. Johnson and M. D. Mosher—© ASBC)

	F	G
1	Trial Number	Gravity (in g/mL)
2	1	1.0402
3	2	1.0400
4	3	1.0399
5	4	1.0405
6	5	1.0401
7		
8		
9		

FIG. 8.14. Positioning the cursor to enter the macro shortcut. (Courtesy S. E. Johnson and M. D. Mosher—© ASBC)

As a side note, the command STDEVP or STDEV.P calculates the standard deviation of the full data set based on the actual set of numbers. This differs slightly from the command STDEV or STDEV.S, which calculates an estimated standard deviation assuming that we are selecting only a sample of all the values. In other words, it assumes that there are more values in the table than were selected. In practice, it's best to determine the standard deviation of the population. As such, the STDEV or STDEV.S calculation provides a larger value for the standard deviation.

For any table of data, we can place the cursor where we want to have the standard deviation reported and enter our shortcut keystroke. The computer will automatically run the macro and enter the value of the standard deviation. This macro will work on any page within the spreadsheet, which allows us to determine the standard deviation quickly on any set of data with just a keystroke.

Finally, if macros are created for an Excel spreadsheet, the spreadsheet must be saved in a format that includes macros. If the spreadsheet is saved in the normal format, the macros will be eliminated. In addition, when a macro-enabled spreadsheet is opened, the software will ask if the macros should be enabled or disabled. It's possible to open and use the spreadsheet with the macros disabled—but then the functions that have been created won't work.

CHAPTER 9

Sensory Analysis

What determines if a beer is successful in the marketplace? In today's world, the sheer number of choices in the beer aisle of your local store can be intimidating. In fact, there are hundreds of different brands, each vying for the consumer's attention. Which one will win?

That question can be answered in multiple ways. The main answer boils down to us—the consumers. Which beer catches our attention first? A beer with packaging that stands out and is easily recognized will likely be one of the brands that we purchase. However, we may never purchase it again if we don't like the beer. This chapter will address some of those concerns: How do we know if the beer matches the style the brewer says it is?

Probably the most important aspect of beer is its taste. Other factors, such as appearance and mouthfeel, impart further sensory experience, but it's ultimately the taste of the beer that decides whether the consumer will buy a specific beer again. Each beer's flavor relies on the presence of hundreds of different chemical compounds. The amounts of all the individual compounds combine to produce the taste of the beer. To complicate matters, two beers could contain exactly the same molecules but have completely different flavors if the quantities of the molecules were different. Even a very small variability in the amounts can have a dramatic effect on the flavor.

These changes can be monitored by chromatographic methods. If the brewery laboratory then carefully examines the quantity of each of the thousand or so individual compounds, it can see the changes. Unfortunately, predicting the flavor of the resulting ratio of compounds will be very difficult. For example, imagine the difficulty in guessing the exact color of paint made by mixing different ratios of blue and yellow—and that's only two compounds.

Instead, the best measure of the flavor of a beer uses **sensory analysis**. Sensory analysis is vital to the production of beer, because it provides the data needed to assess flavor. And by evaluating the data from sensory analysis, the brewer can assess every batch of a particular product to ensure that it has the same flavor. Using humans as sensory instruments, it's possible to operate a successful sensory program with as few as seven minimally trained people.

But sensory analysis isn't simply sitting around a table asking our friends if the beer still tastes good. Doing that doesn't result in any really useful information. A proper sensory analysis program in a brewery is operated

sensory analysis A program that provides the data needed to assess the flavors found in a sample of beer.

Keeping a Sensory Notebook

The first step in developing a sensory analysis program is to record our own tasting experiences. Starting to put words to what we taste and making a conscious effort to notice the flavors in our beer goes a long way toward understanding sensory reactions. Recording what we taste also helps us remember the flavors from a particular brand. When we taste something and give it a name, we are more likely to remember it the next time we taste it. With some practice, we can easily describe a flavor profile for a specific style of beer.

Figure 9.1 shows pages from the authors' own beer journals. Note the level of detail that the authors record as they explore a new beer. Both authors use the **Beer Judge Certification Program (BJCP)** guidelines to compare the beer they are drinking with a particular style. The BJCP guidelines are available online (www.bjcp.org) and are a great tool for any beer drinker who wants to learn more about the myriad of beer styles.

Beer Judge Certification Program (BJCP) An organization that develops and maintains a set of specification ranges for all beer styles; members are certified to serve as judges during competitions.

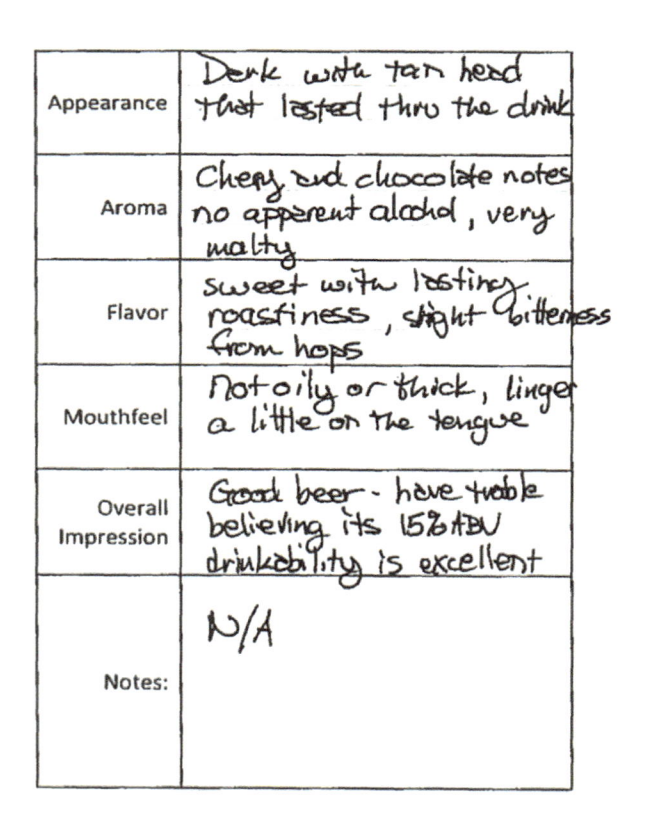

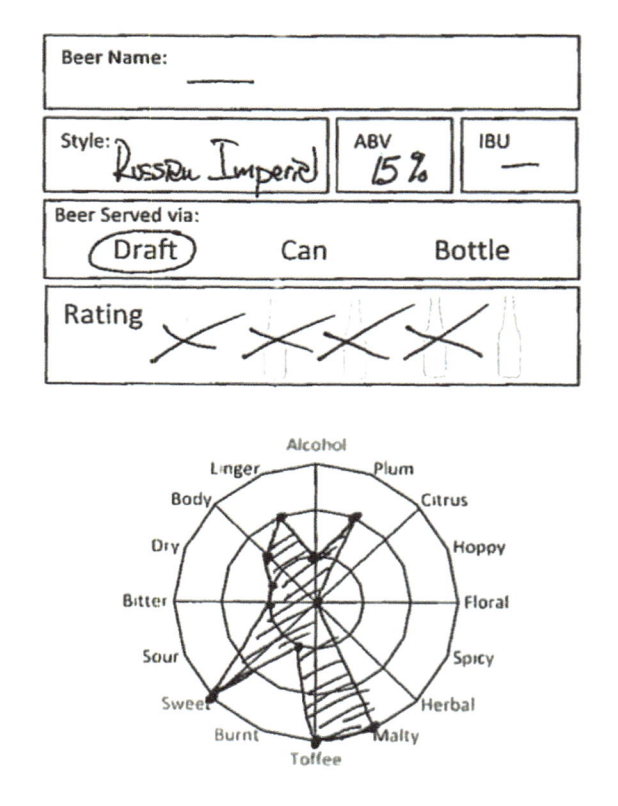

FIG. 9.1. Example of a beer journal record of a Russian Imperial Stout. These pages from the authors' own beer journals denote the sensory experience for a particular beer. The beer brand has been omitted intentionally. (Courtesy S. E. Johnson and M. D. Mosher—© ASBC)

There are many ways to evaluate a beer using sensory analysis, but the authors recommend using the guidelines from the BJCP. Those guidelines focus on five important sensory inputs that fully characterize the sensory experience for a specific beer:

1. **Appearance.** This is the first of the sensory inputs during a tasting. How does the beer look? What color is the foam? Does the beer have nice effervescence?
2. **Aroma.** The glass has to come close to the nose during an evaluation. How does the beer smell? Does it have any off-putting aromas? Does the smell of the beer remind the drinker of anything?
3. **Flavor.** Next, the taste is considered. How does the beer taste? Does it taste the same on the second sip as on the first? How verbose can the description become?
4. **Mouthfeel.** After the beer is tasted, it lingers in the mouth. Is the beer watery or thick on the tongue? Does the CO_2 sting? Is the beer refreshing?
5. **Overall impression.** The final decision now looms. How do you feel about the beer? Would you like to have another of the same beer?

appearance A measure of the look of a beer.

aroma A measure of the smell of a beer.

flavor A measure of the taste of a beer.

mouthfeel A measure of the feel of a beer in the mouth.

overall impression A summary of the beer-tasting experience as a whole.

These factors are discussed in detail in the following sections. However, asking questions that relate to each factor and critically thinking about the beer while tasting it allows us to delve deeper into the sensory world. We begin to educate ourselves about beer sensory when we do this. In fact, the best technique that's available to anyone and obtainable on any budget is becoming educated through questioning, practice, and application.

Appearance: Making a First Impression

Sensory analysis isn't limited to aroma and flavor. The first sense that's engaged when analyzing a beer is vision. Evaluating appearance involves using the eyes as the beer is poured. The formation of the head and the cascade of bubbles offer some measure of the appearance. The color and clarity of the beer contribute, as well. Each style of beer has a defined set of appearance characteristics that should be apparent.

Because our perception of the flavor of something can be modified by its appearance, most sensory analyses that focus on flavor are performed in a room with red lighting. This makes it difficult for an analyst to be swayed by the appearance of the beer. (We will return to this topic later in this section.)

Different color scales are used within the brewing industry to measure the color of a beer. The first color scale implemented successfully was introduced by Joseph Lovibond in the 1880s. As we note in Chapter 10, the color of the beer or wort was determined by comparing it with a set of colored panes of glass. Each pane of glass was assigned a specific value in degrees Lovibond (°L).

The use of the degrees Lovibond scale to measure the color of wort or beer has been supplanted by other methods that don't rely on human

standard reference method (SRM) A numerical rating that describes the color of a beer based on the absorbance of light at 430 nanometers; approximately equal to the color in degrees Lovibond.

European Brewery Convention (EBC) A numerical rating that describes the color of a beer based on the absorbance of light at 430 nanometers; approximately equal to twice the SRM value.

interpretation. One commonly used method is the **standard reference method (SRM)**. Even though use of the SRM is common, the maltster still provides the color of the malts in °L. It should be noted that the SRM color scale for malt doesn't evaluate the color of the malt but rather the color of a wort made from the malt under specific conditions. In other words, a base malt with a color of 4°L refers to the color of the wort that could be made from that malt.

As for beer, the color is most often reported using the SRM color or the **European Brewery Convention (EBC)** color (Fig. 9.2). Both are highly reproducible scales based on spectrophotometric analysis of the beer sample. Using an instrument to determine the color reduces human error as a factor in measurement. Calculating the SRM color requires multiplying the absorbance of 430 nanometers (nm) light by a scaling factor of 12.7. This scaling factor was chosen so that the resulting SRM color values closely align with the degree Lovibond scale. Calculating the EBC color involves multiplying the absorbance of 430 nm light by a scaling factor of 25. The result is that the EBC color is 1.9685 (25 ÷ 12.7 = 1.9685) times the SRM value. For quick calculations, EBC color value is approximately twice the SRM color value. Note that because of the limitations of the spectrometer (which we will discuss in Chapter 11), the amount of error in measuring the color increases when the color is greater than 12.7 SRM color or 25 EBC color.

The SRM color is important both to the beer style and to consumers' perception of flavor. Consumers are quick to associate dark-colored beers with heavier malt flavor, whether that flavor exists or not. For example, a dark India pale ale (IPA) can be perceived as having a richer malt flavor than actually exists. Similarly, two light-colored beers can vary significantly in maltiness. In other words, the appearance can deceive the consumer into thinking that a flavor exists. For this reason, sensory analysis of the flavor of a beer is ideally performed in a room with red lighting. The red lighting obscures the color of the beer so that it can't deceive the analyst into imagining a flavor that doesn't exist. If a room with red lighting isn't available, it's important to consider that color or clarity can influence the data obtained from the test. At a minimum, opaque cups should be used for sensory testing.

Most of the dark color in beer comes from the use of specialty malts in the brewing process. These malts have undergone additional kilning after the original malting process, which increases the Maillard reactions within the malts. Specialty malts can make up a significant portion of some grain

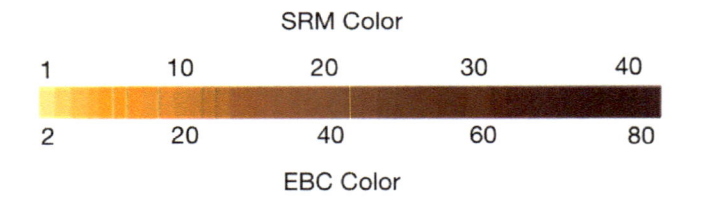

FIG. 9.2. Color scale relating the standard reference method (SRM) and European Brewery Convention (EBC). (Courtesy S. E. Johnson and M. D. Mosher—© ASBC)

bills, depending on the style. The additional kilning they undergo denatures the enzymes needed in the mash. When used with an appropriate amount of base malt, they provide colored compounds and additional flavors that you can't get from the base malt alone.

As noted earlier, the color of a beer results from the color imparted by the malts used to make it. We can estimate the color of the final wort made from the malts by measuring the contribution from each malt or grain added during the brewing process. This measurement is known as the **malt color unit (MCU)** analysis. The calculation is based on the mass of the malt and the volume of the wort that's prepared:

malt color unit (MCU) The amount of color in SRM that a given amount of malt will add to a beer.

$$MCU_{Malt\ 1} = \frac{\text{malt color (°Lovibond)} \times \text{malt mass (pounds)}}{\text{volume of wort (gallons)}}$$

If metric units are being used (which is common outside the United States), the equation becomes:

$$MCU_{Malt\ 1} = \frac{\text{malt color (°Lovibond)} \times \text{malt mass (kilograms)}}{\text{volume of wort (liters)}} \times 0.1198$$

Once MCU values have been determined for all the malts used in the grain bill, the values are added together to provide a total MCU value, which represents the color of the final wort. The value of MCU_{total} is equal to the expected SRM color of the wort. This method works best for SRM colors that are less than 10.5. For darker beers and worts, the MCU method tends not to work well. In fact, the MCU method tends to overestimate the SRM color. To account for this, the equation below can be used to determine the SRM color. It relates the SRM color to 1.492 times the value of MCU raised to the power of 0.686. This modified equation has been shown to improve the agreement between the calculated MCU and SRM for darker-colored worts and beers:

$$MCU_{Total} = MCU_{Malt\ 1} + MCU_{Malt\ 2} + MCU_{Malt\ 3} + \text{etc.}$$

$$\text{if SRM} < 10.5 : \text{SRM} \cong MCU_{Total}$$

$$\text{if SRM} \geq 10.5 : \text{SRM} = 1.492 \times (MCU_{Total}^{0.686})$$

Each specialty malt provides both color and flavor to the beer. Some dark malts will produce a finished beer with a very strong coffee or burned-toast flavor, while others will impart little to no flavor but darken the color of the beer significantly. This is important to consider when calculating the color of a beer using the recipe. A flavor profile for each malt may be more important than the color the malt imparts. Even then, the artistic side of the malt and mashing process will come through, because the true flavor of the finished beer can't be determined until the beer is in the glass.

Most beer styles have been traditionally made with particular sets of malts. If we are trying to replicate a beer style as precisely as possible, we should consider using the same base and specialty malts noted for that style. If we use a different specialty malt to impart the correct color but don't consider the flavors that will be contributed, the flavor profile will be vastly different than what it should be (based on expectations for that style). However, this is where the artistry comes into the brewing process. When we explore different malts and flavors and use style guidelines only as suggestions, the possibilities of color, appearance, and flavor are endless. Some work very well together, but others, not so much.

Another important aspect of appearance is the foam and carbonation. The foam, or head, on a poured beer is worthy of an entire book by itself! In fact, there are many useful books and articles devoted exclusively to foam. For sensory analysis, it's critical to note the color and appearance of the head: Are the bubbles of the foam small and uniform? Are there both large bubbles and small ones? Does the head look like whipped egg whites? How long does the head last? Does it "lace" the glass as the beer is consumed?

The level of carbonation in a beer can influence the appearance. Carbonation level may be more difficult to detect in a darker beer, such as a porter or a stout, or a hazy beer, such as a hefeweizen. But the level of carbonation—the number of bubbles—is another attribute of the appearance of a beer. We expect a pilsner to be highly carbonated, and we expect a cask ale to be lightly carbonated. If we order a pilsner and get one that doesn't meet our expected level of carbonation, we will likely be very disappointed. In addition to affecting appearance, the level of carbonation increases the acidity of the beer. This is the tartness or crispness we sense in a highly carbonated drink.

Keep in mind that the carbonation level doesn't directly relate to the amount of foam or foam stability. Foam and foam stability are related to the types and concentrations of proteins, the concentrations of hop oils, and the level of tannins in the beer. These compounds come from the malts, grains, and hops used to make the beer.

Flavor: Training the Palate

To operate a successful sensory program, the people involved in the analysis must be able to taste a wide variety of flavors and relate what they taste to others. To do so, they will need some form of training to align their palates with the words they will use. Training the palate to analyze the flavors in beer efficiently requires tasting beer. The authors recommend finding many different styles of beer by consulting commercial examples using the BJCP guidelines. These examples fit the styles of beer fairly accurately.

Visiting a local microbrewery may not be the best way to formally train your palate. Although most breweries do a great job of accurately describing the flavors of the beers they sell, their descriptions and assignments of style are sometimes contrary to the actual style of a given beer. In addition, a small microbrewery won't likely assess each beer with a sensory panel to confirm that it fits within a given style guideline. Often, no formal group of experts gathers to decide the style or assess the flavors. So while the beer may be amazing and have tons of flavors that anyone can evaluate, it may not match the style or the flavors may not be typical, characteristic, or even useful for helping hone our skills in flavor detection.

Aroma: A Key Component of Taste

Most of what we experience as taste and flavor comes from the aroma. However, there is still some debate over the exact nature of what can be tasted without the input of aroma. Many of us were taught in elementary or middle school that the tongue can sense only a few flavors: salty, sweet, sour, bitter, and fatty (or umami). Yet recent research has shown that there are many more flavors that can be sensed by taste alone. For example, the holistic medical practice of Ayurveda suggests that there are six taste sensations. In addition to salty, sweet, sour, and bitter, there are pungent (such as the flavors of peppers, onions, or garlic) and astringent sensations. The debate about the number of discrete sensations is still raging.

Also consider this example: When we lick a penny, we end up with a flavor in our mouths that can only be described as metallic. Is this an aroma or a taste? We could argue that the penny doesn't vaporize and end up as an aroma. But metallic isn't one of the taste sensations on the tongue—or is it?

Nevertheless, research has shown that aroma is a key component of taste. Consider another example as confirmation of that fact: If your nose is pinched shut and a strong-tasting candy is put into your mouth, the flavor you taste will be limited to sweet, sour, or salty (depending on the candy). When your nose is released, the full flavor of the candy will be revealed. This occurs because the aroma of the candy moves up into the nose via the back of the throat. That aroma is then detected by the olfactory nerves, and the flavor of strawberries or some other flavor is perceived. This doesn't mean that the tongue isn't important in sensory analysis. In many cases, it's essential to determining the entire full flavor of the beer. We rely heavily on the sensations within our mouths to describe tartness, saltiness, carbonation level, and the mouthfeel of the beer.

The Sensation of Tasting

So, what contributes to the sensation of tasting? What are the limits on what we can taste? Taste buds along the tongue, palate, esophagus, and pharynx respond to the beer and report chemical signals to the brain. Our **taste buds**, shown in Figure 9.3, do the biochemical work that we associate with those signals.

Taste buds are scattered across many places, but most of them are on the tongue. When a food or drink is ingested, the chemical compounds that make up the food or drink are grabbed by tiny hair-like components known as **microvilli** at the surface of the taste bud. A process known as **gustation** then occurs, in which chemicals from food or drink are captured within a taste bud pore and then absorbed into receptor cells in the taste bud. The receptor cells send an electrical signal to the brain via nerves attached to the base of the taste bud. Depending on the chemicals present, the taste bud sends one of a variety of signals. This process occurs very quickly. In fact, the speed of the process is important in human evolution, because if we find

taste bud A receptor organ that contributes to the sensation of tasting; thousands are located along the tongue, palate, pharynx, and larynx.

microvilli Tiny hair-like components on the surface of a taste bud that grab chemicals to taste.

gustation The process in which chemicals from food are captured within a taste bud pore and then absorbed into receptor cells in the taste bud.

ourselves eating something that could harm us or make us sick, we know we must spit it out quickly to protect ourselves.

The most common use for taste without aroma is to determine mouthfeel. A beer such as a pilsner can feel light, like water on the palate. An oatmeal stout can feel oily and heavier in the mouth, like milk. Two identical beers—one served on cask and the other from a tap—can have drastically different mouthfeels because of their carbonation levels.

Aroma and Overall Flavor

Even though taste is very important in sensory analysis, most beer drinkers focus only on the overall flavor of the beer. That's what is perceived when taste and aroma merge. The human body has roughly 400 olfactory receptors. Most are located in the nose, but some are located in the throat and possibly in the stomach. These receptors are responsible for our ability to distinguish any of roughly 1 trillion different aromas. Some people are unable to distinguish between aromas, and most human olfactory systems suffer from sensory thresholds (minimum concentrations of chemicals that cause specific biological responses) and sensory overload (overstimulation of the biological system to the point where it won't respond). Both of these topics will be discussed in sections later in this chapter.

When all of these factors are considered, determining the aroma of a beer can very difficult. The difficulty lies in our ability to distinguish between similar aromas. But with a panel of people who can detect and

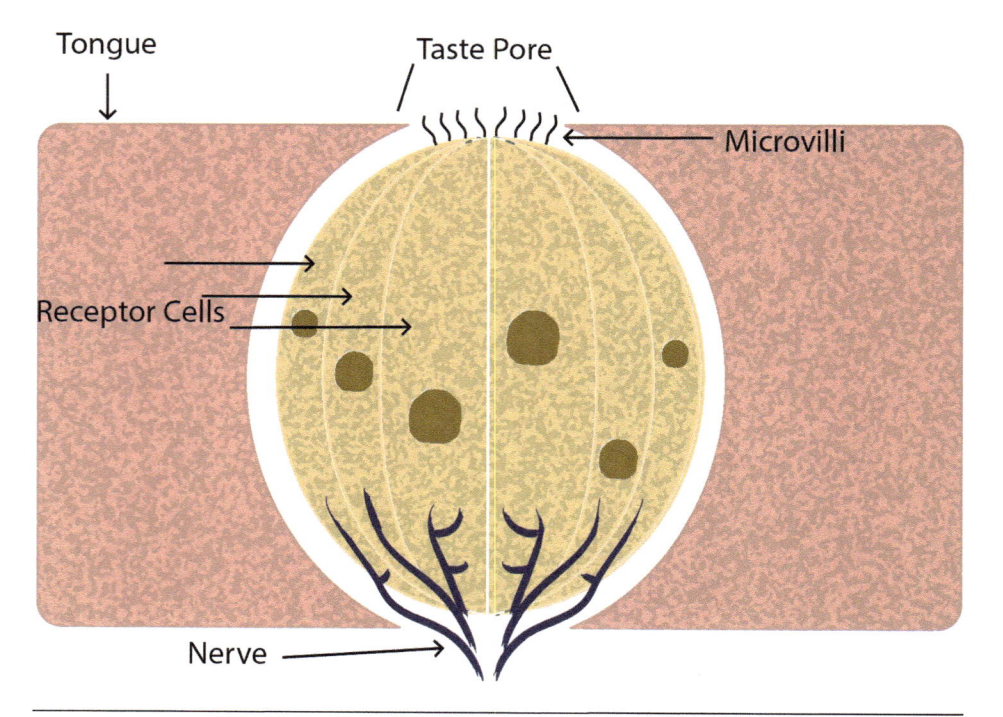

FIG. 9.3. Elements of a taste bud. (Courtesy S. E. Johnson and M. D. Mosher—© ASBC)

identify a variety of flavors and aromas, the distinguishing characteristics of a beer can be determined.

A useful tool for describing the flavors in beer is the flavor wheel shown in Figure 9.4. It provides a set of terms that can be used as specific descriptors for the different flavors in a beer. The individual flavors are organized

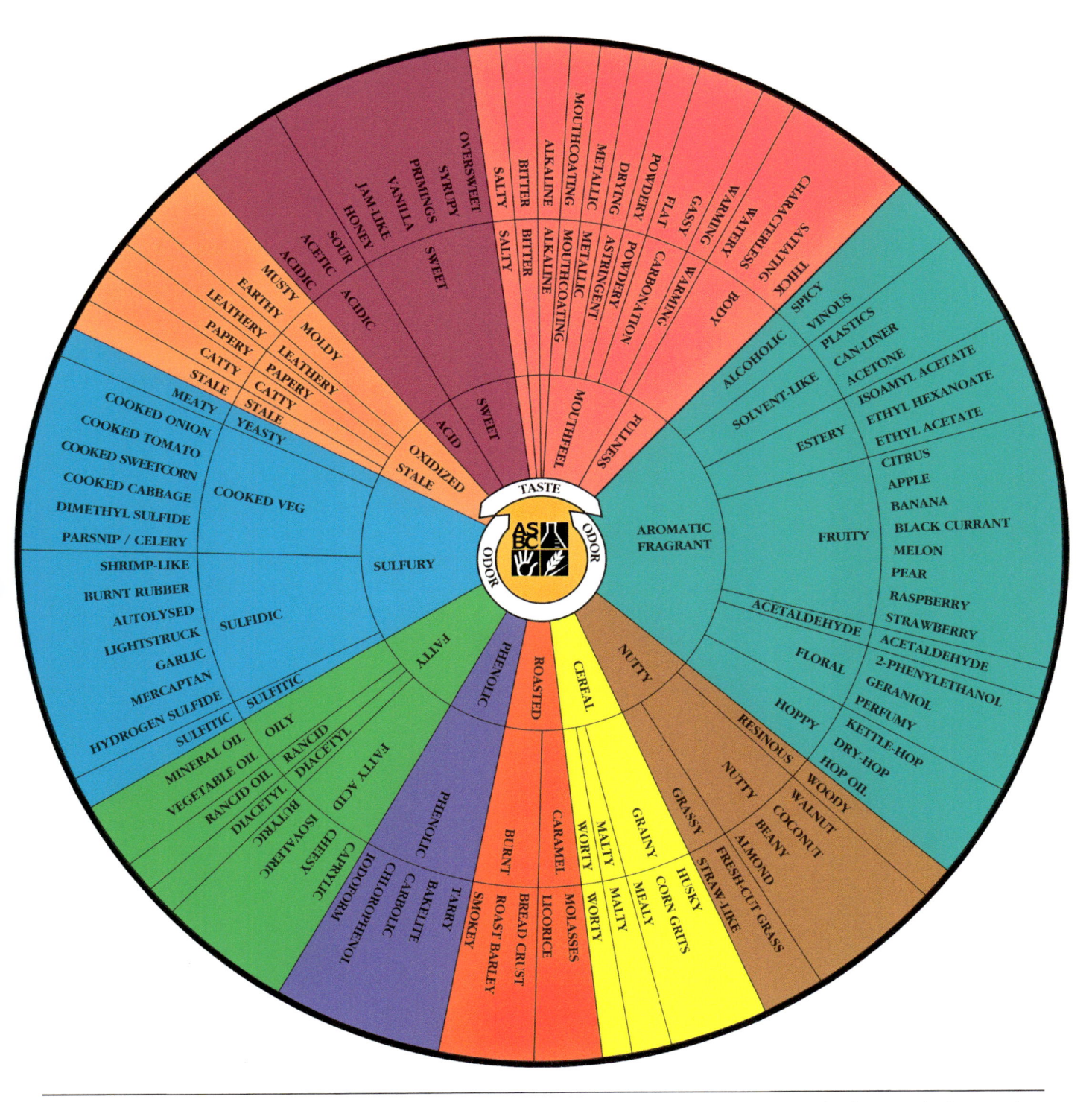

FIG. 9.4. The ASBC Beer Flavor Wheel, which groups flavors into categories and provides key words that can be used to describe a beer. (Courtesy ASBC—Reproduced by permission)

into categories that focus on key similarities, including vegetal, cereal, Maillard, phenolic, fatty, sulfury, stale, acidic, sweet, salty, bitter, mouthfeel, fullness, and aromatic. There are some "tongue-based" flavors on the wheel (sweet, salty, and bitter) and other categories that require use of the olfactory senses.

Everyone in a sensory panel should use the same or similar terminology to describe beer flavors. Using consistent terms will limit confusion from analyst to analyst and allow the panel operator to understand the flavors being sensed. While most people understand the flavors and their differences, putting words to them can be incredibly difficult.

Sensory Thresholds

No two people have identical fingerprints. The same is true about people's ability to smell and taste different compounds. Each of us is sensitive to a different set of chemical compounds found in our food and drink. Some of us are very sensitive to a wide variety of these flavors, while others may be unable to sense these flavors at all. For example, some people are very sensitive to diacetyl (an undesired butter flavor in most styles of beer), while others are unable to detect diacetyl until the concentration of this compound is quite large. This doesn't mean that someone who can't sense a low level of diacetyl can't provide information about the characteristics of a beer. It just means that some people aren't useful for determining the level of diacetyl in beer.

sensory threshold The lowest concentration at which a compound can be sensed by a particular individual.

Each compound has its own **sensory threshold**, which is the lowest concentration at which it can be sensed. Thresholds vary significantly from one chemical compound to the next and reflect how well our taste buds and olfactory nerves are able to respond to specific chemical compounds. For example, acetic acid (a vinegar-like flavor) has a threshold of 200 parts per million (ppm), while vanillin has a threshold of 10 ppb. Our bodies are much more sensitive to vanillin than they are to acetic acid. In Chapter 2, we discovered a wide variety of different off-flavors and their flavor thresholds. As noted, many compounds can be detected by humans in a range of parts per million or parts per billion. For a few compounds, the flavor thresholds are in the range of parts per trillion.

Thus, it's very important to be aware of concentrations when analyzing flavors in beer. For example, diacetyl can be detected in a gas chromatographic analysis of beer (see Chapter 13). However, the presence of this chemical compound in beer doesn't mean that the sample will taste like butter. The concentration that's detected may be lower than the concentration that can be discerned by the human palate. In addition, other flavors in the beer can mask the flavor of an individual compound, such that the flavor threshold seems to be higher. For example, a very rich Baltic porter may have a diacetyl concentration that's well above the flavor threshold but may not be detected by a human sensory panel.

Variations in people's abilities to detect given compounds are common. There will likely be a panel member or two who detect a certain compound below the threshold value that's common for the compound. And as noted earlier, there may be panel members who are insensitive to a flavor until it's 10 times the reported threshold value. There may also be panel members who suffer easily from sensory overload and become blind to a flavor very quickly. In other words, they may be able to detect the compound on the first sip but lose the ability to detect it by the second sip. Even reliable sensory panelists exhibit variations in what they can detect from day to day and even from hour to hour. This is what makes sensory analysis difficult. The variability within a sensory panel means that statistical analysis of the results is required to pull out the correct answers.

Sensory Overload

Sensory overload is an unfortunate problem when trying to determine the flavors in beer. We can get a really good sense of how overload occurs by performing an experiment.

To begin, we select three or four beers of the same style—say, a series of IPAs. We pour three samples of each IPA into separate opaque cups that have identifying marks on the bottoms. Then a friend scrambles the arrangement of the cups and gives us one sample of each beer. We taste each sample and write notes about it in a notebook. We then wait about 15 minutes and repeat the analysis on the next set of samples in a different order without looking at our previous notes. Finally, we repeat the analysis a third time and write notes again. Comparing the notes across the three rounds, we can see that some flavors became easier to detect and more pronounced. Other flavors became muted and, in some cases, were difficult to detect.

The muting of the flavors results from the fact that the alcohol lowers our ability to sense some of the flavors. In addition, some of the flavors become less easily recognized because of the overload to our sensory system. While our taste buds are likely still able to recognize and discern chemical compounds, our olfactory system is becoming overloaded. It needs time to recover.

The sensors that make up the olfactory system essentially get clogged or become less responsive if they are used repeatedly to detect the same chemical compounds. We don't lose all of our sensing and tasting ability from repeated exposure to the same compounds, but as our notes in the experiment will attest, certain flavors become muted. This is important to remember when performing sensory analysis for a brewery. Panel members should not be asked to do too many samples at one time. They will become insensitive to some chemical compounds and less likely to discern the flavors associated with those compounds. Sensory analysis should focus the senses on the task at hand when they are at their peak of performance.

> **sensory overload** Occurs when the body's ability to detect certain flavors is overwhelmed by overstimulation of the taste buds or olfactory system.

How to Set up Your First Sensory Panel

Chapter 15 provides easy-to-use worksheet guides to begin analyses with a sensory panel. Those guides assume that the user has no previous experience.

Sensory analysis should typically be conducted in a room with minimal sensory distractions (Table 9.1). The area should be free of strong aromas, and it should be quiet and well lit by white or red light. Red light will make the colors of beers more difficult to distinguish, which may be helpful if the colors could influence the results. The tests should be conducted at least 2 hours after a meal and not at a time when panelists might be tired. For many breweries, the optimal time for a sensory panel is between 10:00 and 11:00 A.M.

The sensory panelists should approach the tests as analysts, not beer drinkers. It's important that they provide quality scientific feedback, rather than just use sensory as an excuse to have a beer! The beer samples should be free of any identifying marks, such as labels and brand-specific bottles. The best samples are poured into freshly washed and rinsed, opaque glasses that are identical across all the samples. Panelists should see all the beer samples as identical.

TABLE 9.1. Key factors in setting up a sensory panel[a]

Factor	Attributes
Room requirements	Free of distractions.
	No background music.
	Warm and inviting.
	Air conditioning vents should be indirect.
	No bold wall colors.
	Odor free.
Lighting requirements	Soft lighting; avoid use of fluorescent lights.
	White light or red light.
	Cameras are unnoticeable.
	No direct or spot lighting.
Panel seating	Individual seating.
	Comfortable chairs that don't squeak.
	Limited or no view of other panelists.
	Comfortable countertop height.
Sample delivery	Each sample is free of identifying marks.
	Samples are delivered to all panelists at the same time.
	All samples are the same temperature.
	All samples are freshly poured and have the same/similar level of head.
	Containers are free of odors and flavors.
	Containers are clean and dry.
	No more than seven or eight samples in a session for a generic test.
	No more than three or four samples in a session for a specific test.

[a] Courtesy S. E. Johnson and M. D. Mosher—© ASBC.

The panelists should be provided with a simple question to answer about the samples placed in front of them. Examples include: Are any of the samples different? Is beer X more or less malty than beer Y? Does beer X taste exactly like beer Y? Questions that ask for opinions (Which is better?) and broad questions (What are the differences?) require panelists to be highly trained and highly skilled. Questions that ask for opinions should not be used for sensory analysis when the panel is untrained or unskilled. Sensory analysis is a difficult skill to master; therefore, the questions given to panelists should be as straightforward as possible.

Once a specific analysis has been chosen, it's important to choose a sensory analysis technique that will work well with the goals of the analysis. For example, a true-to-brand test is better performed with a paired comparison test than a triangle test. In this case, the question is: Is beer A the same as beer B? A triangle test requires all the samples to be unknown. The panelists in a triangle test are forced or predisposed to find the two samples that are similar. In this case, the results can be skewed to find a difference that may not exist. In other words, the panelists may be forced to overthink the test to give feedback. The results may not be useful if there's no difference between the two beers under consideration. Additional examples of when to use a triangle test versus a paired comparison test are provided at the end of this chapter.

Trusting Your Panel

You want to be sure that the members of your panel are reliable and will report similar results each time they perform an analysis. An easy way to screen panel members for reliability and consistency is to perform the same test 5–10 times. Doing so will reveal each panel member's variability. Performing simple triangle tests or paired comparison tests and recording whether panel members answer the same way on a consistent basis will also indicate variability. It's important for you to be able to trust that the data coming from each panelist—and thus the entire panel—are reliable and useful for a sensory program.

As noted before, if a panel member isn't a reliable analyst for one test, that doesn't necessarily mean that he or she is of no use to the sensory panel. Each person's sensory palate is unique, and in most cases, anyone can be a reliable sensory analyst for some portion of the analysis. However, if a panelist is unreliable for a number of flavors, his or her participation in the analysis may be questioned. In the end, it might be best for the results of the sensory panel to replace that person.

Providing formal training can be a great way to improve your sensory program. Training can be accomplished by identifying samples with known amounts of specific flavor compounds and then providing them to panel members in different concentrations. Alternatively, sending panel members for formal instruction at a company that focuses on sensory training can renew or advance a sensory program incredibly quickly. Not all students in a lecture class learn the same way, and the same is true about sensory training. Getting a fresh view or additional training to grow and enhance a sensory program should always be considered.

Types of Sensory Analysis

When preparing for sensory analysis, it's important to have a specific question in mind—a question that needs to be asked. The next step is to determine a test that will provide the best results to answer that question. Many sensory tests can be performed, but most commonly, a triangle test or paired comparison test is conducted because the setup and reporting are easy. Other analyses, such as duo-trio tests and threshold tests, are provided in the Sensory Analysis Methods of the American Society of Brewing Chemists (ASBC).

Triangle Test

A triangle test is performed to determine whether samples are significantly different or significantly similar (Fig. 9.5). This type of analysis works best when panel members can taste a specific flavor in a spiked sample when compared with two unspiked samples of the same beer. The test is performed by providing three samples of beer to each panelist. Two of the samples should be the same, and one should be different. The main question put to panelists is to identify which two are the same. Advanced triangle testing can be performed by asking panelists to describe why the third sample is different. Remember that advanced sensory analysis requires the panelists to be skilled in flavor analysis.

Paired Comparison Test

Variations of the paired comparison test can be performed for the following purposes:

1. The directional difference test determines how a sensory characteristic differs between two samples (for example, sweeter or less sweet).
2. The paired preference test establishes whether a preference exists between two samples (for example, in consumer tests).
3. Assessor training selects, trains, and perfects panelist sensory analysis.

FIG. 9.5. Two arrangements of the triangle test (left and right). The colors are suggestive of the samples. The triangle test is best performed when all the samples have the same apparent color. Red lighting can be used to obscure minor differences in color. (Courtesy S. E. Johnson and M. D. Mosher—© ASBC)

Each type of analysis compares two samples and determines the similarities and differences between them. The specific type of paired comparison is based on the question posed to the panelists.

Tetrad Test

The tetrad test is a relatively new analysis that's becoming more popular in the world of sensory analysis. The evaluation is performed by providing an analyst with two samples of one beer and two samples of another—in random order. The analyst must then determine which samples are the same by making two sets. Compared with the triangle test, the tetrad test may seem to provide more precise results; however, the probability of randomly guessing the right answer is still 33%. The benefit, however, is that the analyst is more likely to focus intently on the beer, rather than rely on guessing. Using the tetrad test, fewer analysts are required to achieve reliable data. Statistically significant results can be obtained with as few as 25 analysts.

Threshold Test

The threshold test is useful in determining panel members' thresholds of a specific flavor compound. It's performed by setting up a series of beer samples. The flavor compound concentration is increased among the samples, and one or more "blanks" are included in the series (Fig. 9.6). Panelists are asked to determine which samples contain the specific flavor. This type of test provides specific information about each panelist's sensitivity to the given compound and to the panel's overall sensitivity to the specific flavor.

Often, panelists aren't told about their specific sensitivities to flavor and aroma compounds. If they were told, they might develop a bias based on when they were used for a sensory experiment. For instance, suppose that a panel is formed to determine if diacetyl exists in a beer sample, and the panelists all know that they can sense diacetyl at very low levels. The panel will have a bias toward diacetyl. The organizer of the panel should expect to find elevated diacetyl in the sample. However, if the panelists don't know their specific sensitivities, then they won't have any biases and the data from the experiment will be more meaningful.

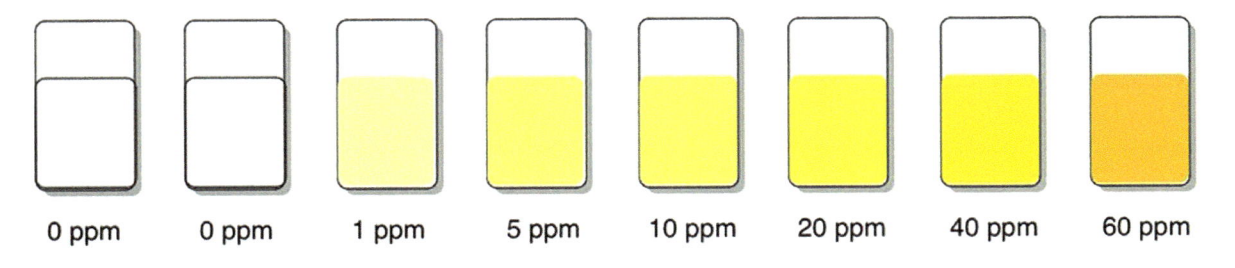

FIG. 9.6. A series of beer samples used in a threshold test. The colors of the samples are suggestive of the concentrations of flavor. Note that there is more than one "blank" sample. (Courtesy S. E. Johnson and M. D. Mosher—© ASBC)

Performing Sensory Analysis

Specific examples of what can be tested for various stages of the brewing process are shown below. These are not limiting data but rather ground-level ideas to grow and expand on as each brewery sees fit.

Wort Sensory Analysis
- Explore flavors of specialty malts using congress wort from those malts.
- Evaluate flavors between suppliers of the same malt using congress worts.
- Evaluate the flavor profile of kettle-soured wort.
- Determine off-flavors, such as stale and green flavors, that might be caused by malt.

Hop Sensory Analysis
- Compare sensory analysis of hop teas to determine aroma and flavor contributions.
- Evaluate old hops for off-flavors or hop oil contribution.
- Evaluate hops from different suppliers or different harvest years.

Beer Sensory Analysis
- Evaluate beer for any off-flavors in the finished product.
- Evaluate beer to compare perceived bitterness versus international bitterness unit (IBU) calculation.
- Evaluate beer to determine if it's true to the brand.
- Evaluate new brands to compare them with existing consumer demands.
- Evaluate beer for yeast health.

It's important to perform multiple sensory analyses that address a single question and to compare the results from each analysis. Making the comparison will reveal if the question is being answered. Also, performing multiple analyses allows the organizer of the sensory program to consider whether the results are consistent from panel to panel and whether the results are statistically significant. For these reasons, a sensory program should be fluid and able to change and adapt based on what panel members are available and what analyses are needed during beer production. Analyses should also be repeated multiple times to learn which analysis works best given the schedule and resources available.

CHAPTER 10
Colorimetric Analysis

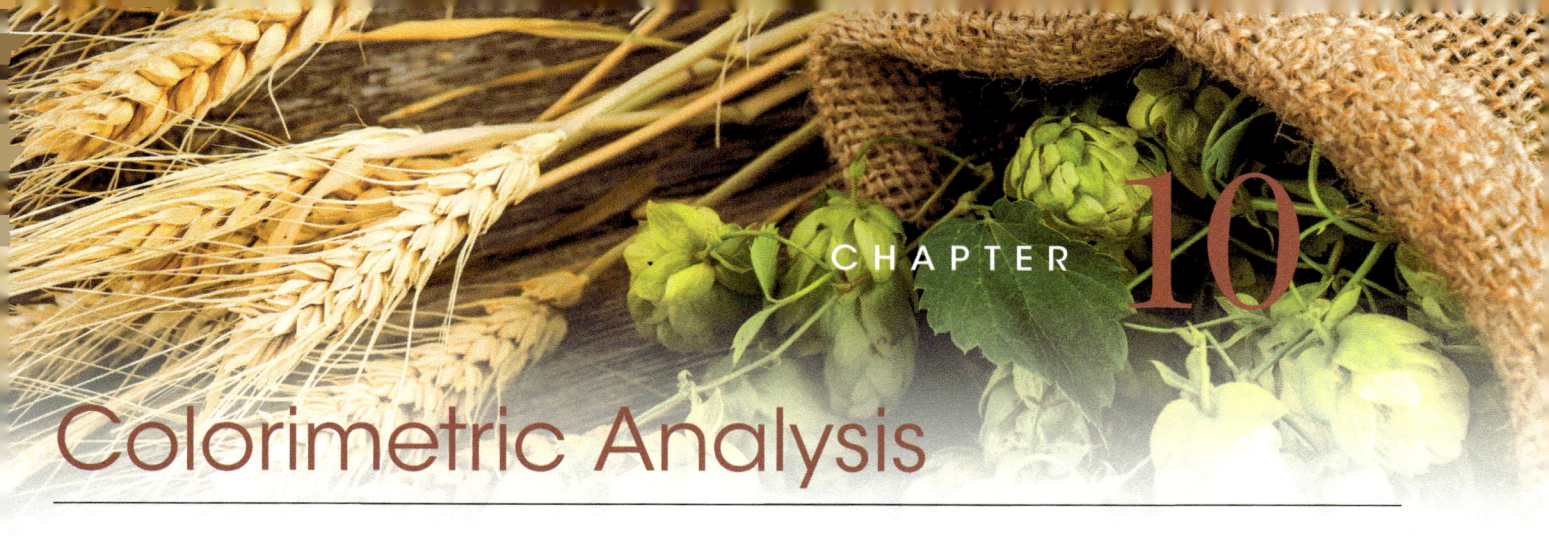

In the early 1700s, scientists began working on the theory that would govern much of our current analytical instrumentation. The first was Frenchman Pierre Bouguer, who in 1729 used experiments and reasoning to show that the intensity of light decreases in a geometric progression as it passes through a transparent object. In 1760, German scientist Johann Heinrich Lambert developed a more exact mathematical relationship explaining this principle. He found that the intensity of light absorbed by an item as light passes through it is directly proportional to the thickness of the item. This became known as the **Lambert absorption law**. As modern chemical thought developed during the next 100 years, no additional work advanced this idea.

In the mid-1800s, August Beer, a mathematics professor in Bonn, Germany, built on the Lambert absorption law. Although he didn't refer to Lambert's work, he likely knew about it, given that he didn't claim he had discovered a new relationship between light intensity and absorptive materials. In 1852, Beer presented his research results, which demonstrated that the reduction in the intensity of transmitted light was directly proportional to the concentration of a solution. Just a couple of months later, French scientist Felix Bernard published a very similar set of results that related absorptive capacity and concentration. However, the resulting law became known as the **Beer-Lambert law** (Fig. 10.1), giving credit for the discovery to the scientist who published his work first. Bouguer is sometimes noted as a contributor to the Beer-Lambert law, but Bernard is rarely noted.

The work these scientists performed provided the equation that we discussed previously (see Chapter 8). More importantly, their work ushered in a new way to analyze solutions in chemistry. Although the first record of the law coined as "Beer's law" wasn't made until the late 1880s, the work was so profound that scientists across the globe instantly implemented it in their laboratories. In fact in 1854, Frenchman Louis Jules Duboscq, a prolific and popular instrument maker, created one of the first commercially available instruments capable of comparing the intensities of two solutions. After improving the design in 1870, Duboscq's company sold the **Duboscq colorimeter** to almost every scientific laboratory (Fig. 10.2). The device compared the color intensities of two solutions by adjusting the depths of the solutions. Unfortunately, Duboscq's company manufactured these instruments as works of art. They were extremely accurate and precise but fairly expensive. By the 1880s, other optics companies in Germany and

Lambert absorption law Describes how objects absorb light.

Beer-Lambert law Provides the direct relationship of molar absorptivity, path length, and concentration to the absorbance of a solution.

Duboscq colorimeter A microscope-like device that compares the intensity of light in the sample to a standard by adjusting the path length of each sample.

FIG. 10.1. The Beer-Lambert law. This mathematical expression relates how the absorbance of a solution is directly related to a constant (ε) for the analyte; the thickness (b, in centimeters) of the solution; and the concentration (c, in molarity) of the analyte in the solution. The absorbance increases as the path length increases and as the concentration increases. (Courtesy S. E. Johnson and M. D. Mosher—© ASBC)

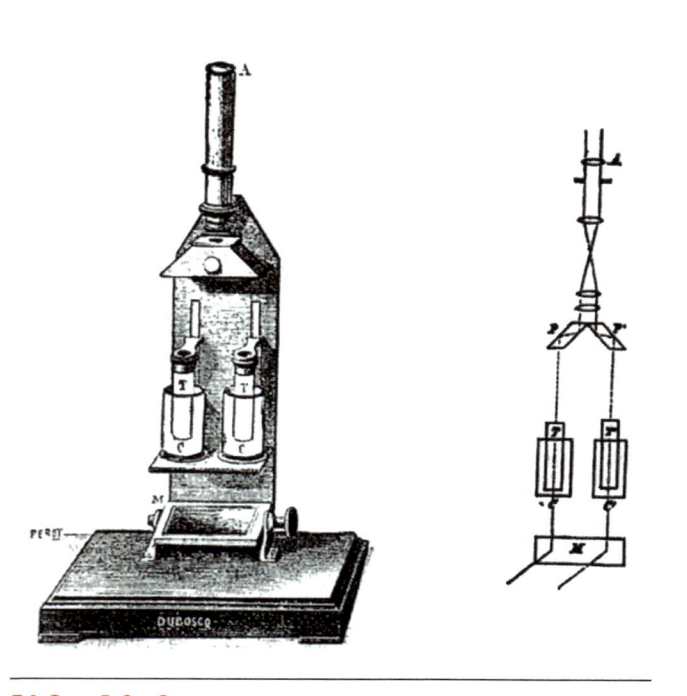

FIG. 10.2. The Duboscq colorimeter. Two solutions, a standard and a sample, are compared simultaneously. The depths of the solutions are adjusted until they are equal in intensity. The result is related to the concentration of the sample. (Courtesy *Chemical News*, London, Jan. 21, 1870)

FIG. 10.3. The Lovibond tintometer. The beer sample was placed on the shelf behind the open rectangle and the user compared the color to the glass plates, which could be rotated. (Courtesy The Tintometer, Ltd.—Reproduced by permission.)

Britain had overtaken the market with less expensive but still accurate and precise instruments.

For example, Joseph Lovibond, an English brewer, noted while working in his father's brewery that the color of a beer was related to its quality (a topic about which we still have plenty of questions). Lovibond wanted to find a way to identify the color of a beer more accurately. A colorimeter could do the job, but to produce accurate results, it required using standard solutions and having some skill. Recognizing that the device could be simplified by using panes of colored glass, Lovibond started a company called Tintometer, Ltd., and it marketed the **Lovibond comparator** for sale in 1885 (Fig. 10.3). The user simply placed a sample of beer in a glass jar of a known size and then compared the color against a rotating disk of glass plates. Given the success of this device, the company still manufactures high-quality, easy-to-use colorimeters and comparators.

Humans can be used to measure the color intensity of a solution to determine its concentration. In fact, the primary operating feature of both the Duboscq colorimeter and the Lovibond comparator is the user. Specifically, both devices rely on the user's eyes.

Lovibond comparator
A device that uses a set of standards to determine the color of a beer by comparison with shaded panes of glass, plastic, or other material.

Eyes Are the Instrument

Our eyes contain two important structures, rods and cones, that allow us to see shades and colors in the everyday world. Both absorb light that enters the eye. Cones make up about 10% of the receptors in the eye and can discern differences in an object's color and intensity of color. Cones can perceive red, blue, and green. Rods are much more abundant and allow us to see in very dark environments; they are sensitive only to light, not to specific wavelengths of light. That's why it is harder to perceive color at night.

For most people, all the colors in the visible spectrum from red to violet are observable. Unfortunately, some of us are unable to tell the difference between certain colors because of defects in the cones within our eyes. There are three main forms of this deficiency: deuteranopia (green color blindness), protanopia (red color blindness), and tritanopia (blue-yellow color blindness). Someone with one of these conditions isn't just blind to a color; rather, he or she has reduced ability to distinguish between colors that have a component of that color. For example, a protanopic person has difficulty differentiating objects that are red. He or she also has trouble distinguishing between colors such as purple and blue, because purple is made up of blue and red. The result is that purple and blue look very similar.

This deficiency occurs in people of all races and both sexes. Studies have shown that in preschoolers, the deficiency is most prevalent among Caucasians (~6%). Men are much more likely to be color blind than women, because the gene responsible for the most common forms of color blindness, red and green color blindness, is located on the X chromosome. Since women have two of these chromosomes and men have only one, it's likely that a woman will have at least one correctly functioning gene. In addition to genetic causes, certain maladies and diseases, such as Parkinson's disease, can cause color blindness. Color blindness can also be a side effect of some medications.

Even though color blindness makes it difficult to distinguish between colors, a person with this condition often has no difficulty noticing the difference between colors when they are side by side. He or she can tell the difference between the two but is unable to tell which color is which. Distinguishing between different intensities of the same color is only rarely impacted by color blindness. One of the most common influences on this ability occurs when the lens of the eye becomes cloudy, a condition known as cataracts. In that case, the perception of intensity is different for each eye. Of course, some medications can also reduce the ability to discern shades of the same color.

Many colorimeters are designed to use only one eye in making color determinations. Or for a device such as the Lovibond comparator, the user can simply close one eye. Doing so eliminates any impact for a user whose eyes are not matched completely.

People as Instruments

Since the main instrument used in colorimetry is the human eye, people are involved heavily in determining the result of a measurement. That means that all the variability in a measurement rests in the perception of the user of the colorimeter. **Visual perception**—our individual ability to interpret our environment based on visual stimuli—drives the accuracy and precision of the colorimeter.

visual perception The ability to interpret the environment based on visual stimuli; a measure of how a particular sample is observed.

What is perception? Perception involves taking in information using the senses (in this case, the eyes) and then translating that information into an interpretation. For a colored object, this means that the user's eyes need to visually represent the object he or she is observing. Then the chemical signal needs to be interpreted by the brain in a way that makes sense. Issues can arise from the ability of the eyes to see specific colors and to determine the intensities of those colors. How the brain receives the information and then translates that information also impacts our ability to interpret what we see.

The variability in visual perception between people can be quite significant, but the variability within a single person from measurement to measurement isn't nearly as great. A quick survey of laboratory personnel might reveal that each person has a different name for the color of a particular beer. That discrepancy might stem from the individual's labeling of the color or variability in his or her perception of that color. The brewery laboratory can reduce the variability in colorimetry measurements by using trained personnel.

In addition, the lab should require that all colorimetry experiments for a particular analysis be performed by the same person or group of people. For example, the morning analyses can be performed by one lab worker and the afternoon analyses by another. Doing so will ensure that the readings are relatively consistent across a single analysis. In contrast, if all the workers in the lab assist with the morning colorimetry measurements, their variability in perception will reduce the accuracy of the measurements.

To reduce variability, a person assigned to perform colorimetry measurements should take frequent breaks from the task. These should not be coffee breaks but instead involve performing other tasks that are less visually intensive. Our eyes get tired over time, which causes our perception to diminish. As a result, there is a decrease in accuracy over the course of measurements when extended colorimetry experiments are performed by the same person. Initially, there is a small increase in accuracy, as the person gets used to the measurement and hones his or her skill, but the eyes eventually get tired and the person's ability to distinguish between intensities of color declines. At that point, the accuracy of the measurements begins to suffer.

To maintain accuracy during the analyses, the brewery laboratory should insert frequent known **standard samples** into the schedule of samples. In addition, the identities of the samples and the inserted standards should be obscured from the person using the colorimeter. As noted earlier, perception is very important in this measurement. If the user knows the identity of the sample or knows that the next sample is actually a standard, he or she will subconsciously perceive the answer to the measurement as a defined value. For example, if a user knows that the **standard reference method (SRM)** 10 color standard is the next sample, he or she will be predisposed to match that sample with the SRM 10 color standard and write that value in the data table. But if the user doesn't know that the next sample is a standard, he or she will write down the value that he or she observes.

standard samples A set of solutions with known different concentrations.

standard reference method (SRM) A numerical rating based on the intensity of color determined using a laboratory method of analysis.

Testing for Presence Versus Testing Against a Standard

Two different measurements use colorimetry: **testing for the presence of a color** and **testing against a standard color**. We have hinted that both measurements are related to the intensity of the color, rather than the specific color. Both focus solely on the intensity of a particular color.

The first measurement, testing for the presence of a color, is the hardest of the two to perform accurately. The user evaluates whether or not the specific color exists. Perception and the ability to determine if a color exists govern an individual's ability to observe the color. In addition to being person dependent, this test is color dependent. Some colors are easier to recognize than others.

For example, studies have shown that light with a wavelength of 555 nanometers (nm) is the easiest to observe (Table 10.1). This is a bright-green color. Detecting the presence of this color, even at a very small intensity, is much easier than detecting the presence of any other color. Yellow is the next-easiest color to observe. Because yellow and green are easy to observe, many cities and towns paint their emergency response vehicles a greenish-yellow. Red, the third color on a stoplight, is one of the least visible colors. Not only is it hard to see, but very dilute solutions of red are also sometimes

testing for the presence of a color An evaluation to determine whether a color exists.

testing against a standard color An evaluation to determine the intensity of a color.

TABLE 10.1. Colors in the visible region[a]

Colors that are easy to observe		Colors that are difficult to observe	
Green	Wavelength: 555 nm	Violet	Wavelength: 380 nm
Yellow	Wavelength: 590 nm	Red	Wavelength: 670 nm

[a] Courtesy S. E. Johnson and M. D. Mosher—© ASBC.

FIG. 10.4. Hydrion indicator paper used to determine the pH of a solution. Comparing the color of the exposed paper to the chart on the container is a version of colorimetry. (Courtesy Micro Essential Laboratory—Reproduced by permission.)

hard to distinguish from colorless solutions. Given these issues, many cities and towns have added flashing yellow lights to their emergency response vehicles, which makes them easier to see.

Unfortunately, red is one of the most common colors in chemical analyses. For instance, indicator paper used to measure the pH of a sample turns red when it's dipped in a solution that's acidic (Fig. 10.4). For indicator paper, the scale of colors goes from a greenish-yellow near pH 6 to an orange near pH 3 and to a dark red near pH 0. Luckily, the center of the pH scale on indicator paper is green. Notice in Figure 10.4 that the colors representing the low and the high pH values are visually harder to distinguish than those in the center of the pH range. Comparing an exposed pH paper to a set of standards is testing against a standard color—the second type of measurement that uses colorimetry. This is much easier to do than perceiving if a color is present.

Red is also observed when titrations are performed using a phenolphthalein indicator. In this case, adding a basic solution to an acidic solution produces a pinkish-red color at approximately pH 8 (Fig. 10.5). The appearance of the color signals the end of the experiment. Thus, perceiving the slight presence of red determines the accuracy of the experiment. If too much base is added, the red becomes more intense, indicating that the end of the experiment was missed.

Luckily, most experiments that require the user to determine the intensity of a color involve comparing the color to a standard. The Lovibond comparator and the Duboscq colorimeter are two examples of this type of colorimetry. The user places a sample in the instrument and then refers to a standard color to match the intensity of the color. Not only is this technique

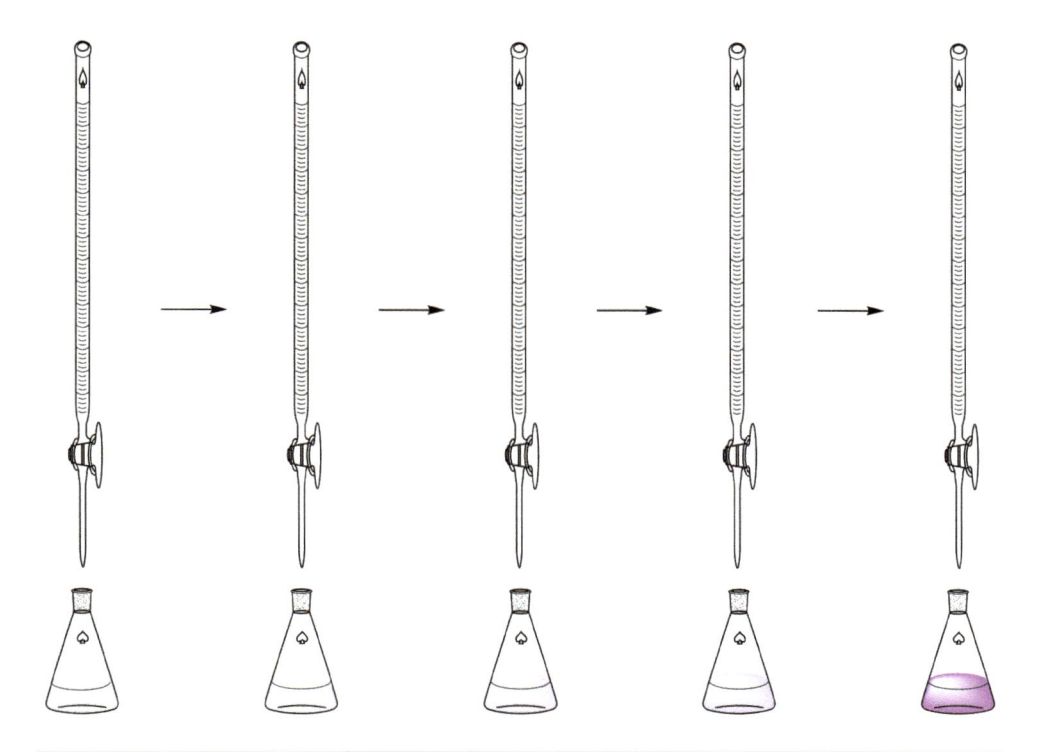

FIG. 10.5. Titration of a solution using a phenolphthalein indicator. The solution goes from colorless to pinkish-red as the solution in the burette is slowly added to the Erlenmeyer flask. Can you identify which flask (from left to right) first shows the faint pinkish-red color? (The second image from the left.) (Courtesy S. E. Johnson and M. D. Mosher—© ASBC)

more common among the laboratory experiments performed in the brewery laboratory, but it's also easier to perform and results in significantly better precision and accuracy.

Making a Reference Chart

A colorimeter or comparator can be purchased from a wide variety of commercial sources and will come with specific standards or directions for preparing standards. In addition, it will provide as precise a method as possible to obtain the answer to the analysis. Even though the colorimeter and the comparator are among the least expensive instruments used in the brewery laboratory, the cost for a small microbrewery may be prohibitive. Reagents (if needed) must also be purchased, and funds will be needed to maintain the instrument.

However, the instrument can be made quite easily by laboratory personnel for almost no cost. Doing so requires having the essential parts available and/or prepared using as precise a technique as possible. A homemade colorimeter/comparator includes a test tube rack and a set of test tubes with well-defined and consistent diameters. Flat-bottomed test tubes work best, but curved-bottomed test tubes can be used. The rest of the homemade colorimeter/comparator includes the sample to be analyzed and a set of standard solutions that span the range of values found in the analysis. Let's look at these in greater detail.

Colorimeter Setup

The test tubes used for the standards and samples should be thoroughly cleaned and free (inside and out) of any hard water spots, scratches, and dirt/smudges. Conducting this preparation step may take a little longer than anticipated, but cleaning, drying, and polishing the test tubes will pay off. Scratched, etched, or imperfect test tubes should be rejected.

The analysis is performed using the written method, part of which will include preparing standard samples. In most cases, brewery lab personnel make a standard with a known concentration. That concentrated sample is then diluted numerous times to prepare a set of standards with different but known concentrations. For example, a 100 parts per million (ppm) sample of an analyte can be prepared by dissolving 100 milligrams (mg) of the sample in 1 liter (L) of solvent. Remember from our earlier discussion of ppm (see Chapter 2) that the 100 ppm sample is simply 100 parts by mass of the analyte dissolved in 1 million parts by mass of the solution. Since 1 L of solvent has an approximate mass of 1,000,000 mg, the 100 ppm solution is very easily prepared using a balance and a 1 L volumetric flask.

A 10 ppm standard can be made by taking 1 milliliter (mL) of that 100 ppm solution and diluting it to 10 mL (a 1:10 dilution). A 5 ppm standard can be made by taking 0.5 mL of the 100 ppm solution and diluting it to 10 mL or by taking 5 mL of the 10 ppm solution and diluting it to 10 mL. Each dilution is governed by the **dilution equation**:

$$C_1 V_1 = C_2 V_2$$

where C_1 and C_2 are the concentrations and V_1 and V_2 are the volumes of the two solutions. Remember that C_1 and C_2 may be units of any concentration. For example, to make 10 mL of a 3 ppm solution from a stock 100 ppm solution, the equation would be as follows:

$$C_1 V_1 = C_2 V_2$$

$$(100 \text{ ppm}) \times V_1 = (3 \text{ ppm}) \times (10 \text{ mL})$$

$$(100 \text{ ppm}) \times V_1 = 3 \text{ ppm} \cdot \text{mL}$$

$$V_1 = \frac{30 \text{ ppm} \cdot \text{mL}}{100 \text{ ppm}}$$

$$V_1 = 0.3 \text{ mL}$$

We would measure out 0.3 mL of the 100 ppm solution using a volumetric pipette, place it into a 10 mL volumetric flask, and then dilute it to the mark to make the 3 ppm solution.

Care should be taken when making standard solutions. Any deviation or error in measuring the volumes of the samples will cause an error in the final

dilution equation Used to calculate how to dilute a solution to obtain a particular concentration.

result. Volumes should be measured using accurately calibrated glassware, such as volumetric pipettes and volumetric flasks. Care should also be taken (especially when going from a large concentration to a small concentration) to ensure that any error in measuring the volumes will cause only a small error in the final concentration. For example, in the previous example, if we had accidentally added an extra drop of liquid (~0.03 mL) when we obtained the 0.3 mL of the 100 ppm solution, our final concentration after dilution to 10 mL would be 3.3 ppm. A 10% error in the first measurement seems like a very large mistake, but it isn't; it's just a small drop. And that large an error results in a standard that really isn't the standard we want. That means we should try to keep the volumes as large as possible so that being off by a small amount, such as a drop of liquid, does not produce a large error in the final concentration.

Using our previous example, a more accurate 3 ppm standard could be made by first making a 10 ppm sample. We would take 10 mL of the 100 ppm stock solution using a volumetric pipette and dilute it to 100 mL. This would give 100 mL of the 10 ppm standard:

$$C_1V_1 = C_2V_2$$

$$(100 \text{ ppm}) \times V_1 = (10 \text{ ppm}) \times (100 \text{ mL})$$

$$V_1 = 10 \text{ mL}$$

Then we would use this solution to create the 3 ppm solution. Using the dilution equation ($C_1V_1 = C_2V_2$), we would find that we need to obtain 3 mL of the 10 ppm solution using a volumetric pipette, add it to a 10 mL volumetric flask, and then dilute to the mark:

$$C_1V_1 = C_2V_2$$

$$(10 \text{ ppm}) \times V_1 = (3 \text{ ppm}) \times (10 \text{ mL})$$

$$V_1 = 3 \text{ mL}$$

This would make the same 3 ppm solution. However, if an extra drop (~0.03 mL) accidentally were added in the final step of the process to make the solution, the result would be a 3.03 ppm solution. This is only a 1% error from what we think the standard should be, but it's still off because our standard isn't the concentration we wanted to make. However, the error isn't as critical as when we made the 3 ppm standard from the 100 ppm stock solution.

Once a set of standards have been prepared, they are treated the same as the samples when the analysis is conducted (heating steps, cooling steps, filtration, etc.). Then when the color has fully developed in all the samples and standards, the test tube rack is loaded with the standards in sequential order, from lowest concentration to highest concentration.

We should also prepare a standard with no analyte present—that is, a standard with a concentration of 0 ppm. This standard is known as a **blank**. The blank can be distilled water or some other compound that lacks the analyte. The blank contains all the chemicals and has been treated the same as the other standards and samples. As the colorizing agents are added, they

blank A sample that lacks the analyte in question.

should not develop into the color that indicates the presence of the analyte. In some cases, the blank remains clear, and in other cases, it has a slight color. In still others, it develops a color, but the color is a different hue than the standards. In any case, we know that the blank (the standard with 0 ppm concentration of the analyte) does not have any of the analyte.

The standards and the samples are then treated with the reagents according to the method being used to test for the concentration of the analyte. A set amount of each (for example, 10 mL) is transferred to the cleaned and dried test tubes. Then a test tube rack is loaded with the test tubes containing the standards in one row in order from least concentrated to most concentrated. A test tube of one of the samples is then placed in the second row of the rack. Then the lab workers position themselves above the test tube rack so they look down on the tubes. The workers move the sample from one position to another until the intensity of the color is determined to match one of the standards (Fig. 10.6).

It's quite possible that the depths of the solutions in the test tubes make the intensities of the solutions appear very similar. When that happens, the amount of standard and sample added to each tube can be adjusted. In some cases, the volume in the test tube may need to be reduced so that the lab worker can see through the sample. In other cases, the volume may need to

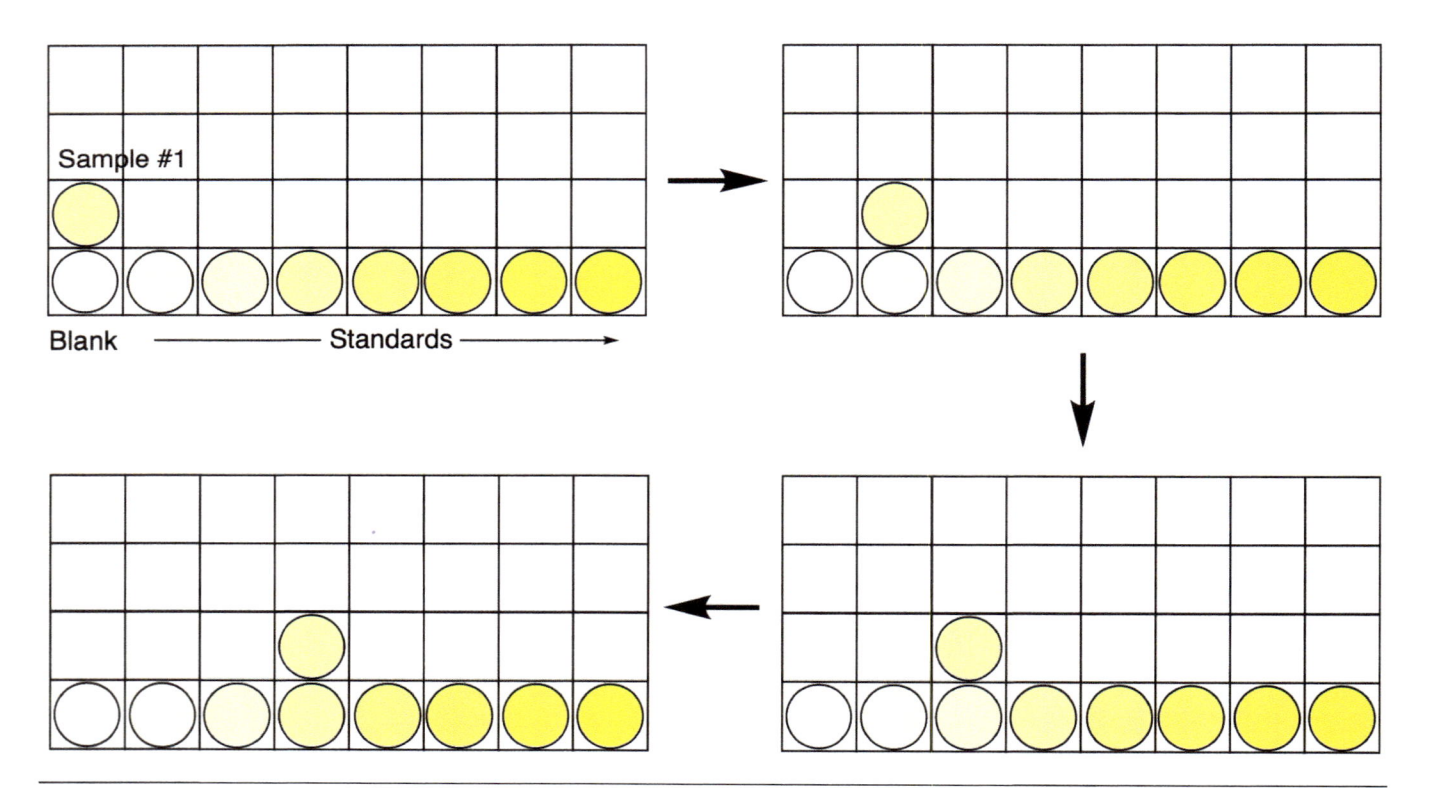

FIG. 10.6. Colorimetry using test tubes in a test tube rack. The blank and standard test tubes are placed in order in one row of the test tube rack. Then the sample test tube is placed in the next row of the test tube rack and the intensity compared with the intensity of the color in the standard row. The sample is moved along the test tube rack until the intensity is determined to match the intensity of the standard. The concentration of the analyte in the sample is then determined to be equal to the concentration in the standard row. (Courtesy S. E. Johnson and M. D. Mosher— © ASBC)

be increased to increase the intensity of the color that's observed. As long as the amount of solution in each test tube is identical, the specific volume in each is immaterial.

As a word of caution, it's tempting to observe the samples by peering through the sides of the test tubes. This should be avoided. While it's possible to compare the intensities by holding the tubes up and comparing them side by side, the more accurate method of peering down into the tubes from the top gives the best results. Adjusting the volumes in the tubes until the distinction between samples and standards can be made provides the most accurate results.

In most cases, the intensity of the sample lies between two standards. The sample has less intensity than one sample but more than the next lower concentrated standard. For such a case, the brewery laboratory should have a recorded method that describes how to report the value of the concentration. One method is to estimate the concentration as best as possible. For example, if the standards are 2 ppm and 4 ppm and the sample appears to be halfway between the two, the user can estimate 3 ppm for that sample. If the standards are 4 ppm and 5 ppm and the sample is halfway between the two, the user might record 4.5 ppm for the sample. Instead, the brewery laboratory may institute the policy to report the concentration as being between the two values. For example, if the sample intensity is between the 4 ppm and 5 ppm standards, the sample can be reported as "less than 5 ppm" or "4 ppm–5 ppm." Reporting the value this way is more accurate than estimating it, but doing so causes some issues when the lab is trying to plot the results.

Given the chance of having samples that don't exactly match a standard, it's very useful to create standard concentrations that are very close to each other. This improves the precision of the results. Unfortunately, in most cases, color intensities are only subtly different between samples with similar concentrations. This makes distinguishing between two standards very difficult. For example, in Figure 10.6, it's very difficult to tell the intensity difference between the fourth and fifth standards.

Minimum and Maximum Concentrations

The minimum and maximum concentrations of the standards should already have been determined by the method of analysis. In most cases, the minimum involves use of the blank as "no concentration of the analyte." The maximum may be less than the actual maximum concentration that could exist in a sample. This is quite common and results from the maximum color intensity that can be developed by the method. For example, the maximum concentration of an analyte might be only 10 ppm because beyond that concentration, the color intensity doesn't increase significantly.

When this occurs, the sample may need to be treated so that it can be evaluated in the range allowed by the method. This can be easily accomplished by diluting the sample with distilled water before starting the experiment. If there's a chance that the sample might be more concentrated than the standards, the dilution and the sample are prepared. The dilution can be by half or more, depending on the concentration anticipated to be in the sample. For example, 50 mL of the sample can be diluted with 50 mL of distilled water to make the solution used in the analysis. Let's assume that the results of the analysis reveal that the diluted sample has a concentration of the analyte equal to 4 ppm. The original sample will then have twice that concentration, or 8 ppm of the analyte. A word of caution: If the concentration is likely more than twice the largest standard, the error in measuring the concentration can increase.

The minimum concentration can also cause an issue. If the method indicates that the usable standard range is from 1 to 10 ppm, then samples that have less than 1 ppm of the analyte won't provide useful results. For example, in the analysis of diacetyl in finished beer, one method creates standards in the range of 1–10 ppm. But if the samples have less than 1 ppm and the user can clearly identify that the color is present in the sample test tube, then the reported value of the concentration should be less than 1 ppm. In such a case, the exact concentration can't be determined.

Colorimetric Methods

The *ASBC Methods of Analysis* contain many methods that can be adapted for use in colorimetry. In addition, a number of other methods of analysis are published exclusively as colorimetry methods that are not specific to beer. If the brewery laboratory is interested in expanding its services to include water analysis, for example, it can find published methods that use colorimetry as the instrumental method. Among the ASBC methods, however, the majority of the methods rely on other instruments, even if they are fundamentally colorimetric methods. The reason for this move from colorimetry to other methods to determine the intensity of color in a solution is that instruments are more easily able to discern slight differences in intensity compared with the human eye. In addition, the wide ranges in color and turbidity of beer can make it difficult to adapt colorimetry methods to the analysis of certain analytes.

Any method that bases the detection of an analyte on a color change to colorimetry can be easily adapted by increasing the number of standards. However, the stability of the intensity of the color in the standards must be considered. In some cases, the standards should be prepared every hour, because the intensity of the color changes after the standards are made. In other cases, the standards can survive 1 week or more if stored in amber-colored glass bottles in the refrigerator. For example, if the procedure indicates that color development continues beyond the test, the standards should be replaced regularly throughout the day. For those analyses for which the standards are replaced regularly (especially when standards need to be replaced hourly), moving to permanently colored solutions (as done by Lovibond when he created the colorimeter) or to a different instrumental method will save time and improve precision and accuracy. If doing so isn't possible, each standard should be photographed as its color is fully developed. Then each sample should be photographed under the same lighting conditions as its color intensity becomes fully developed. The analysis involves comparing the photograph of the sample to the photographs of the standards. Given the availability of computers and digital cameras, performing this additional step doesn't diminish the usefulness of colorimetry as a method of analysis.

ASBC Method **Beer 10A** was adapted from a method that was traditionally a colorimetric method based on many beers being yellow. The original method required the use of beer standards with known SRM values. Alternatively, water samples containing brown food coloring in amounts that match SRM values of color were used. Then a beer sample was compared with the standards to determine its SRM. Alternatively, a color chart or a set of colorized glass panes (as in the Lovibond apparatus) can be used to provide known SRM values. Determining the SRM or European Brewery Convention (EBC) color may not need to be as precise in some applications; in those cases, the colorimeter method is quick and easy to perform and provides good data that can inform the brewery.

ASBC Method **Beer 25D** provides a useful example of the conversion of another instrumental method to colorimetry. The method was created to prepare a series of diacetyl standards and samples that develop color based on the concentration of the diacetyl. The intensity of the color is then determined by an instrument, rather than the eyes. However, if the number of standards were increased slightly, the method could be adapted for use in colorimetry. Unfortunately, the standards have a finite lifetime, so they must be made regularly. Brewery lab personnel will have to decide if the results from using Beer 25D or any other analysis provide reliable data to be used for regular analysis. Using a commercial colorimeter or developing a homemade colorimeter that uses a series of colored glass plates could be helpful.

ASBC Method **Beer 26** is directly applicable to colorimetry, although technically, this method is used to determine the turbidity (cloudiness) of a sample. Turbidity in finished beer is determined by comparison with a set of standards based on various concentrations of formazin suspensions in water. The standards take 24 hours to prepare and are good for about 24 hours before they should be replaced. Thus, the analysis is suitable for colorimetry. In this case, the turbidity of the standards is compared with the samples. While adding a few standards is helpful, the precision of Beer 26 as a colorimetric method isn't as good as the instrumental version of the method. However, if the brewery is concerned only with whether a particular sample is less than some cutoff value of turbidity, the colorimetric method may provide the best and fastest option for the laboratory. Again, it may be helpful to use a commercial colorimeter that uses **formazin turbidity units (FTUs)** or to create a homemade colorimeter using colored glass, acrylic panes, or painted glass that matches liquid FTU standards.

Beer 10A An ASBC method used to determine the color of beer.

Beer 25D An ASBC method used to determine the presence of diacetyl.

Beer 26 An ASBC method used to determine the turbidity of beer.

formazin turbidity units (FTUs) Used to measure the clarity of beer based on the chemical formazin.

CHAPTER 11
Spectrophotometric Analysis

Spectrophotometry is the branch of scientific study that focuses on the interaction of light with compounds. In Chapter 10, we covered another laboratory technique very similar to spectrophotometry known as colorimetry. However, as we will discuss in this chapter, there are some significant differences between spectrophotometry and colorimetry. The basis of those differences lies in the definition of the word "spectroscopy."

Spectroscopy is the field of study that encompasses both spectrophotometry and colorimetry. The word comes from the Latin *specere*, meaning "to look at," and the Greek *skopia*, meaning "to see." Spectroscopy is thus the study of how matter and light interact. Spectrophotometry is the quantitative analysis of light as it's either reflected or transmitted through matter. As we learned in Chapter 10, colorimetry requires that the light be within the visible spectrum. In other words, spectrophotometry involves measuring the amount of light from across the entire electromagnetic spectrum. This means that some spectrophotometric techniques include the interactions of X-rays, infrared rays, radio waves, and microwaves with matter.

The most common spectrophotometric technique in most analytical laboratories (including the brewery laboratory) involves measuring the interaction of ultraviolet and visible light with matter. This seems very similar to colorimetry but is quite different in terms of the technique that's used. Colorimetry uses the human eye to perform the analysis. Although doing so has both advantages and disadvantages, the technique has limitations because it relies on human observation. For example, colorimetry works best when the solution is yellow or green. In comparison, spectrophotometry uses an instrument to measure the intensity of light at a specific wavelength. The limitations of the instrument's measuring ability are governed by the design of the instrument and its sensitivity to specific wavelengths of light. So if a laboratory wants to measure the intensity of a particular wavelength, it needs only to ensure that the instrument is tuned to that particular wavelength.

spectroscopy The study of how matter and light interact.

What Is a Wavelength?

A **wavelength** is essentially a single "color" of light. That color could be within the visible spectrum of electromagnetic radiation, or it could be invisible to the human eye. For example, a wavelength of light in the infrared region could be

wavelength A single "color" of electromagnetic radiation, or light; inversely related to frequency.

electromagnetic spectrum The range of energy that can be transmitted through space.

photon A discrete packet of energy.

frequency The number of times an electromagnetic wave passes a given point per unit of time; inversely related to wavelength.

speed of light 3×10^8 m/s; the velocity of electromagnetic radiation in a vacuum.

used to measure the amount of alcohol in a sample, because alcohol can absorb the specific wavelength of infrared radiation. Infrared radiation, like ultraviolet light or visible light, is a region within the electromagnetic spectrum.

The **electromagnetic spectrum** is referred to as that because of the nature of energy. Energy is made up of an oscillating electric field and an oscillating magnetic field, which are perpendicular to each other. In 1900, physicist Max Planck suggested that any energy could occur in a discrete unit or packet. This idea was suggested again by Albert Einstein in 1905. By 1928, American physicist Arthur Compton had used the word **photon** to describe the discrete packet of energy, and that word is still used today.

Every photon of light has a specific wavelength and a specific **frequency**. The two, wavelength and frequency, are actually indirectly related:

$$c = \lambda \times \nu$$

where c is the **speed of light** (3×10^8 meters per second [m/s]); λ is the wavelength of the light in meters; and ν is the frequency of the light in hertz (Hz, or 1/s). This equation is simple and straightforward but also reveals some very interesting information.

First, the product of the wavelength times the frequency can be used to determine the speed of light. Although the speed of light is slightly variable depending on the medium in which it's traveling, in practice, we can assume that the speed of light is equivalent to its speed in a vacuum, where it's 3×10^8 m/s. This is the fastest that anything can go. And if you are thinking that a rocketship in outer space can go this fast (or faster), you are wrong. Thanks to Einstein and his work on understanding general relativity, we know that as an object (such as a rocketship) approaches the speed of light, its mass increases. Thus, more energy is needed to make it go faster. This cycle continues until the amount of energy required becomes infinite. In other words, matter doesn't travel at or even close to the speed of light. Only energy can do that.

Second, the wavelength and the frequency of a photon are indirectly related. As the wavelength increases, the frequency decreases. Light with

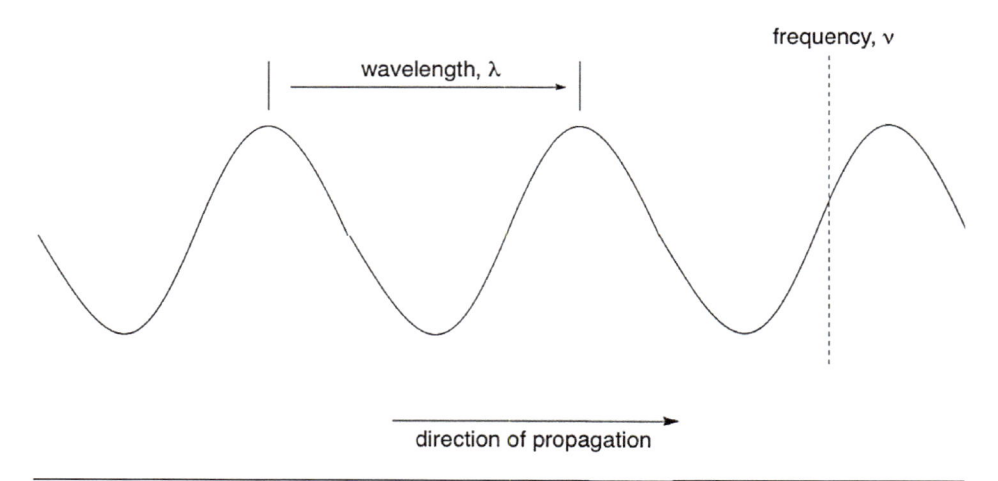

FIG. 11.1. The relationship between wavelength and frequency. The wavelength is the distance from the top of a wave to the top of the next wave in meters. The frequency is how many wave crests pass a given point per second. (Courtesy S. E. Johnson and M. D. Mosher—© ASBC)

a wavelength of 500 nanometers (nm) (500 × 10⁻⁹ m) has a frequency of 6 × 10¹⁴ Hz, if we assume the speed of light is 3 × 10⁸ m/s. If the wavelength is 600 nm (600 × 10⁻⁹ m), the frequency is 5 × 10¹⁴ Hz. Because electromagnetic energy propagates, or moves, in an oscillating field, the definitions of wavelength and frequency can be visually applied to the resulting sine wave (Fig. 11.1).

Infrared, Visible, Ultraviolet, and Beyond

The work of Planck and Einstein further revealed that the energy of a photon is directly proportional to the frequency of the photon. Substituting that relationship with the speed of light equation illustrated that the energy was indirectly related to the wavelength of the photon:

$$E = h \times \nu$$

$$E = \frac{h \times c}{\lambda}$$

where E is the **energy** of the photon in **joules (J)**. The joule is named after James Prescott Joule, a nineteenth-century English physicist who illustrated the relationship between heat and energy. The joule is actually the same as the units kg·m²·s⁻², which arises from the units in the equation used to determine the quantity. You may be more familiar with the energy term "calorie" used in the United States. To convert between the two, we can use this equivalence:

$$1 \text{ calorie} = 4{,}184 \text{ joules}$$

When calculating energy, we find that the energy associated with the wavelength of a 500 nm photon is 3.98×10^{-19} J. This doesn't seem like a lot of energy (and it isn't), but that's the energy of a single photon. In fact, we often think of the energy associated with 1 mole of photons. Remember that a mole is just a counting number equal to 6.02×10^{23} things. In other words, if we had 1 mole of photons with a wavelength of 500 nm, we would have 239,000 joules. And when numbers like this get this large, we often describe them in kilojoules (kJ). One mole of photons with a wavelength of 500 nm is associated with 239 kJ of energy.

To put it simply, light is energy. In fact, energy can be transmitted in essentially every possible wavelength, and the specific wavelength indicates the exact amount of energy that's transmitted. That energy propagates through space until it interacts with an object, such as a molecule.

If the energy impacts a molecule, it can be absorbed by the molecule. The molecule then experiences an increase in energy. We say that the molecule moves from a **ground state**, where it has the lowest possible energy, to an **excited state**, with higher energy. For light in the visible range, the molecule absorbs the energy and promotes an electron to a higher energy level. For light in the infrared region of the electromagnetic spectrum, the energy causes bonds within the molecule to become excited. In fact, each region of the electromagnetic spectrum is associated with different changes within a molecule (Table 11.1).

energy The power within a system that can be transferred.

joule (J) A unit of energy measured in kg·m²·s⁻².

ground state The lowest possible energy state of a molecule.

excited state Any energy state of a molecule that's higher than the ground state.

Once formed in a molecule, the excited state eventually collapses back to the ground state. That collapse can occur by a number of different means based on the specific excitation that occurs. For example, molecules experience vibrations under normal circumstances. Think of the atoms of a molecule being held together with springs. At room temperature, the molecule has enough energy to wiggle those springs. This is the ground state. When energy within the infrared region strikes the molecule, the amplitude of those wiggling springs increases. The molecule is now in its excited state. The molecule could collapse back to the ground state by releasing the exact amount of energy that was absorbed. This would return the vibrations back to their normal state. However, the molecule could release the excited state energy by transferring the energy to its surroundings as heat.

Within the visible region, the electron is promoted to a higher energy level when the molecule absorbs the photon. Then the molecule collapses back toward the ground state. This means that the electron can return back to where it came from in one step, or it may step down to other energy levels along the way. Think of an apartment building in which a resident goes from the lobby to the fifth floor. (The electron is promoted to a higher energy level.) The resident could return to the lobby by going there directly, or he or she could stop at any floor along the way. Each collapsing transition (from the fifth floor to the fourth, the fifth to the third, the fifth to the lobby, the fourth to the first, and so on) is accompanied by the loss of the corresponding amount of energy. In other words, as the excited state drops back to the ground state, a set of photons with discrete energies is released. Moreover, the photons can be released in any direction. Most of the time, they don't travel in the same direction as the original incoming photon. The result is that the original beam of light appears to have lost some of the specific photons that were absorbed.

How long does the excited state exist? The answer depends on the specific energy that's absorbed and the resulting change that the molecule undergoes. However, most excited states don't exist very long. In the **ultraviolet-visible spectroscopy (UV-visible spectroscopy**, or **UV-vis)** excited state, the lifetime is on the order of femtoseconds (1×10^{-15} s). The collapse

ultraviolet-visible spectroscopy (UV-vis)
The study of how ultraviolet-visible light interacts with a solution.

TABLE 11.1. Molecular transitions and the energies that cause them[a]

Region	Wavelengths	Type of transition
γ-ray	Less than 10^{-12} m	Nuclear excitation
X-ray	1.0×10^{-12} to 1.0×10^{-8} m	Core electron excitation
Ultraviolet	1.0×10^{-8} m to 3.1×10^{-7} m	Valence electron excitation
Visible	3.1×10^{-7} m to 7.1×10^{-7} m	Valence electron excitation
Near-infrared	7.1×10^{-7} m to 2.5×10^{-6} m	Molecular vibration
Mid- and far-infrared	2.5×10^{-6} m to 1.0×10^{-3} m	Molecular vibration
Microwave	1.0×10^{-3} m to 0.01 m	Molecular rotation
Radio wave	0.01 m to 1.0×10^{5} m	Nuclear spin excitation

[a] Courtesy S. E. Johnson and M. D. Mosher—© ASBC.

back to the ground state is almost immediate. Some of the longer excited states occur in nuclear magnetic resonance spectroscopy and in phosphorescence, where they can exist for multiple seconds.

Relationship to Colorimetric Methods

UV-vis spectroscopy is quite useful in the analytical laboratory. It uses an instrument to detect even the smallest changes in the absorbance of light passing through a sample and is more precise than the human eye. Besides the obvious differences between using a human as an instrument versus using a computer-based instrument, how do the colorimeter and spectrometer differ? Here are the key differences:

- The instrument can distinguish between colors that are very close together. Most UV-vis instruments can tell the difference in color (i.e., wavelength) between photons as close as 5 nm apart. The human eye has difficulty discerning between colors this close in wavelength.
- The instrument can distinguish between very similar intensities of color in the sample. The human eye requires a much larger difference in intensity to perform that analysis.
- The instrument can be tuned to evaluate wavelengths outside the range that the normal human eye can see. Whereas the human eye is particularly good with light near the green color of the spectrum, the instrument can be adjusted to be equally good across almost all wavelengths—even those invisible to the human eye.
- The instrumental method is usually slower and much less portable than using the colorimeter. Time must be allocated for the instrument to perform the measurement. The colorimeter (especially a portable unit with glass plates) can easily provide the measurement quickly.
- The instrument will outperform the colorimetry user in accuracy of measurement over time, because the instrument doesn't get tired.

General Instrument Design

Figure 11.2 illustrates the general design of a spectrometer. The photons of interest are generated from a source. If the spectrometer is measuring interactions in the visible region of the electromagnetic field, the source is

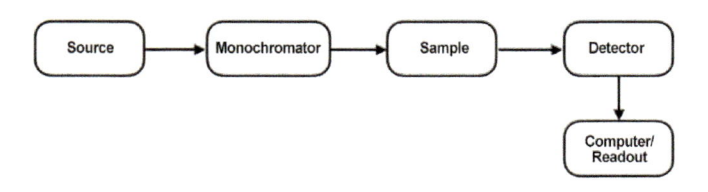

FIG. 11.2. General design of a spectrometer. There are many options in the layout of the parts of the spectrometer, including one in which the sample and monochromator are switched. (Courtesy S. E. Johnson and M. D. Mosher—© ASBC)

a tungsten lamp. The photons pass through a wavelength selector known as a monochromator. The specific wavelengths are then directed through the sample containing the analyte of interest. When the photon wavelength matches the wavelength that's absorbed by the analyte, the analyte becomes excited. Almost instantly, the excited analyte collapses back to the ground state and emits wavelengths of light that correspond to the wavelength that was absorbed or to portions of that wavelength, if the collapse involves many steps. Most of the emitted photons are directed everywhere except in the original direction of the beam of light. The original beam continues through the sample (now missing some of the photons) and strikes the detector. The detector converts the response into an electrical signal based on the number of photons it detects. That signal is sent to a computer, and the intensity is recorded. The readout may be a graph of wavelength versus intensity or a simple number readout.

To locate the sample within the beam of photons, it's necessary to use a sample holder. For most spectroscopic methods, this is known as a **cuvette**. A cuvette is a carefully made container that's designed to fit specifically into the sample chamber within the instrument. Square-bottomed cuvettes are most common, but some instruments use round-bottomed cuvettes that look a lot like small test tubes. If the instrument uses round cuvettes, it's best to keep them in a drawer separate from the test tubes in the lab. Cuvettes are different and more expensive than test tubes. Using a test tube instead of the specially designed cuvette can result in measurement issues.

Cuvettes are made of many different materials. Plastic cuvettes are the least expensive. They can be purchased in bulk and are considered disposable. But unfortunately, plastic absorbs light in the ultraviolet region, so plastic cuvettes are not useful for many analyses that measure absorbance below about 300 nm. Glass cuvettes are a little more expensive and a little more robust than plastic cuvettes. They are also much easier to clean. Whereas a nonaqueous sample can dissolve, soften, or melt plastic cuvettes, glass cuvettes will hold up. However, glass also absorbs light below approximately 300 nm, so glass cuvettes are not useful for ultraviolet measurements. Cuvettes can also be made of quartz. Quartz cuvettes are easy to clean, just as robust as glass cuvettes, and able to hold aqueous and nonaqueous samples. Quartz does not absorb ultraviolet light, which makes quartz cuvettes perfect for analyses that use wavelengths below 300 nm.

Cuvettes should be treated with care. When they are cleaned, they should not be scrubbed with anything that might scratch the surface where the beam of light passes. Any chemical residue on the cuvette surface should be removed by rinsing and washing, as that residue will affect the results of the analysis. Finally, the transparency of cuvettes (especially plastic cuvettes) should be visually inspected. Plastic cuvettes that show cloudiness should be replaced.

The beam of light that's passed through the sample should be generated by a **source**. The source can be a lightbulb or other object that generates energy in the region of the spectrum in which we are interested. The most common source for visible light is a tungsten lightbulb. For wavelengths in the ultraviolet region, a deuterium lamp is the best choice. If the spectrometer is used to analyze infrared wavelengths, one of the sources that can be used is a globar: a small, bar-shaped piece of silicon carbide that when heated emits energy in the infrared region of the electromagnetic spectrum.

---

cuvette A sample holder with a known path length.

source A lightbulb or other object that generates energy in the region of the spectrum of interest.

Like the lightbulbs in our homes, all sources have finite lifetimes, and because the sources are made to very high standards, they can be expensive. Luckily, they are very easy to replace in the instrument if they do burn out.

The **detector** in the UV-vis spectrometer converts analyte information into a computer signal. There are a number of types of detectors:

- A commonly used type of detector is the **photocathode**. This interesting piece of technology is the simplest of the different types of detectors. It converts photons that strike it into an electrical signal. The theory of the process, known as the photoelectric effect, was described by Einstein and others. In this effect, electrons on the surface of a metal can be ejected from the metal when a photon of sufficient energy strikes the surface. Those electrons then strike an anode and are read as a potential (that is, a signal).
- A more sensitive and versatile detector is the **photomultiplier tube**. It works similarly to the photocathode, but as the electrons are first ejected, they are directed toward another material that ejects additional electrons. This cascade continues until a large number of electrons finally strike the anode. The result is a greatly enhanced electrical signal; a single photon can produce 10^6 to 10^8 electrons at the anode.
- A much simpler detector, a **photodiode**, is made from solid-state materials. Essentially, two semiconductors are fused together; one is electron rich and the other is electron poor. The result is a diode with a given voltage between the two materials. If a voltage is applied to the diode, the potential for electrons to move between the materials is enhanced. Then when a photon strikes the material, the electrons are promoted within the material and flow from one semiconductor to the other. The result is an electrical signal.
- A **diode array** is another type of detector often used in UV-vis spectroscopy. This detector has a series of photodiodes aligned in an array. The purpose of the array is to provide readings at every wavelength instantaneously.

The instrument then "sees" that energy and records the intensity of the signal at each wavelength. That information can be displayed in numerous ways—most commonly as a specific absorbance value at a given wavelength or as a plot of the absorbance values at every wavelength. A **spectrum** (*plural: spectra*) is a plot of the absorbance (*y*-axis) versus the wavelength (*x*-axis) (or frequency) across all the wavelengths within the region being explored. As noted before, the *y*-axis can be best displayed as the absorbance. However, it may be more useful to display the *y*-axis as **transmittance** (*T*) or **percent transmittance** (*%T*). The relationship between the absorbance (*A*) and the transmittance is given by this equation:

$$\%T = 100 \times T$$

$$A = -\log(T)$$

where *T* is the transmittance. The percent transmittance is the transmittance multiplied by 100. The effect of the equation is to illustrate an inverse

detector The part of an instrument that converts analyte information into a computer signal.

photocathode A detector that converts photons into an electrical signal.

photomultiplier tube A detector that cascades electrons to result in an enhanced electrical signal.

photodiode A detector that measures a change in voltage when it's struck by a photon.

diode array A detector made up of a series of photodiodes aligned in an array.

spectrum (*plural:* **spectra**) A plot of the wavelength of electromagnetic radiation versus its intensity as it interacts with a sample.

transmittance (*T*) The intensity of electromagnetic radiation that passes through a sample; related to absorbance via the equation: A = –log(*T*).

percent transmittance (*%T*) One hundred times the transmittance.

relationship between absorbance and percent transmittance. While peaks hang down from the top of the spectrum when the y-axis is in percent transmittance, they point up when the y-axis has units of absorbance. Peaks point down in a spectrum for which the y-axis is given in units of percent transmittance.

Fixed or Single Wavelength

The key pieces within a **fixed wavelength spectrometer** include those listed in the previous section. Older instruments, which are still very useful to the brewery laboratory, use a simple **monochromator**, which can select specific wavelengths of light for analysis. The monochromator can be a prism that separates the original beam of light into a rainbow of the wavelengths found in the beam. This is necessary because the detector in the instrument can't differentiate between wavelengths.

The beam of light is directed to the sample using mirrors and slits that force the beam into a single path. Then the beam passes through the sample, where some of the wavelengths are absorbed. The remnants of the original beam then enter the prism and are split into different wavelengths of light. Those wavelengths fan out into the instrument. A slit next to the prism allows only a small number of wavelengths to escape and hit the detector. The prism can be rotated until the selected wavelength passes through the slit (Fig. 11.3).

A fixed wavelength spectrometer has a display that reports the chosen wavelength and the absorbance of the sample at that wavelength. Older instruments used a needle and gauge to report the absorbance. Newer instruments report the absorbance using a digital display.

The advantage of using this type of instrument is that once the wavelength has been selected and the instrument tuned for zero absorbance and full absorbance, all the user has to do is place the sample in the cuvette. The value of the absorbance is read directly. This makes the process quick and easy to perform. The results are just as precise and accurate as those obtained using more expensive computer-based instruments. The disadvantage of using a single wavelength instrument is that only one wavelength can be analyzed at a time. If multiple wavelengths are needed, the instrument must be adjusted to each wavelength and re-blanked (re-zeroed). Thus, switching between wavelengths during an analysis greatly increases the time required to perform the experiment.

fixed wavelength spectrometer A spectrometer that can measure only the absorbance of a single wavelength at a time.

monochromator A device that selects and transmits a single wavelength of electromagnetic radiation.

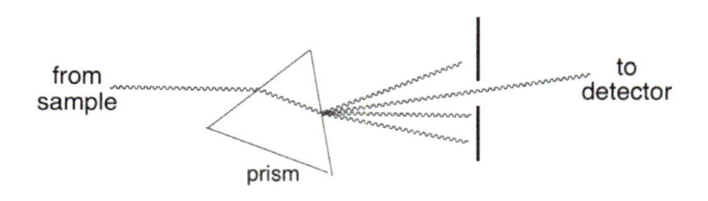

FIG. 11.3. The light from the sample enters the prism and is split into a rainbow of individual wavelengths. A slit allows only one wavelength at a time to continue on to the detector. By rotating the prism, different wavelengths can be selected. (Courtesy S. E. Johnson and M. D. Mosher—© ASBC)

Continuous Wave

A **continuous wave spectrometer** results in the generation of a spectrum that appears one wavelength at a time as the prism is rotated. The advantage of this layout is that it's very easy to perform maintenance on the instrument. The only moving part is the motor that controls the rotation of the prism. One disadvantage is that the motor and its motion must be calibrated to specific wavelengths. This isn't difficult, as the motor is usually a variable speed motor, but the rotation needs to be checked periodically. The other disadvantage is that the spectrum takes time to be collected; the time is based on the speed of the motor. Unfortunately, if the motor moves too quickly, the quality of the data collected by the detector isn't as good.

One modification of this arrangement is the fixed wavelength spectrometer, described earlier. The parts and everything else about the fixed wavelength spectrometer are essentially identical to the continuous wave spectrometer, but the instrument can't draw the entire spectrum. Instead, the fixed wavelength spectrometer displays the absorbance at a given wavelength. The user selects the wavelength and adjusts the baseline of the measurement to give zero absorbance when the blank is in the sample chamber. When the sample is in the instrument, the instrument reports the absorbance at that wavelength. For obvious reasons, the fixed wavelength instrument is relatively inexpensive. Plenty of them can be found in the resale market.

An upgrade to the spectrometer involves the use of a **diffraction grating**, instead of a prism (Fig. 11.4). A diffraction grating is a piece of plastic, glass, or other material with carefully cut and closely spaced grooves. The grooves make the surface of the material act like a prism. The grooves on a CD are close enough that the surface acts like a diffraction grating. But there are two major differences between a diffraction grating and a prism. First, a diffraction grating produces multiple rainbows of varying intensity, so the

> **continuous wave spectrometer** A spectrometer that produces a spectrum across a wide range of wavelengths by rotating a monochromator.

> **diffraction grating** A monochromator with carefully cut and closely spaced grooves; acts like a prism.

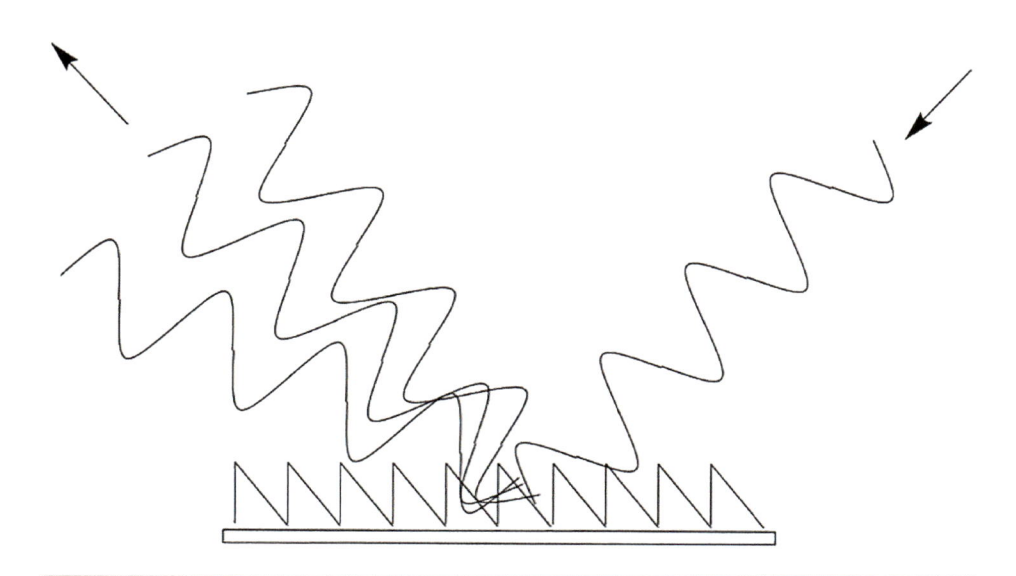

FIG. 11.4. Diffraction grating. The incoming photons strike the diffraction grating and are reflected in different directions as a function of their wavelengths. The resulting rainbow can be directed to the detector. (Courtesy S. E. Johnson and M. D. Mosher—© ASBC)

instrument is often tuned only to the most intense set of wavelengths. Second, if the light is reflected off the surface of the grating, the beam of light doesn't lose any intensity, which would result from passing through a prism. For the prism, this loss is minimal, but it does exist.

In the continuous wave instrument, the monochromator is attached to a motor. The motor moves the monochromator slowly, so that every wavelength is exposed to the detector. This results in the detector response being recorded as a function of the wavelength of light—a spectrum. The instrument can be adjusted to report the absorbance of the sample at a single wavelength, but as with the single wavelength instrument, being able to display the entire spectrum at once allows the user to find and explore the different signals that a sample produces.

The advantages of using a continuous wave instrument are the ability to obtain the entire spectrum at one time, the ability to use the instrument as a single wavelength instrument as needed, and the relatively low cost compared with the most advanced versions of the instrument. One of the disadvantages of using a continuous wave instrument is that the entire spectrum isn't needed for most measurements. In addition, because the instrument rotates the monochromator to produce the spectrum, the speed of the process to develop the entire spectrum isn't very fast. Some settings can require as long as 30 seconds to display the entire spectrum; for most analyses that require only a single absorbance reading, the wait time for the instrument can be limiting.

A modification of the continuous wave instrument allows for detecting all the different wavelengths of light at one time without rotating the monochromator. Doing this requires using a diode array detector. This detector is essentially a string of diodes arranged side by side, which the rainbow of the beam of light will hit after passing through the monochromator. Thus, the speed at which the instrument produces measurements increases dramatically, because there is no motor turning the monochromator.

Fourier Transform

The most advanced version of the spectrometer is the Fourier transform (FT) instrument. This instrument doesn't use a monochromator, and it uses a different type of detector than the single and continuous wave instruments. Specifically, the beam of light is passed through a device called an **interferometer**, which creates an interference pattern in the beam of light by moving a mirror that reflects the light (Fig. 11.5). That pattern is then passed through the sample and to the detector. The result is an interference pattern of the sample as a plot of the intensity of the light versus the position of the moving mirror. That plot is called an **interferogram**.

A computer converts the interferogram into a spectrum using FT mathematics. The complexity of the math is beyond the scope of this text, but it works well to convert the interferogram into an entire spectrum quickly. The instrument usually repeats the collection of the interferogram multiple times, and those individual interferograms are added together to give an averaged interferogram. Averaging reduces the noise in the measurement and allows the signals from the sample to be more apparent. Very precise measurements are obtained of the intensities of the different wavelengths.

interferometer A device that creates an interference pattern in a beam of light by moving a mirror.

interferogram A plot of the intensity of the light versus the position of the moving mirror in an interferometer.

The advantages of using an FT spectrometer are its speed in collecting a spectrum and its ability to signal average by adding together multiple interferograms. The power of signal averaging increases the ability to accurately and precisely measure the intensity of a weak signal. One disadvantage of using an FT spectrometer is the high cost of the instrument compared with other versions. In addition, the interferogram is created by moving parts that must be moved very accurately. This increases the need for maintenance of the instrument. Although replacing the motor on a monochromator isn't cheap, it's considerably less expensive than replacing an interferometer. Luckily, the huge advantages to using an FT spectrometer far outweigh these disadvantages, as shown in Table 11.2.

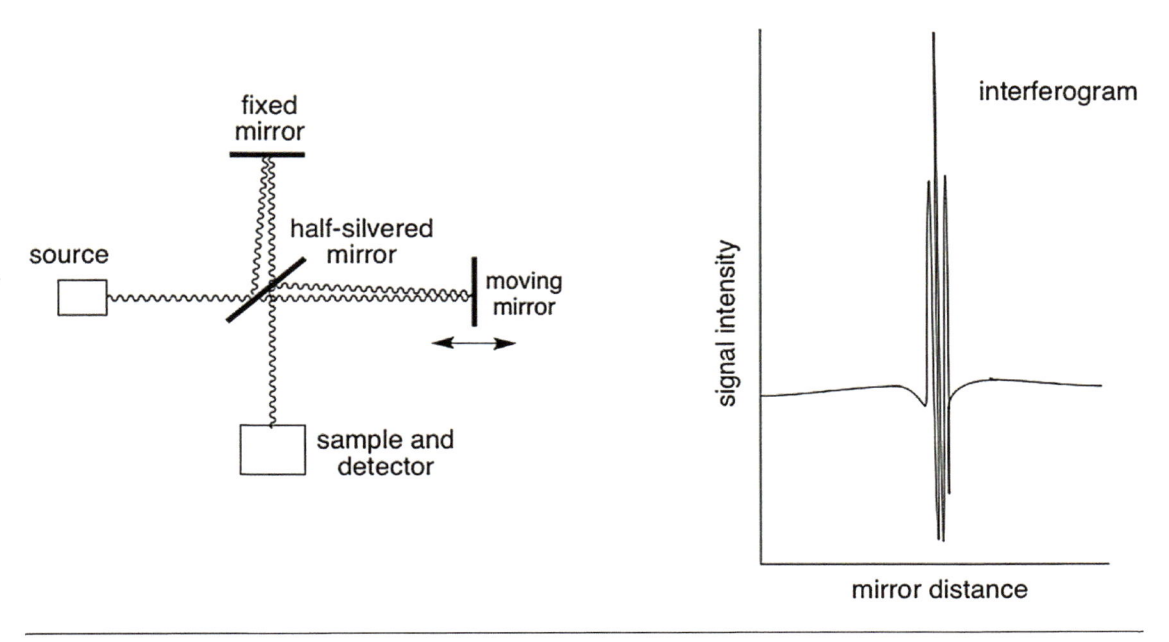

FIG. 11.5. An interferometer produces an interferogram at the detector. (Courtesy S. E. Johnson and M. D. Mosher—© ASBC)

TABLE 11.2. Pros and cons of using different types of spectrometers[a]

Instrument	Advantages	Disadvantages
Fixed wavelength	Inexpensive	Measures only one wavelength at a time
	Rapid analysis at one wavelength	Must be reset for each wavelength
	Computer control not needed	
Continuous wavelength	Moderately priced	Time required to display entire spectrum
	Can operate at fixed wavelength or full spectrum	Very sensitive to selecting wavelengths
	Large bench "footprint"	Requires blanking across all wavelengths used (can be computer controlled)
Fourier transform (FT)	Signal averaging to reduce noise	Expensive
	Rapid full spectrum or single wavelength	Interferometer requires maintenance to maintain accuracy of wavelength selection

[a]Courtesy S. E. Johnson and M. D. Mosher—© ASBC.

Infrared Analysis

Simple molecules, such as a hydrochloric acid (HCl) molecule, can be thought of as 2 atoms attached by a spring. The spring oscillates based on the energy of the system, such that the distances between the 2 atoms changes in a regular pattern. In other words, the bond vibrates. This can be extended to molecules with more than 2 atoms. In those cases, many different vibrations take place.

In fact, each atom within a molecule can move in any of three directions along the *x*-axis, the *y*-axis, and the *z*-axis. Atoms can also move diagonally, which can be reduced to thinking about the movement as a combination of movement along two or all three axes. For example, movement diagonally within the *x*–*y* plane can be thought of as movement along the *x*-axis and the *y*-axis at the same time. The result is that every atom has three **degrees of freedom**.

When atoms are attached to make a molecule, the total number of degrees of freedom is three times the number of atoms. For most molecules, three of the resulting modes of freedom are rotations (spinning of the molecule) and three are translations (movement of the molecule). If the molecule is linear, there are only two rotations. The remaining modes stem from **vibrations**, in which the atoms are moving different distances relative to each other. The formula for a linear molecule is:

$$(3 \times N) - 5 = \text{vibrational modes}$$

The formula for a nonlinear molecule is:

$$(3 \times N) - 6 = \text{vibrational modes}$$

To determine the number of vibrational modes, we need to know the shapes of the molecules. The HCl module we discussed earlier is linear and will have only one vibrational mode. Water (H_2O), a nonlinear molecule, will have three vibrational modes. Ethanol (C_2H_6O) will have 21 vibrational modes, and glucose ($C_6H_{12}O_6$) will have 66 vibrational modes.

Each vibrational mode for a molecule can absorb a specific wavelength of infrared energy but only if that vibration results in a change in the center of mass for the molecule. Some vibrations don't adjust the center of mass and are therefore considered infrared inactive. The result is a vibration that doesn't absorb a wavelength of infrared energy. For example, carbon dioxide (CO_2), a linear molecule, has a vibrational mode that's invisible (Fig. 11.6).

Each specific vibrational mode that's infrared active will absorb a specific wavelength of infrared energy. Thus, the number of vibrational modes indicates the number of peaks we can expect to find in an **infrared spectrum** of the molecule. For CO_2, we should expect to see only two peaks—one at 15.0 μm and another at 4.26 μm—in the infrared spectrum for this molecule.

The large number of vibrational modes means that the infrared spectrum for a mixture of compounds tends to be very crowded. Luckily, modes that are similar absorb similar amounts of energy, even if they are in different molecules. For example, ethanol and propanol are similar molecules. Structurally, they differ only in 1 carbon atom. And while their physical properties

---

degrees of freedom The ability of a molecule to move in three dimensions along the *x*-axis, the *y*-axis, and the *z*-axis.

vibrations Movement between atoms within a molecule in which the distances between the atoms change.

infrared spectrum A spectrum generated by the amount of infrared light absorbed versus the wavelength of that light.

are pretty different, their infrared spectra are somewhat similar because they have very similar structures. In fact, it can be difficult to tell using infrared spectroscopy if a sample of ethanol is contaminated with propanol.

There are three major regions within the infrared (IR) spectrum. As shown in Table 11.3, they are the near-IR (NIR), midrange-IR, and far-IR regions. The most commonly explored region of the infrared spectrum is the midrange-IR or mid-IR, although the near-IR is also used quite often. Absorptions of infrared energy within the mid-IR region tend to be very well defined, and they are often used to characterize the type of molecule. For example, a mid-IR spectrum can easily determine if the molecule of interest is an alcohol or an ester. The far-IR region can also have very well-defined absorptions and very sharp peaks in the printed spectrum. Unfortunately, not a lot of chemical studies have been conducted in the far-IR region, so its utility really hasn't been defined. The near-IR region, on the other hand, has been very well studied.

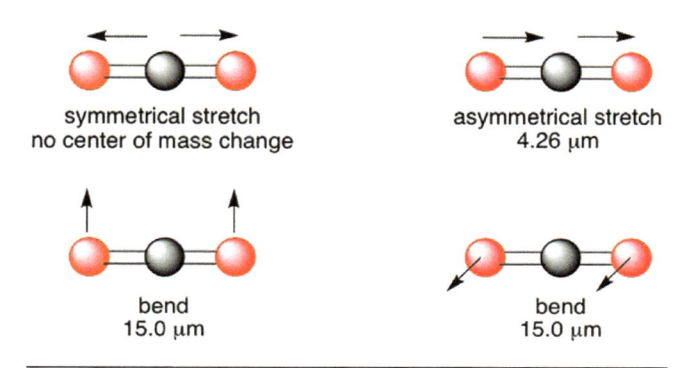

FIG. 11.6. Vibrational modes for carbon dioxide molecules. The symmetrical stretch in carbon dioxide doesn't result in the change in the center of mass and so is infrared inactive (upper left). The values shown for the other modes are the wavelengths of infrared energy for those particular vibrations. The two bending modes are the same (i.e., they have the same bend and are only perpendicular to each other), so they absorb the same wavelength of infrared energy. (Courtesy S. E. Johnson and M. D. Mosher—© ASBC)

TABLE 11.3. Infrared (IR) regions and some practical uses in brewing laboratory analyses[a]

IR region	Wavelengths	Uses in brewing laboratory analyses
Near-IR (NIR)	780–2,500 nm	Minimal interference from water
		Alcohol measurement
		CO_2 measurement
Midrange-IR	2,500–25,000 nm	Identifies more organic functional groups than NIR
Far-IR	2–1,000 μm	Not many uses

[a] Courtesy S. E. Johnson and M. D. Mosher—© ASBC.

Absorptions of energy in the near-IR region aren't direct absorptions by the molecules. Instead, they are known as overtones or harmonics of actual absorptions. The nature of an overtone can be illustrated by plucking a string on a guitar. Once the string starts to vibrate, much smaller vibrations begin in other strings that correspond to harmonics. In the near-IR region, the overtones are much less intense than in the mid-IR region. In addition, many of the overtones overlap and make broad regions of absorption. Thus, measurements of analytes in the near-IR region require deconvoluting multiple wavelengths to obtain usable data.

Pros and Cons of Infrared Analysis

The advantage of performing an analysis using infrared spectroscopy is that some detailed information can be obtained about the structure of an analyte. And if the analyte in question has a unique structure, it can usually be teased out of a mixture of compounds. Another advantage is that sample preparation for infrared analysis is usually very simple. Solutions can be placed in chambers that are invisible to the infrared wavelengths. For these samples, little sample preparation tends to be involved. The absorption wavelengths can be determined just by pouring a liquid sample into the sample chamber.

The disadvantages of using infrared spectroscopy for analysis are few, but they can be significant:

1. The sample chamber must be invisible to infrared energy. For some regions in the wavelength range, that means using a sodium chloride container. A sample chamber made from this material is very sensitive to the presence of water (sodium chloride is soluble in water), which severely limits which samples can be analyzed. There are alternatives to sodium chloride chambers, but they are expensive and problematic. For example, a holder made out of a zinc-selenide crystal is invisible to infrared energy. However, this particular crystal isn't stable below pH 5, and so it really isn't useful for analyzing beer samples.
2. Measuring the percentage alcohol by volume can be performed using infrared spectroscopy. This is commonly done within the near-IR band, where the absorbance at multiple wavelengths is evaluated. A formula is then applied to the results, and the percentage alcohol is estimated. Because of the similarity between the structures of various analytes, the evaluation of a sample must be based on the assumption that other similar compounds don't exist in the sample. For example, when determining the percentage alcohol by volume, the user must assume that other compounds (methanol, propanol, isopropanol, and so on) don't exist in the sample (or exist at levels so small that they are insignificant to the overall measurement).
3. Infrared spectroscopy instrumentation is expensive. The mid-IR instruments tend to be the least expensive; the near-IR instruments can be very expensive. Of course, the cost of the instrument is entirely related to its level of precision—something the brewery laboratory must consider.

Using Infrared Analysis in Brewing

As noted earlier, the most common use for infrared spectroscopy in the brewery laboratory is to determine the percentage alcohol by volume. Anton Paar manufactures an instrument that measures alcohol using near-infrared (NIR) spectroscopy. An infrared spectrometer can also be used to explore other analytes. For example, infrared spectroscopy can be used to assess levels of carbon dioxide in gases and to determine the concentrations of proteins, carbohydrates, and fats. Additionally, infrared analysis and other spectroscopic methods can be performed inline, allowing measurements of alcohol or CO_2 to be made inside a vessel or in a transfer line without withdrawing a sample. However, because of the assumptions discussed earlier, confirmation and correlation to actual values require using a secondary method that's more specific.

UV-Vis Analysis

As noted earlier in this chapter, spectroscopy is quite useful in the brewery laboratory. In fact, after a pH meter, the next instrument to add to a brewery laboratory should be a UV-vis spectrometer. There are UV-vis analyses for malt, wort, hops, beer, and various other substances that make this instrument quite useful. Typical experimentation using a UV-vis spectrometer involves preparing a series of standards and samples (including a blank) according to the method directions. Finally, each standard, the blank, and each sample are measured on the instrument. Typically, within the visible region of the spectrum, analytes have very broad absorbances that correspond to the promotion of electrons to higher energy levels. The most common use of the data returned from the instrument involves recording the value of the absorbance at a particular wavelength.

To ensure that our standards and samples provide the most accurate and precise results, we must make sure that the absorbance is taken from the very apex of a peak in the UV-vis spectrum (Fig. 11.7). This location in the spectrum should have the most responsive changes when the concentrations of the standards or samples change. The method should report the wavelength at the top of the important peak in the spectrum; however, it's best to confirm that the wavelength is actually at the top on each individual instrument. A small alignment error may move the peak away, such that the user thinks the instrument is reading the top of the peak when it isn't. If that's the case, a small adjustment to the wavelength might be necessary.

In some cases, there are multiple peaks within the spectrum (Fig. 11.8). The method should have identified the most important peak or peaks in the spectrum. If we are developing our own method or want to see the results if we use another peak in the spectrum, we can select the wavelength that corresponds to the top of another peak. It's important not to select the side of a peak or the valleys between peaks. These locations provide greatly reduced or nonexistent responses to the difference between standards. Thus, these wavelengths don't correlate well to the concentration of an analyte.

As with any set of standards, it's best to prepare a standard curve. We explored standard curves earlier in this text (see Chapter 8). In short, the resulting standard curve should span the range of all the samples. When plotted as a function of the absorbance at a given wavelength versus the concentration of the individual standards, the standard curve should result in a straight line. This provides good evidence that the standards follow the Beer-Lambert law.

Finally, as noted previously, it's important to keep the instrument within the most responsive area. That means the concentrations of standards and samples should be less than the maximum values the manufacturer has set

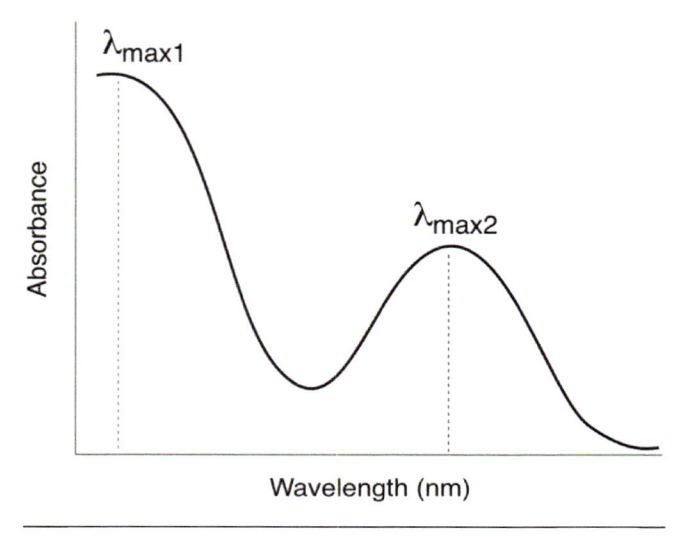

FIG. 11.7. UV-vis spectrum. Each peak is characterized by a wavelength (known as λ_{max}) and an absorbance at that wavelength. (Courtesy S. E. Johnson and M. D. Mosher—© ASBC)

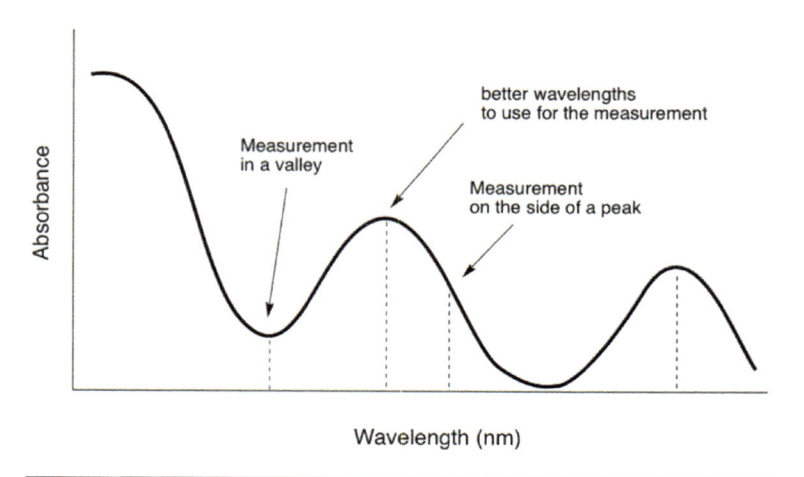

FIG. 11.8. Where to measure the absorbance within a spectrum. Valleys and the sides of a signal should never be used to measure the absorbance. The best location is the top of a peak. (Courtesy S. E. Johnson and M. D. Mosher—© ASBC)

as guidelines. In fact, a much better response is often seen at the lower concentration ranges. In some cases, we may need to dilute our samples to fall within that lower range. In other cases, we can use the samples without diluting them.

Pros and Cons of UV-Vis Analysis

The advantages of using UV-vis for an instrumental method are numerous. One advantage is that UV-vis analyses can be obtained using a relatively inexpensive instrument. Also, methods are available that can measure almost any analyte in a sample. This is done either by using the existing ability of the analyte to absorb specific wavelengths or by adding chemical reagents that convert the analyte into colored compounds. The extensive list of available methods is a tremendous advantage to the brewery laboratory. Most key analytes can be done with only one instrument.

The disadvantages of using UV-vis reside in the fact that not all analytes can absorb specific wavelengths within the ultraviolet-visible region. For example, determining the percentage alcohol by volume can't be accomplished using UV-vis spectroscopy, because ethanol doesn't absorb light within these regions. In addition, the sensitivity of the method can sometimes be less than ideal. If the laboratory needs to measure the concentration of an analyte at a level below what the instrument can measure, the method isn't useful.

Using UV-Vis Analysis in Brewing

Many of the analytical methods used in brewing don't require adding chemical colorizing agents. Within the brewery laboratory, this can include analyses such as the evaluation of standard research method (SRM) color, the concentration of iso-alpha acids, and the measurement of the turbidity of a sample. These values can be determined simply by measuring the absorbance of a sample at a specific wavelength using the UV-vis spectrometer. The colorimeter also works for those methods that develop a color within the visible region of the spectrum. Most of these analyses involve very little sample preparation. Thus, performing these methods of analysis tends to be rather quick.

Other methods require modification of the sample by adding chemical reagents that cause a color to develop. Although the addition and development of color take a little time, the measurement of the analyte can be fairly straightforward. With a well-developed method of analysis, the process can be routine. The methods of analysis provided in Chapter 15 include numerous examples of UV-vis spectroscopy.

CHAPTER 12

Chromatography

Chromatography is an instrumental technique that many scientists consider to be their greatest ally. Chromatography is the process used to separate mixtures of chemical compounds. After the compounds have been separated, they are detected by the instrument. This allows for far superior accuracy and precision when analyzing a mixture compared with a spectrophotometric analysis, which measures the entire mixture at once. By itself, chromatography is a quantitative method, but it becomes an incredibly useful qualitative analysis tool when coupled with a detector such as a UV-visible (UV-vis) spectrophotometer, mass spectrometer, or other instrument that can relate information about the structure of a chemical compound.

The theory of chromatography began with the work of Christian Friedrich Schönbein, a German–Swiss chemist, and one of his students, Friedrich Goppelsröder, at the University of Basel in Switzerland. In 1861, they published the results of their studies on capillary action. In these studies, a solution containing a dye was dripped onto filter paper. The researchers noted that the water in the solution seemed to travel farther than the dye did. In 1903, Mikhail Tsvet, a Russian–Italian botanist, used this observation to develop a process to separate plant pigments. He extracted the pigments from plants and passed the extract through a column of ground limestone. He noted that the different pigments moved at different rates as they moved down the column. It wasn't until the mid-1940s that Archer John Porter Martin and Richard Laurence Millington Synge described the theory behind the process. Their work won them the Nobel Prize in Chemistry in 1952. By that time, chromatography had become a useful tool in the chemist's toolbox.

Chromatography is taught in most introductory chemistry courses. In fact, some elementary school projects use chromatography to explain chemical theories. The most common methods of chromatography used today are thin-layer chromatography (TLC) and paper chromatography. Both methods use paper or plates of adsorbent material instead of a column to separate a mixture of compounds. Columns are used in other methods, including flash chromatography, gas chromatography, and liquid chromatography.

How do these techniques work? The mixture of compounds is placed into a **mobile phase**, which moves through the chromatography system. The

chromatography The process used to separate a mixture of compounds.

mobile phase The portion of the chromatography system that moves during an analysis.

stationary phase The fixed portion of the chromatography system that doesn't move.

mobile phase then passes through an adsorbent known as the **stationary phase** (because it doesn't move). As the mixture of compounds moves along, the compounds interact with both the mobile phase and the stationary phase. If the compounds interact strongly with the stationary phase, they slow down. If the compounds interact strongly with the mobile phase, they move quickly through the adsorbent. Because different chemical compounds have different levels of interaction with the stationary and mobile phases, the mixture is separated.

There are many unique forms of chromatography based on the different types of interactions that are possible. This allows compounds to be separated by mass, by boiling point, by biological interaction, by polarity, and by many, many more criteria. Most of these specific methods won't be explored in this text. Instead, we will focus on the interactions that are based on the polarity of the molecules.

Polar Versus Nonpolar

polarity A measure of the distribution of electrons within a molecule.

nonpolar A molecule in which the electrons are evenly distributed.

polar A molecule in which the electrons are unevenly distributed.

dipole moment The field that results from the separation of charges in a molecule because of an uneven distribution of electrons.

electronegativity The measure of how electrons are polarized in a bond within the molecule.

Polarity is a measure of the distribution of electrons within a molecule. In some cases, the electrons are evenly distributed in the molecule. Such a molecule is considered nonpolarized, or **nonpolar** for short. In other cases, a region or regions within the molecule that contain greater electron density take on a partial negative charge because of the charge on the electron. This means that other regions of the molecule are missing some of the electron density and take on a partial positive charge because of the exposed nuclei of the molecule. Two poles result within the molecule. We say that the molecule has a dipole. Such a molecule is polarized, or **polar**.

The dipole produces an electric field that stretches from one end of the molecule to the other (Fig. 12.1). This field is known as the **dipole moment** of the molecule. As the amount of separation and the amount of the partial charge both increase, the dipole moment increases. The electric field allows the molecules to interact.

How do the electrons in a molecule distribute themselves? The dispersion of the electrons is caused by a phenomenon known as **electronegativity**: the measure of how electrons are polarized in a bond within the molecule. An atom that is highly electronegative has a nucleus with a large number of protons surrounded by electrons that are held close to the atom. This arrangement means that the nuclear charge is felt strongly by the electrons. If a bond is made using those electrons, the electrons will more likely be associated with the highly charged nucleus than with the atom at the other end of the bond. For example, assume that a fluorine atom shares a bond with a tin atom. The electrons will be polarized toward the fluorine atom more than the tin atom, because the fluorine atom uses electrons that are held close to its nucleus while the tin atom uses electrons that are much farther away from its nucleus.

Linus Pauling, an American chemist, created a scale of electronegativity values in 1932 that we can use to determine if a bond is polarized (Fig. 12.2). If the difference between the values is less than 0.4 units, we say that the bond is nonpolarized. (The bond lacks a dipole moment.) If the difference is greater than 0.4 units, we say that the bond is polarized. (It has a dipole moment.) For example, if we compare the electronegativity values of fluorine and tin, we see that fluorine has a value of 4.0 and tin has a value of 1.8. The difference between the two is 2.2, so the tin–fluorine bond is polarized. Similarly, if we compare a hydrogen atom and a carbon atom, we see that the bond between them is nonpolar. (Hydrogen is 2.1 and carbon is 2.5, so the difference is 0.4.) A quick look at Figure 12.2 reveals that electronegativity values increase as we move up and to the right on the periodic table.

What does this mean for our molecules of interest? If a molecule has only nonpolar bonds, then the entire molecule lacks a dipole moment. This molecule is nonpolar. Such a molecule will interact with other molecules without using the dipole moment and associated electric field. If the molecule is polar, then it possesses a dipole moment. Such a molecule will interact with other molecules via its associated electric field—an interaction known as an **intermolecular force of attraction**.

Intermolecular forces should not be confused with intramolecular forces. An intramolecular force of attraction is the force that holds a

intermolecular force of attraction An attraction between molecules that holds them closer together.

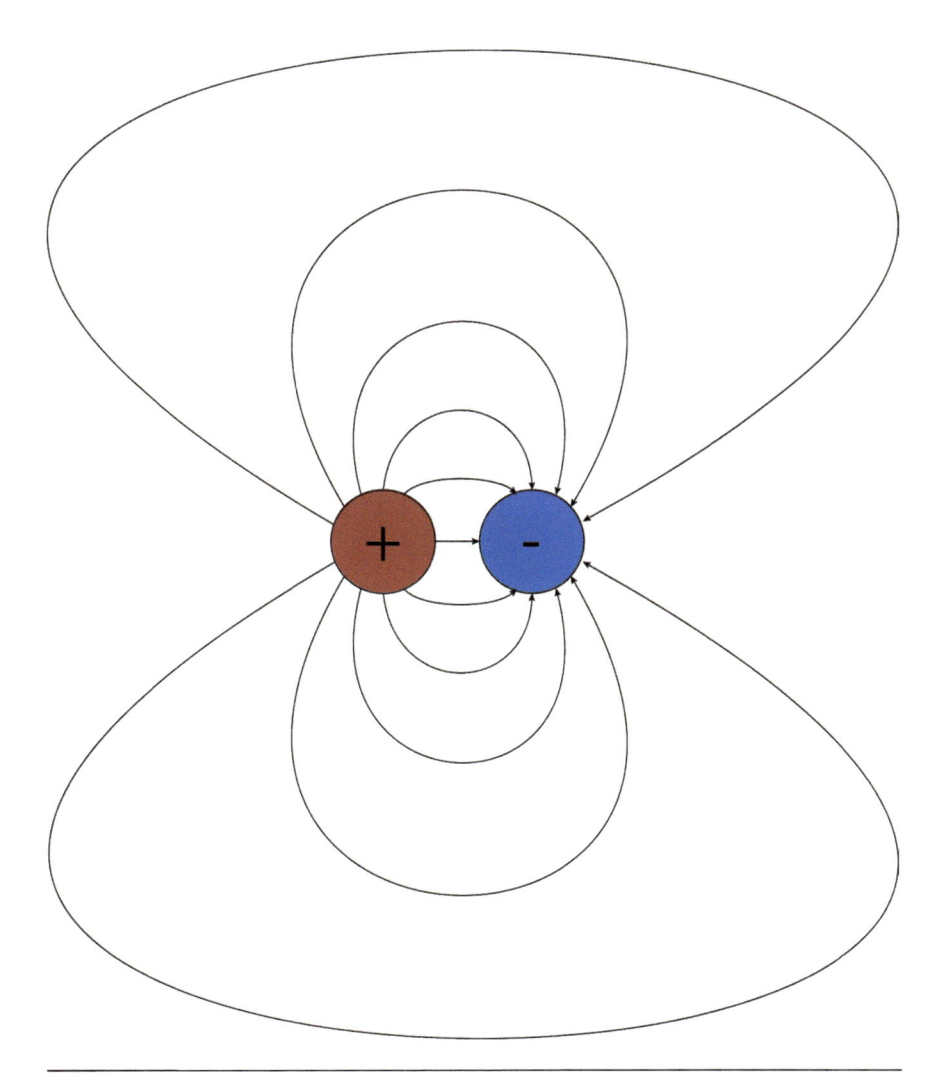

FIG. 12.1. Electrical field set up by the separation of charge within a molecule. (Courtesy S. E. Johnson and M. D. Mosher—© ASBC)

molecule together. These are the bonds within a molecule, and they are considerably stronger than intermolecular forces of attraction. There are three main intermolecular forces between molecules: London dispersion forces, dipole–dipole forces, and hydrogen bonding.

London Dispersion Forces, Dipole–Dipole Forces, and Hydrogen Bonding

To fully understand intermolecular forces, we must remember that a solution often contains more atoms and molecules than can be comprehended. Each of the molecules in a solution can interact with each of the molecules immediately around it, whether they are molecules of the solvent or molecules of other dissolved compounds. When we look at a solution, the trends we observe are a composite of the behaviors of all the molecules in solution, rather than a single effect imparted by an individual molecule. If we consider all the molecules together, the interactions between molecules can result in reactions, affect the boiling points of the substances, and affect the viscosity of the solution, among other interactions. Predicting these interactions will help us evaluate how a mixture can be separated using chromatography. Understanding the interactions will allow us to determine the order in which compounds in a mixture will be separated during analysis.

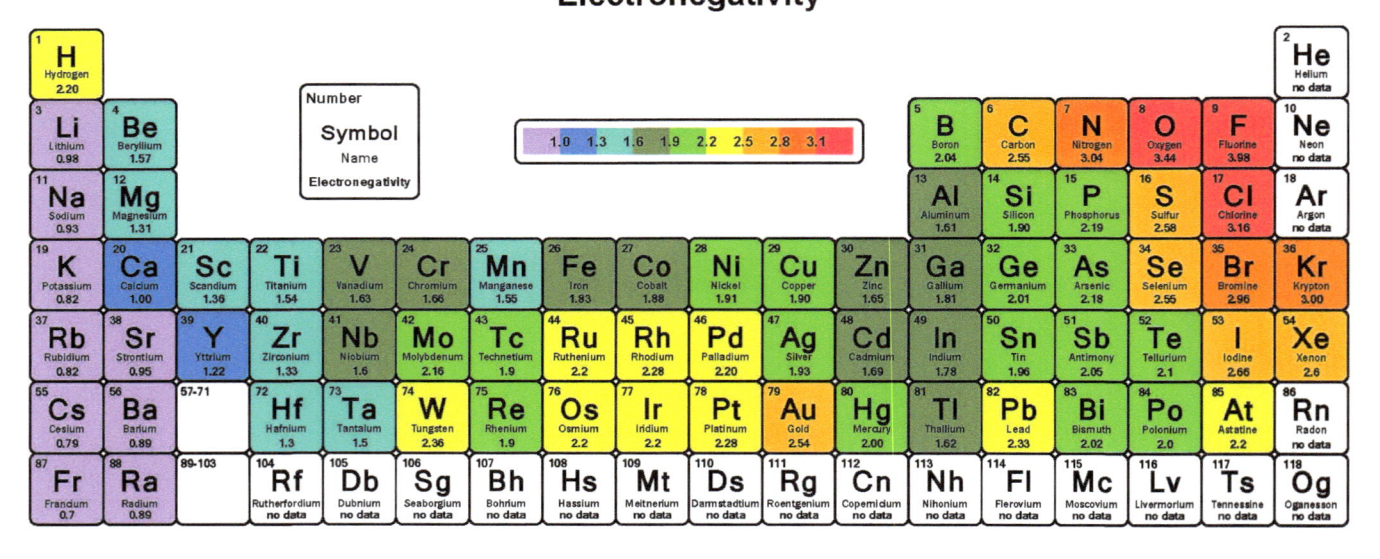

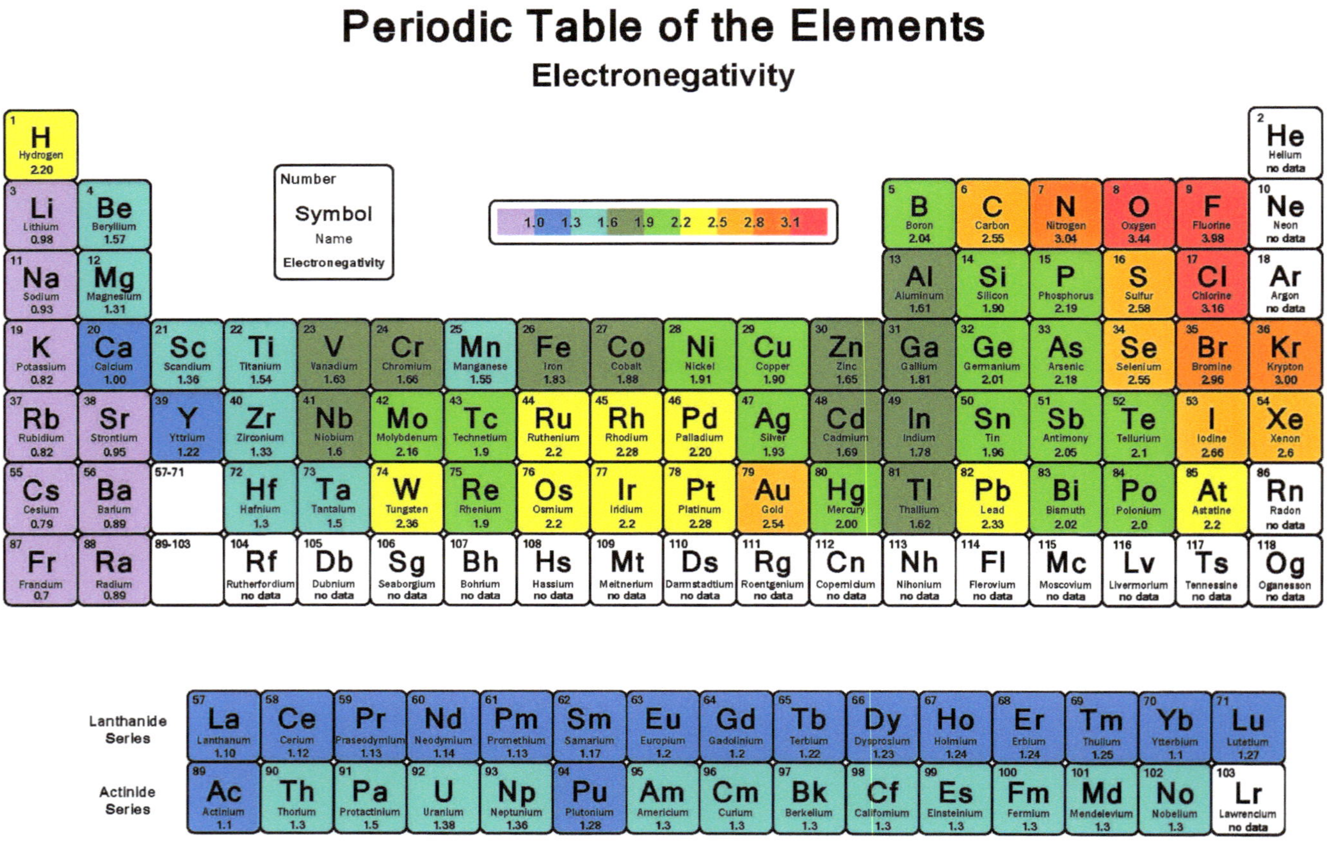

FIG. 12.2. Electronegativities of selected elements on the periodic table. The largest electronegativities are located in the top-right corner of the periodic table. (Courtesy ScienceNotes.org—© Todd Helmenstine)

To assess the interactions, we need to determine if the molecule in question is polar or nonpolar. Making this determination requires more than simply examining the electronegativity values of the atoms in the molecule. We also need to know the shape of the molecule. Luckily, a small 2-atom molecule will always be linear. For a linear molecule, the difference in electronegativity between the atoms will determine the bond dipole. For example, an oxygen (O_2) molecule is made up of 2 of the same atom. The difference in electronegativity between these atoms is 0, and there is no bond dipole. Without a bond dipole, there is no molecular dipole and no dipole moment for the molecule. The molecule is nonpolar.

We can evaluate a molecule such as carbon dioxide (CO_2) in the same way. The structure of this molecule is O=C=O. It is linear. The atoms are different, and the molecule possesses 2 bond dipoles between the O and C atoms. Each dipole is directed toward an oxygen atom—the most electronegative atom of the pair. The bond dipoles are equal in magnitude but point in opposite directions. Because of this, no side of the molecule has a partial positive or negative charge. The result is no net molecular dipole moment. The molecule is nonpolar.

Nonpolar molecules can interact with other nonpolar molecules by aligning next to each other so that they have the maximum number of contacts. When this happens, a very weak intermolecular force of attraction arises. That force is called the **London dispersion force** and was named after Fritz London, a German chemist who immigrated to the United States before World War II. Because electrons within a molecule are in constant motion, there are brief moments when the electrons are distributed nonsymmetrically (Fig. 12.3). This fleeting arrangement of electrons results in an instantaneous dipole within the molecule. That dipole causes adjacent molecules to undergo a similar electron distribution, and when that happens, the two electric fields from the molecules can interact (Fig. 12.4). The interaction is a positive force of attraction between the two molecules. Because of the fleeting nature of the nonsymmetrical distribution, the force of attraction is also very fleeting. Given this, London dispersion forces are the weakest of the intermolecular forces.

London dispersion force The weakest intermolecular force that results from electron movement in a molecule.

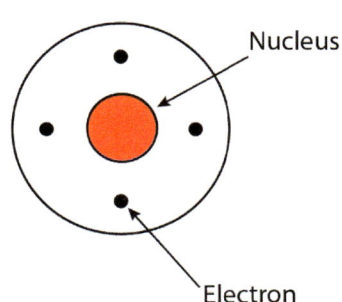

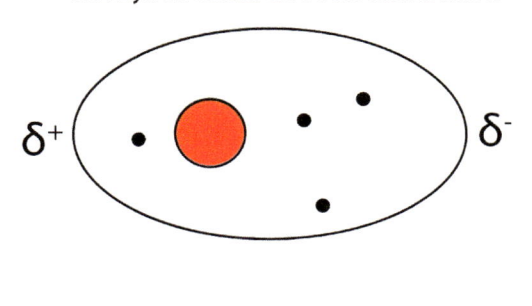

FIG. 12.3. The random nonsymmetrical distribution of electrons results in the increase in electron density on one side of the molecule versus the other side, producing London dispersion forces. (Courtesy S. E. Johnson and M. D. Mosher—© ASBC)

dipole–dipole interaction
A medium-strength intermolecular force between molecules with a dipole moment.

If a molecule has polarized bonds, there's a chance it could be a polar molecule. That chance is based on whether the molecule's bond dipoles add together or cancel out each other. As discussed earlier, the bond dipoles will cancel when the molecule is symmetrical. This implies that we must know the structure and the shape of the molecule to determine if its bond dipoles are symmetrical. In general terms, organic molecules may be symmetrical, but chances are they are not. In fact, the vast majority of organic molecules that contain more than just C and H atoms are polar. For example, consider a molecule of methanol (CH_3OH). Figure 12.5 shows the shape of this molecule. There are two bond dipoles in a methanol molecule, and both point toward the oxygen atom. Because the dipoles are at an angle to each other (because of their shape), they don't cancel out each other. The result is a molecular dipole moment. The methanol molecule is polar.

Unlike nonpolar molecules, polar molecules have a permanent molecular dipole. They readily orient themselves into a pattern where the partial negative end of one molecule is closer to the partial positive end of another, much like the structures shown in Figure 12.4. The interaction between two polar molecules is called **dipole–dipole interaction**. As the name implies, the dipoles of the two molecules are key factors in this force. Because the dipoles are permanent within the molecules, dipole–dipole interaction is much stronger than dispersion forces. These molecules will still experience the London dispersion force, but the effect will be minimal compared with that of dipole–dipole interaction. Figure 12.6 illustrates the dipole–dipole interaction between molecules of methanol.

The force of attraction between molecules can be strong enough to keep them together longer as the temperature increases. This results in increases in the boiling point and the melting point of polar compounds versus

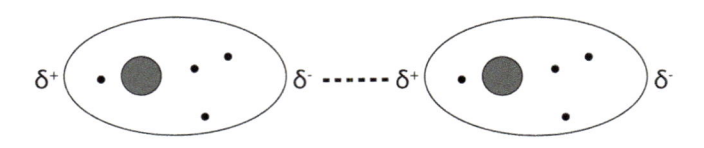

FIG. 12.4. The propagation of nonsymmetrical electron distribution from one molecule to another results in a weak force of attraction. (Courtesy S. E. Johnson and M. D. Mosher—© ASBC)

shape of methanol   bond dipoles   molecular dipole

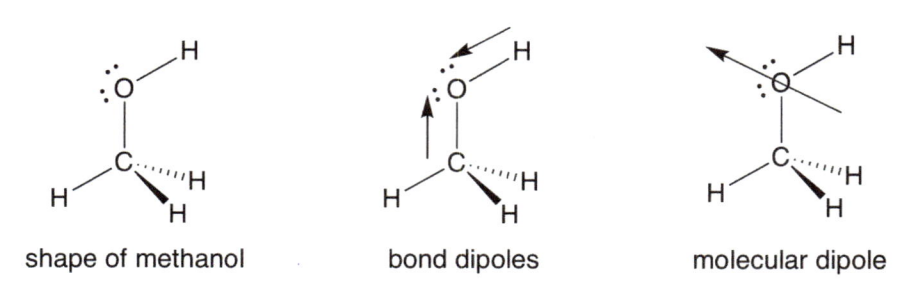

FIG. 12.5. The shape of methanol, its bond dipoles, and the overall molecular dipole. Because methanol has an overall dipole, it is polar. (Courtesy S. E. Johnson and M. D. Mosher—© ASBC)

similar nonpolar compounds. For example, consider propane, acetaldehyde, and ethanol:

- Propane has the lowest boiling point, because the molecule is composed only of carbon and hydrogen atoms. There are no bond dipoles in the molecule, so the molecule must be nonpolar. Only London dispersion forces hold adjacent molecules together in propane, and this weak force is easily broken as the temperature increases.
- The acetaldehyde molecule is composed of C, H, and O. It's a nonsymmetrical molecule, so there is a bond dipole between the C and the O and a net molecular dipole. The molecule is polar. This molecule has dipole–dipole interactions with adjacent acetaldehyde molecules. The boiling point increases because of that.
- Interestingly, the ethanol molecule is very similar to the acetaldehyde molecule and has a very similar molar mass. However, ethanol has a very high boiling point compared with acetaldehyde and propane. The ethanol molecule has dipole–dipole interactions, like the methanol molecule (see Fig. 12.6). But that alone isn't enough to account for the significant difference in boiling point. The ethanol molecule must have another force of attraction that's stronger than the dipole–dipole interaction.

Ethanol has a higher boiling point because of a unique type of dipole–dipole intermolecular force known as **hydrogen bonding**. Hydrogen bonding occurs when a hydrogen atom is bonded to a very electronegative atom, such as nitrogen, oxygen, or fluorine. Because of this polarization of the bond, the hydrogen atom exhibits a large partial positive charge to the neighboring atom. The large deficiency in electron density on the hydrogen atom increases the chance that a nitrogen, oxygen, or fluorine atom on an adjacent molecule will form a partial bond to the hydrogen atom. In other words, the hydrogen bond forms between an NH, OH, or FH on one molecule and an N, O, or F on an adjacent molecule. Figure 12.7 illustrates hydrogen bonding in methanol.

hydrogen bonding A strong intermolecular force between molecules containing an oxygen, nitrogen, or fluorine atom and a hydrogen atom attached to one of those atoms.

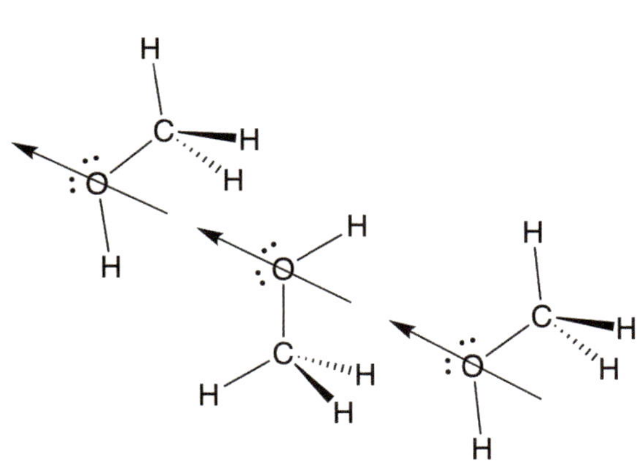

FIG. 12.6. Overlap of the molecular dipoles in methanol molecules results in a dipole–dipole interaction. (Courtesy S. E. Johnson and M. D. Mosher—© ASBC)

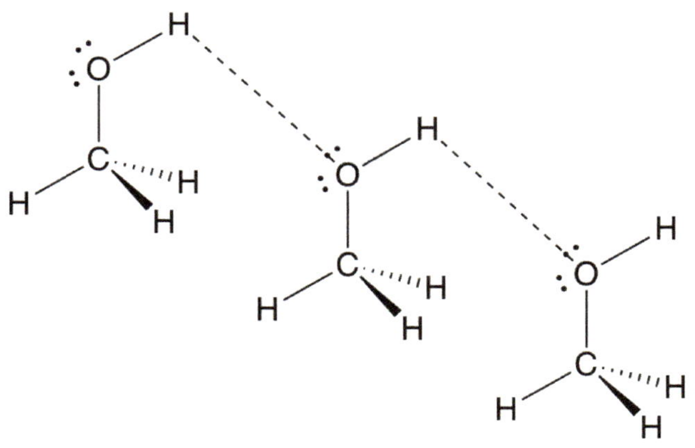

FIG. 12.7. Hydrogen bonding in methanol. The hydrogen bond is a special dipole–dipole interaction that shows a partial bond between adjacent molecules. (Courtesy S. E. Johnson and M. D. Mosher—© ASBC)

The hydrogen atom is partially bonded to both molecules in a hydrogen bond. In a sense, the hydrogen atom appears to be a "bridge" between the two molecules. The bond isn't a true bond, but the force of attraction between the two molecules is significant. Comparing the boiling points of acetaldehyde and ethanol demonstrates this point (Table 12.1).

While intermolecular forces are directly related to the polarities of molecules and their boiling and melting points, they are also related directly to how molecules are separated in chromatography. The interaction, however, is between the molecule and the mobile phase and the molecule and the stationary phase. If a compound has strong forces of attraction to the stationary phase, it will move more slowly through the chromatography system and take longer to reach the end. Similarly, if the compound has strong forces of attraction to the mobile phase, it will move quickly through the chromatography system.

Mobile and Stationary Phases

Suppose we are interested in analyzing the hop oils found in a beer sample. There are likely thousands of compounds within hop oils in addition to ethanol, residual sugars, and water. Many of the compounds, such as water and ethanol, have much higher concentrations than the oils we want to analyze. However, we happen to know that many essential oil compounds are composed mainly of carbon and hydrogen atoms, which means they will be much

TABLE 12.1. Relationships among shape, mass, and boiling point[a]

Molecule	Shape	Mass[b]	Boiling Point	Polar
Propane $CH_3CH_2CH_3$		44.10 g/mol	−42.00°C	No
Acetaldehyde CH_3CHO		44.05 g/mol	20.20°C	Yes
Ethanol CH_3CH_2OH		46.07 g/mol	78.24°C	Yes

[a]Courtesy S. E. Johnson and M. D. Mosher—© ASBC.
[b]g/mol = grams per mole.

more nonpolar than compounds such as ethanol (Fig. 12.8). We should be able to separate the mixture into the individual compounds. The question is, how can we physically make this happen?

Often, an initial step in performing the separation is to extract the molecules of interest from the base material or matrix. Performing this extraction will get most of the unwanted compounds out of the way. For example, many hydrocarbons can be extracted from beer by adding a solvent such as hexane and then swirling or shaking the mixture. After the mixture is left to sit, the hexane will rise to the top, where it can be removed. Hexane (C_6H_{14}) is a nonpolar compound. Each of the nonpolar compounds in the beer will have a much stronger attraction to hexane than to water; they will dissolve in the hexane after the treatment. If the hexane is then evaporated, the residue that remains will result from those nonpolar compounds in beer. This will allow separating the nonpolar compounds from molecules such as water and ethanol.

To separate the compounds via chromatography, the mixture is injected into the instrument. The compounds are carried through the instrument by the mobile phase, which passes through or by the stationary phase as it moves. This movement gives the molecules in the sample the chance to interact with the mobile phase or the stationary phase. The more time that a compound spends attracted to the stationary phase, the longer it will take to **elute** (or pass through the stationary phase), because the mobile phase is what must ultimately bring the compounds to the detector.

elute The process of moving analytes through a chromatography system.

Some analyses require performing a reaction on the mixture of compounds before using chromatography. Doing so is necessary to make the compounds "seen" by the detector. In addition to UV-vis, types of detectors include the flame-ionization detector (FID), mass spectrometer (MS), refractometer, and electron capture detector (ECD). Specific laboratory methods detail the types of columns and detectors that are required. However, many methods require a specific detector, and if the brewery lab doesn't have that detector, the method may not be able to be used in the lab's system. Some detectors are more sensitive to certain compounds. For example, it might be ideal to use an ECD to detect vicinal diketones (VDKs), although they can be detected by other detectors, such as an FID or MS.

Substantial time and research have been devoted to developing suitable methods for analyzing different compounds. However, there may be some analytes that haven't been explored. In those cases, we can use the chromatography system and develop the method ourselves. Unfortunately, method

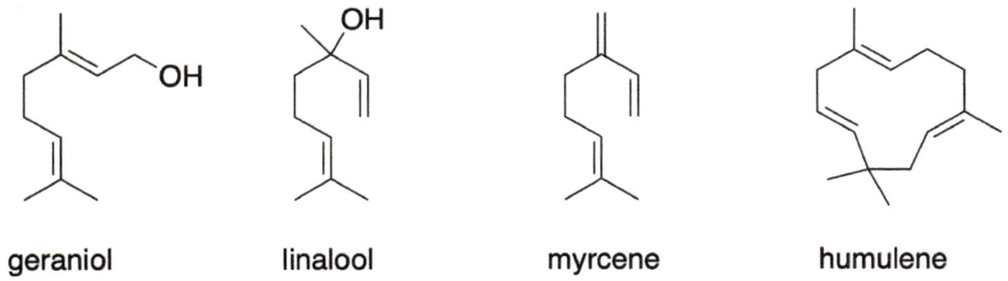

geraniol   linalool   myrcene   humulene

FIG. 12.8. Some of the hop essential oils that may be found in beer. (Courtesy S. E. Johnson and M. D. Mosher—© ASBC)

development is often a "Let's try it and see what happens" approach, which means revising and retrying the method until it works. It can take days or weeks to arrive at the best method.

The most important part of the chromatography system is the column. That's where the stationary phase is located and where the mobile phase must pass through to interact. The best choice for a column used for an analysis must meet these criteria:

- Doesn't react with compounds of interest
- Can withstand the mobile phase without dissolving
- Separates compounds of interest
- Is cost effective
- Performs the analysis in reasonable time
- Doesn't degrade quickly
- Provides reproducible results

The most important part of developing a new method is determining the best column to use. There can be polar, nonpolar, chiral, and various other stationary phases. And within each of these types, there are at least 100 subtypes. In liquid chromatography, commonly used stationary phases include silica gel and reverse-phase columns. Silica gel is a very polar stationary phase and works best if the mobile phase is relatively nonpolar. Reverse-phase columns are nonpolar and work best when the mobile phase is polar. Specific reverse-phase columns include C-18 and C-6 phases. Each of these codes is actually a count of the number of carbon atoms attached to a silica gel column. Like a tail, the long carbon chain points into the mobile phase and allows nonpolar compounds to interact. In gas chromatography, stationary phases can be packed into the column or adhered directly to the column, or silica gel can be used as a coating on the column and the stationary phase is then adhered to the silica gel. There are also many specialty stationary phases, such as the cyanopropyl silyl stationary phase (CN). This is a reverse-phase column that has a large amount of nonpolar character.

Unfortunately, we can't predict the exact order or time it might take for a compound to go through the column. Thus, if we are trying to develop a new method, it may take some time to find the correct mix of column and other conditions to perform the separation of the analyte in the most efficient manner. Luckily, as noted earlier, many experiments have already been published for the extractions and analyses of various compounds of interest in beer—some of which can be found in Chapter 15.

General Instrument Design

thin-layer chromatography (TLC) The application of chromatography to a thin plate containing the stationary phase and a pool of liquid that moves upward through the plate by capillary action.

All chromatographic instruments typically separate mixtures into their component compounds based on polarity. There are three main industrial versions of applied chromatography:

1. **Thin-layer chromatography (TLC)** is by far the cheapest application of chromatography and is a quick and relatively inexpensive process.

However, TLC isn't very useful for anything other than detecting a compound for a qualitative analysis. There are instruments that can read a TLC plate after it's developed, but they are expensive and less useful in the brewery laboratory than other methods of analysis. With appropriate method development, TLC can be used as an alternative method for some analyses. But TLC isn't practical for measuring subpart per million quantities of compounds.

2. **Gas chromatography (GC)** is unique, as it analyzes the components of a mixture in their gaseous states. This means that the mixture to be separated must be composed of **volatile compounds**: compounds capable of being converted into gases at temperatures less than 250°C. GC is often used for **headspace analysis**—a process that analyzes the compounds in the vapors above a sample under normal conditions.

3. **High-performance liquid chromatography (HPLC)** is widely used as a chromatography analysis system. Its greatest advantage is that nonvolatile compounds can be separated by using liquid to elute the compounds through the column. A disadvantage to using HPLC can be the cost of the instrument, but some GC systems can be just as expensive. Another disadvantage lies in the operating cost for HPLC. Given the solvents it uses and the wastes it generates, the cost of operating an HPLC often outpaces that of operating a GC.

gas chromatography (GC) A system that uses gas as the mobile phase.

volatile compound A chemical with a low boiling point.

headspace analysis The evaluation of an analyte by testing the vapors above the sample.

high-performance liquid chromatography (HPLC) A system that uses pressure to push a liquid through the system.

Identifying compounds using chromatographic methods can be challenging. If the detector can't determine the identity of a compound, the user must evaluate the sample to determine the identity. This can be done using the **retention time (RT)** of the analyte. The RT is the amount of time that the individual compound spends inside the chromatography system. A particular compound should reproducibly spend the same amount of time every time it runs through the system. Thus, if the user injects a known compound into the system by itself, the recorded RT value can be used to determine if the analyte exists in the mixture. In addition, if a known amount of that known compound is injected, the size of the signal that results can be used to determine the concentration of that compound in the mixture. For a mixture with a large number of unknown compounds, it could take a while to identify and quantify each compound.

retention time (RT) The amount of time an analyte spends in the chromatography system during separation.

Thin-Layer Chromatography

TLC is a great introduction to chromatography for any level of scientist. Although few methods utilize TLC in the brewing industry, understanding how TLC works can help the user understand GC and HPLC instruments, which are more commonly used. The instrument design is straightforward. TLC plates can be prepared on various backings, also called supports. Aluminum-, glass-, and plastic-backed sheets are common supports. The advantages and disadvantages of using each backing material are outlined in Table 12.2. The process that occurs during chromatography is very easy to observe on a TLC plate, especially if the compounds being separated are colored. TLC analyses are relatively inexpensive compared with the costs of using other instruments.

TABLE 12.2. Advantages and disadvantages of thin-layer chromatography backing materials[a]

Material	Advantages	Disadvantages
Glass	Rigid	Thick
	High chemical and heat resistance	Breaking hazard
		High weight
	Transparent	Takes up more room
Aluminum	Thin	Low chemical resistance
	Low weight	Not transparent
	Not fragile	
	Can be cut with scissors	
Plastic	Thin	A bit transparent
	Not fragile	Low heat stability
	Can be cut with scissors	Stationary phase can crack if bent
	Lower weight than glass	

[a] Courtesy S. E. Johnson and M. D. Mosher—© ASBC.

eluent The liquid used as the mobile phase in TLC.

retention factor (R_f) The ratio of the time an analyte spends versus the time that an unretained compound spends in the chromatography system.

peak broadening The effect of diffusion during a chromatographic process that causes the analyte to spread out.

Most TLC plates can be ordered with the stationary phase already adhered to the support. Alternatively, there are published methods for preparing your own TLC plates from raw materials.

To perform a TLC analysis of a mixture, the first step is to dissolve the mixture in a small amount of a volatile liquid. Using a thin, hollow tube, a small amount of the resulting solution is then spotted onto the TLC plate about 15% away from the edge of the plate. Capillary action transfers the mixture from the tube to the plate. The plate is then placed vertically into a solvent system (the mobile phase), where capillary action carries the solvent up the plate. In TLC, the mobile phase is often referred to as the **eluent**. The compounds in the mixture are carried up the plate by the mobile phase at different rates according to their degrees of interaction with the stationary phase. If the substance is placed too low on the plate, it will dissolve into the eluent and render the analysis useless.

After the eluent reaches the top of the plate, the plate is removed and placed on the benchtop to air dry. The compounds on the plate are then visualized with an ultraviolet lamp (if they are able to absorb ultraviolet light) or sprayed with a chemical that changes color after reacting with the compounds. An example is shown in Figure 12.9.

The most common data reported from a TLC analysis is the **retention factor (R_f)**. The R_f value for a specific compound will remain the same no matter what the starting material is, as long as the stationary phase, mobile phase, and temperature of the analysis are the same. Using the R_f is a good presumptive test to determine whether a compound is present in the solution.

Some compounds exhibit **peak broadening** in TLC, as well as in GC and HPLC analyses. The source of peak broadening is diffusion within the

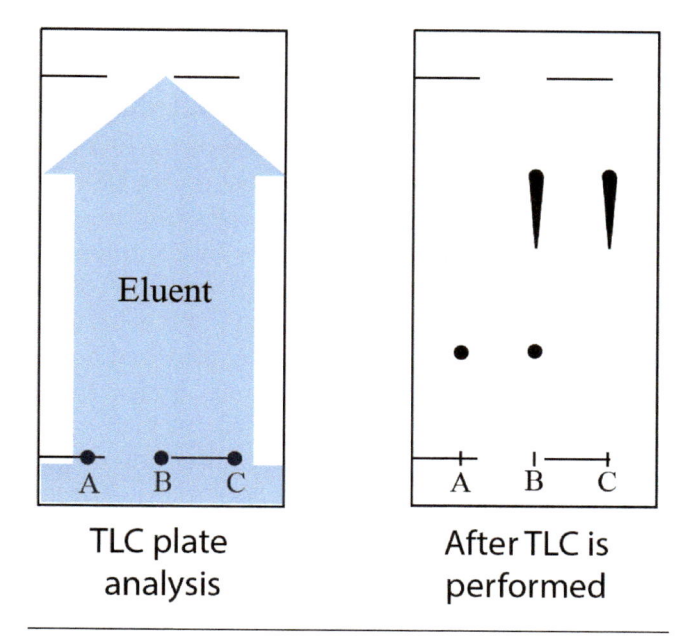

FIG. 12.9. TLC plates before (left) and after (right) analysis is complete. Lane B contains a mixture of the compounds found in lane A and lane C. The compound in lane C exhibits tailing on the plate. (Courtesy S. E. Johnson and M. D. Mosher—© ASBC)

stationary phase. This diffusion can be observed by placing a drop of food coloring on a plate that has a pool of water in the bottom. The food coloring will slowly expand across the plate (which remains stationary). The same diffusion occurs with compounds as they are analyzed with chromatography. As the compound interacts within the mobile and stationary phases, it diffuses in all directions. As the compound moves up the TLC plate, the spot gets bigger. This can be observed in GC and HPLC, as well. The peaks in these analyses become broader as the compound spends time in the system; this can't be prevented. Using smaller drops when putting the sample on the plate will help lessen this effect.

Tailing is another effect that can be seen on the TLC plate. Tailing is caused by the path the compound takes through the stationary phase, and it isn't always uniform. As some of the compound takes a quicker path up the plate, the rest of the compound is left behind. Sometimes, tailing can be resolved by making minor adjustments in the solvent system; in other cases, it's just an unavoidable result of using TLC for that specific analysis.

tailing The retention of an analyte behind its main mass as it moves through the chromatography system.

Gas Chromatography

GC is a widely used instrumental method for separating volatile compounds (Fig. 12.10). A GC analysis begins with loading a sample into a syringe. The syringe is then inserted into the **injection port** and the sample is introduced. The injection port is kept at a very hot temperature (200–250°C) so the entire sample is immediately vaporized.

injection port The location that the sample is added to the chromatography instrument.

GC column A tube containing the stationary phase used in gas chromatography.

carrier gas The mobile phase in a gas chromatography analysis.

column oven A heater used to warm the gas chromatography column.

The vapors are then carried through the **GC column** via a **carrier gas** (usually, hydrogen or helium). The carrier gas is the mobile phase in a GC analysis. The column in the GC is loaded with the necessary stationary phase for the analysis. This column is situated inside an insulated **column oven** that has a controllable temperature, often ranging from room temperature to more than 400°C. Using the oven helps keep the vapors in a gaseous state. Using it can also speed up the analysis, as higher temperatures reduce the degrees of interaction of the compounds with the stationary phase. Figure 12.11 shows the inside of a GC oven that has two GC columns.

GC columns are hollow tubes of varying lengths that have been packed with an internal stationary phase. The stationary phases for GC are much more diverse than those for TLC, because GC is much more efficient at separating and analyzing compounds both qualitatively and quantitatively. Therefore, manufacturers are especially interested in producing GC columns with stationary phases specific to different analytes (Fig. 12.12). The same stationary phases could be adhered onto a TLC plate, but it would be

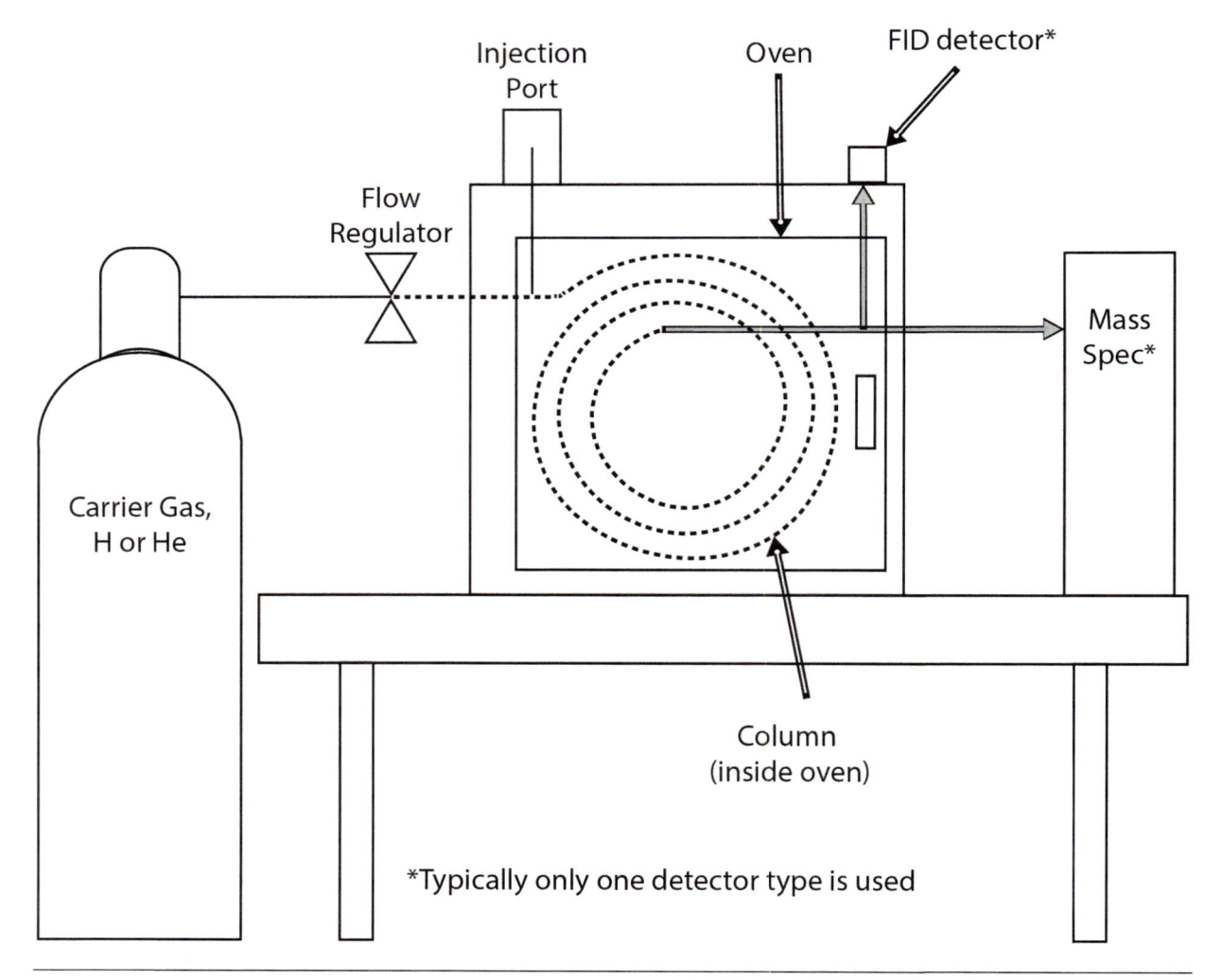

FIG. 12.10. Parts of a gas chromatography instrument. (Courtesy S. E. Johnson and M. D. Mosher— © ASBC)

impractical to perform many analyses using this method. The stationary phases within a GC column are typically made of a porous, solid material that has been adhered within the long, hollow tube. The tube can be made of stainless steel, copper, or glass. The most common column is a glass capillary column that takes on the appearance of copper. The tubing is then wound up so that lengths up to 100 meters can be made to fit into the GC oven.

Depending on the compounds being separated, different columns might be needed for one analysis versus another. Being able to make these changes requires having someone on staff who knows how to replace the columns. There are more complicated stationary phases, but they won't be explored in this text. In short, these stationary phases can separate certain compounds in ways that are unique to specific columns. If a method requires an expensive or unique column, it's likely because the compounds of interest are difficult to separate without the specialized environment that the column provides.

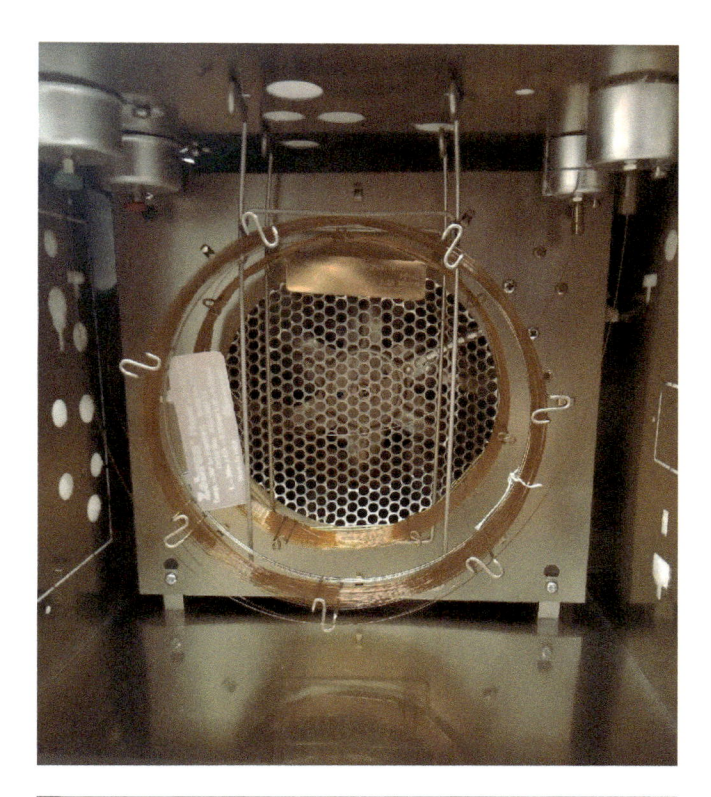

FIG. 12.11. Two GC columns in a GC oven. The injector ports are the silver "cans" in the top left of the oven, and the detector ports are the "cans" in the top right of the oven. The columns are copper-colored, capillary glass columns that run from the injector to the detector. The columns in this oven are 30 meters long. (Image of Zebron™ ZB-WAX reproduced by permission of Phenomenex. Photo courtesy S. E. Johnson and M. D. Mosher—© ASBC.)

Once the sample has passed through the column and been separated into its individual analytes, each molecule will be detected. Quite often, gaseous compounds that have been separated are analyzed with a mass spectrometer, or MS. This type of instrument is commonly referred to as a GC-MS. An MS reports the molar mass of each component in the mixture and the masses of pieces of each compound. It does this by fragmenting the molecule into charged particles, detecting each of the fragments, and reporting them in the ratios in which they exist after fragmentation. Each molecule will fragment in a predictable fashion. This means that the MS spectrum for each molecule is as unique as a human fingerprint. Coupled with a database of known compounds, the MS spectra of a sample can be matched to the spectra of a known compound. Using this information, the MS detector can suggest potential structures of the molecules.

Electron capture detectors, or ECDs, detect the different analytes via a current created by excited electrons in a molecule. When a molecule leaves the column, it's battered with excited electrons, known as β-radiation, typically generated by radioactive nickel-63 within the detector. When a molecule is hit by β-radiation, its own electrons become dislodged. These dislodged electrons are then detected as a change in the current at an anode within the device. In other words, as analytes leave the column, the ECD reports a

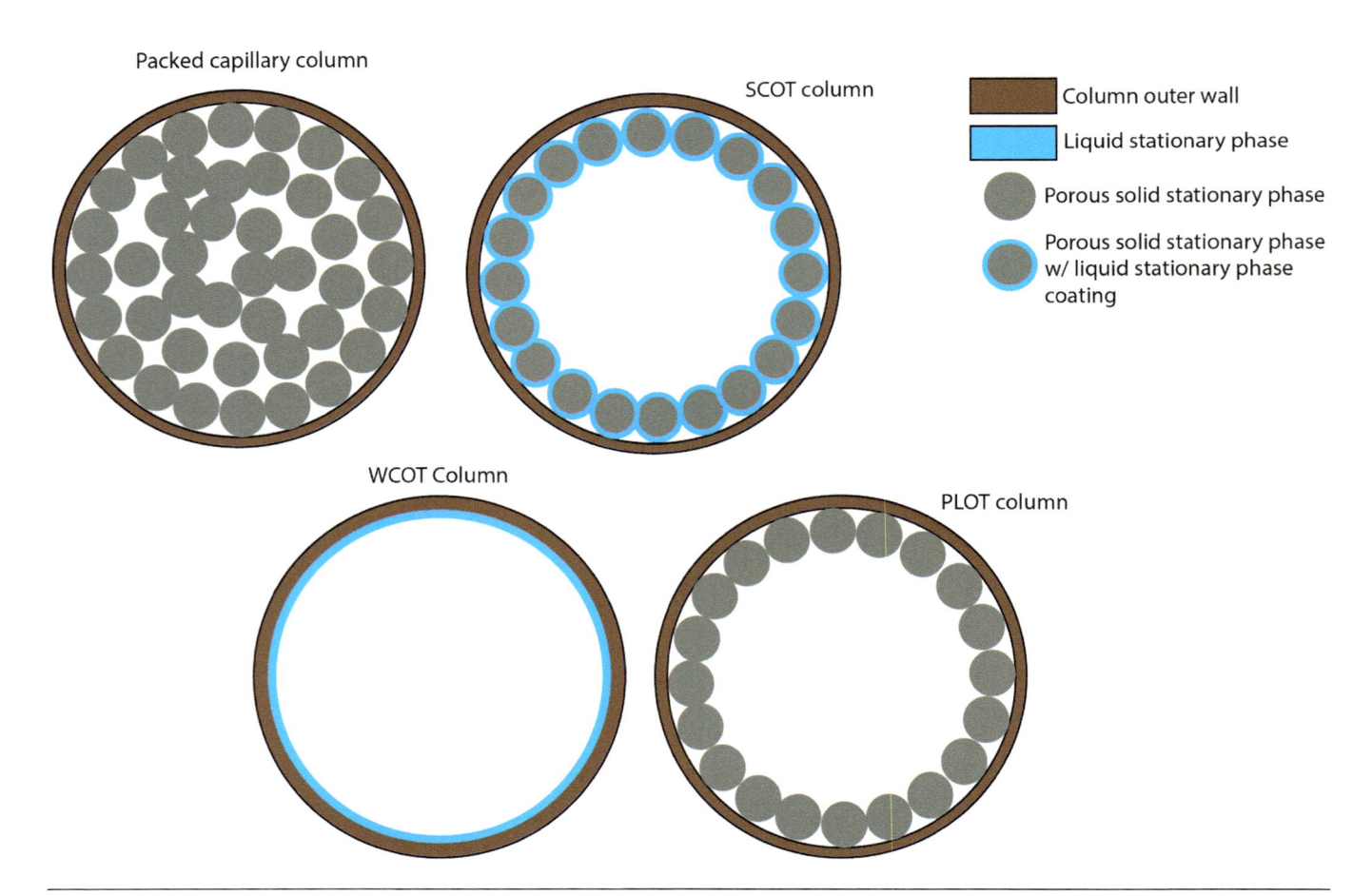

FIG. 12.12. Cross section of GC columns and types of supports. (Courtesy S. E. Johnson and M. D. Mosher—© ASBC)

change in electrical current. ECD detectors are commonly used to monitor diacetyl because of the sensitivity of this detection method.

Another detector used in GC analysis is the flame-ionization detector, or FID. This detector ignites the molecules as they exit the GC column. As hydrocarbons are burned, they produce ions that are detected by the FID. The quantity of ions produced is proportional to the concentration of the compound leaving the GC column. The advantages and disadvantages of using MS, FID, and ECD detectors are summarized in Table 12.3.

The other important consideration when using GC (or HPLC) for the quantitative analysis of a sample is the **internal standard**: a compound that the user has added in a known concentration to the sample. This additional compound is chosen so that it has an RT that doesn't match that of any other compound in the sample. The internal standard is added in a known quantity that's roughly similar to the expected range of the analytes to be measured. Doing so ensures that the area of the peak on the chromatogram can be related to the concentration of the compound; then the quantities of the other compounds in the sample can be determined. Using an internal standard is necessary, because even with extremely accurate automated sample injection, the areas of the peaks representing different compounds can vary greatly. This variation can be easily eliminated by using an internal standard.

GC is a great instrument for many chromatography analyses. The downside is that GC works best when used to analyze compounds that are volatile or semivolatile. This limitation can be overcome by modifying the sample with a chemical reaction. The reaction is usually designed to add large

internal standard A compound added to a sample in a known amount; used to measure the quantity of the analyte in the sample.

TABLE 12.3. Advantages and disadvantages of types of gas chromatography detectors[a]

Detector	Advantages	Disadvantages
Mass spectrometer (MS)	Provides information about molecule structure Large databanks available to identify analytes Can detect analytes for which no standards are available	More expensive Requires more expertise to analyze Requires use of a vacuum pump More involved maintenance than other detectors
Flame-ionization detector (FID)	Very high sensitivity to combustible/ionizable compounds Resulting spectra is simple to analyze if a pure standard is available Fairly robust	Uses hydrogen gas to support flame Requires an open flame Sensitivity varies with each analyte
Electron capture detector (ECD)	Highly sensitive Resulting spectra is simple to analyze if a pure standard is available Good at detecting halogenated compounds	More expensive Limited range of detection Sensitivity is related to type of carrier gas used

[a] Courtesy S. E. Johnson and M. D. Mosher—© ASBC.

hydrocarbon groups to an –OH, –C=O, –NH, or other susceptible reaction site within a molecule. The result significantly reduces the polarity of the molecules. With a reduction in polarity, the boiling point of the molecules drops. The main risk involved in modifying the sample is assuming that the reaction will go to completion and all the compounds in the sample will react at the same rate. If the reaction conditions are not met, the reaction can have lower yields, which will skew the data for the GC analysis. If the only chromatographic instrument available is a GC, this may be the only way to perform some analyses.

High-Performance Liquid Chromatography

HPLC can be set up to separate almost any mixture of nonvolatile compounds. HPLC involves a pump that pushes liquid at relatively high pressure through a column packed with the stationary phase. The sample is injected inline and then carried through the stationary phase to the detector (Fig. 12.13).

As for GC, a plethora of columns can be used for HPLC to get the job done. However, unlike GC, the mobile phase in HPLC analysis is much more flexible. The solvent that makes up the mobile phase in HPLC analysis can be mixed using the pump in the instrument. In addition, the pump can change the amount of each solvent as the analysis is being performed. This is known as a **gradient elution** or **gradient HPLC**. In GC analysis, a similar scenario occurs when the oven temperature is altered, but this doesn't have the same amount of control that's available in HPLC. By altering the mobile phase throughout the analysis, it's possible to move certain compounds along the stationary phase with one mobile phase and then switch to another that moves other compounds. The speed of the mobile phase can also be increased to speed up the HPLC analysis. Doing so involves starting with a mobile phase that will separate the compounds but not move them very far

gradient elution or **gradient HPLC** The change in the polarity of the mobile phase during a chromatographic separation.

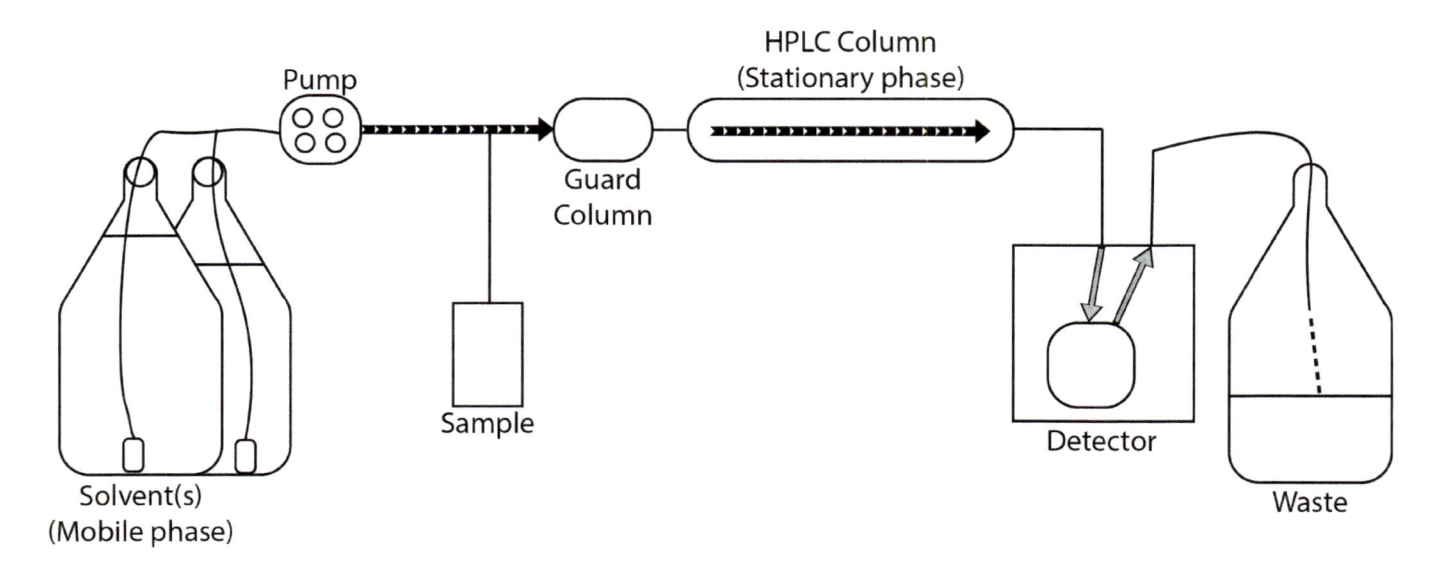

FIG. 12.13. Parts of a high-performance liquid chromatography (HPLC) instrument. (Courtesy S. E. Johnson and M. D. Mosher—© ASBC)

on the column and then switching to another mobile phase that moves all the compounds out of the instrument more quickly. An example of the effect of gradient elution on RT is illustrated in Figure 12.14.

HPLC works similarly to GC: The sample is injected into an injector and carried to the column using the mobile phase. Once in the column, the compounds are separated and passed through to the detector. HPLC offers a wide variety of detectors (more than will be discussed in this text). The

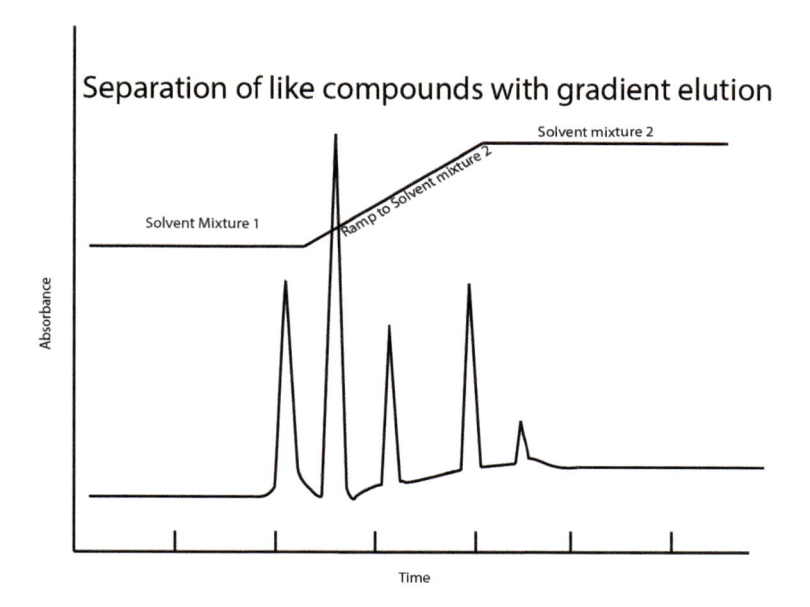

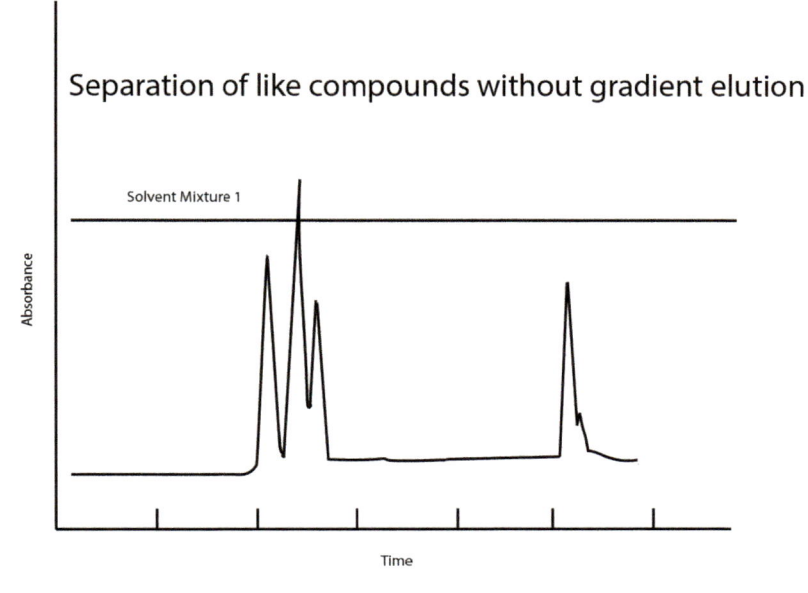

FIG. 12.14. HPLC gradient elution effect. The sample contains five analytes. In the chromatogram without gradient elution (bottom), the compounds are not well separated, and the total time for the separation is quite long. The chromatogram of the gradient elution (top) illustrates that the analytes are separated and have moved through the system more quickly. (Courtesy S. E. Johnson and M. D. Mosher—© ASBC)

detector must be capable of analyzing the desired compounds of interest. The detector could be a spectrometer, MS spectrometer, or more complicated or delicate detection instrument. Knowing what detector an analysis requires is important for the same reasons stated earlier about GC detectors.

In summary, chromatography is an amazing analytical tool for any laboratory. The importance of separating a mixture into individual compounds and then analyzing those compounds both qualitatively and quantitatively can't be underestimated. Many analyses simply can't be done without these instruments. The disadvantages of using them are that the initial cost is high and that they usually require a well-qualified scientist for operation and maintenance. However, when it comes to the quality of a beer, there's almost nothing that an HPLC or GC can't tell us.

We don't recommend jumping straight into chromatography when starting a brewery laboratory, because the concepts and equipment require a lot of knowledge for efficient use and operation. As a brewery grows, however, using chromatography will allow very accurate analysis and detection and help ensure that the quality of the beers produced is always up to the brand's standard.

CHAPTER 13

Other Instruments in the Brewery Laboratory

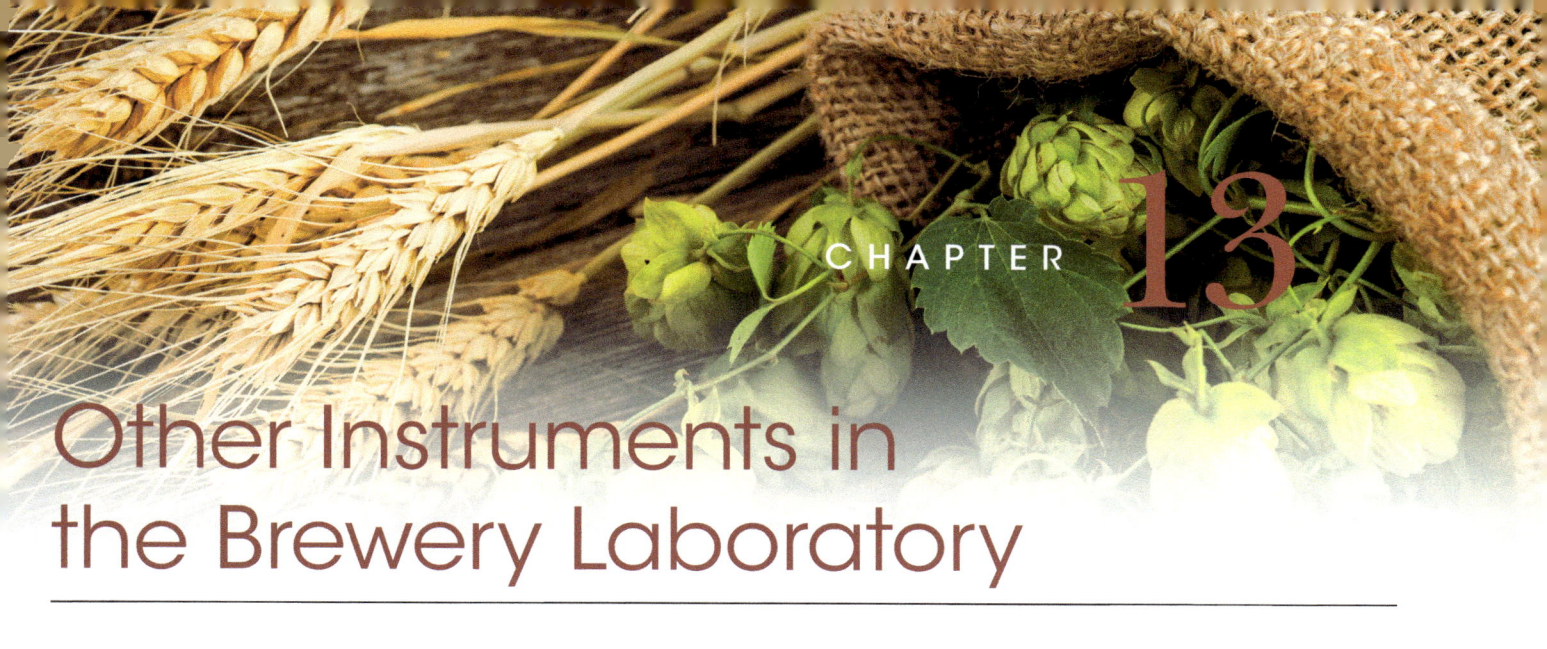

Spectroscopic and chromatographic instruments have a variety of uses in the brewery laboratory. In fact, a majority of analyte-specific methods require using these instruments. Yet other instruments and techniques are just as valuable, even though they aren't as diverse in terms of their capabilities. Many are essential to accurately analyzing the product stream. These other instruments don't rely on the principles and theories that we have already uncovered. Instead, they tend to focus on measuring physical properties of a sample. Because they analyze the sample in bulk, they are not analyte specific. Our exploration of these instruments will focus on the techniques that are unique to the brewing industry. We will discover how these instruments are used in the brewery laboratory and how they are important to beer quality.

Density Meters

In Chapter 1, we discussed the concept of **specific gravity** as an estimate of the sugar concentration in wort. As it relates to the brewery laboratory, the specific gravity of a solution is the density of the solution divided by the density of pure water—the solvent used to make the solution. At sea level and a temperature of 4°C (39°F), the density of pure water is 1,000 kilograms per cubic meter (kg/m^3) (or 1 gram per milliliter [g/mL]). Under these conditions, the density of a solution and its specific gravity are equivalent.

We also noted earlier, however, that the density of a solution doesn't explicitly determine the concentration of sugars. Specific gravity is a measure of the effect of all the compounds dissolved in the solution on the density of the solution. In other words, multiple solutions can have the same density but be vastly different in makeup. For example, one could be a sugar solution, another could be a salt solution, and another could be a solution that's a mixture of both salt and sugar. In the brewery, however, the major solutes within a solution that might be measured are sugars, ethanol, or a mixture of the two. If we limit our analysis to solutions that contain either

specific gravity The density of a solution divided by the density of pure water.

sugar or ethanol and avoid the mixtures, the density of the solution can serve as a good estimate of the concentration of sugars. Of course, many different solutes can be present in relatively large amounts in the brewery. Thus, any measure of the density of a solution should be an estimate of a concentration only. The brewery laboratory should never rely on a density measurement as an accurate report of the concentration of a particular solute.

Although this issue alone should discourage the use of density to determine the concentration of anything other than a sample containing only sugars or only alcohol, an additional issue complicates the measurement of the density. The temperature of the solution results in changes to the density. For example, at 25°C (77°F), the density of water is 0.997 g/mL (997 kg/m^3). The difference from the density at 4°C isn't very significant, and the difference is slight enough that there is little deviation in the estimate of the sugar concentration. For example, wort with a specific gravity of 1,060 kg/m^3 has a density of 1,056 kg/m^3 (1,060 kg/m^3 ÷ 997 kg/m^3) at 25°C. The change in temperature could translate into an error of about 0.5% alcohol by volume (ABV) in the finished beer if we used the density of wort alone to perform the calculation of ethanol in a sample. This is a large error and worth correcting, but it's relatively small for such a large temperature difference. Yet if we measure the density of the same wort when it's 71°C (160°F) and the density of pure water is only 977 kg/m^3, the density will be 1,035 kg/m^3. This could translate to a significant error of 3.3% ABV in the finished beer.

Many breweries brew high-gravity beers and dilute them to arrive at the reported % ABV for a particular brand. In addition, the temperature of the wort often changes throughout the brewing process. This is why specific gravity is measured and reported instead of density. Specific gravity is more accurate, because it's a ratio of the densities of the solution and the solvent at a specific temperature. It results in a more accurate estimate of the sugar concentration in a solution. The specific gravity of a solution won't change as the temperature changes, because the value of specific gravity is a relative measurement.

density meter An instrument that measures the density of a liquid.

densitometer An instrument that measures the optical density of a sample but doesn't physically measure the density of a solution.

A **density meter**, or densimeter, is an instrument that measures the density of a liquid. Many densimeters include a thermometer so the instrument, or user, can calculate and report the specific gravity of the solution. Densimeters should not be confused with **densitometers** (although the two are sometimes used interchangeably), which measure the optical density of a sample but don't physically measure the density of a solution. The optical density of a sample may be able to be related to the density of a solution, but the relationship will need to be based on some external calculation.

Many different types of densimeters are used in the brewery laboratory. In this section, we will explore three of the most common: hydrometers, gravimetric density meters, and Coriolis density meters, also known as mass-flow meters.

Hydrometers

hydrometer A simple instrument that can measure the density of the liquid in which it's placed based on the instrument's buoyancy.

A **hydrometer** is a simple instrument that can measure the density of the liquid in which it's placed (Fig. 13.1). It has been calibrated by the manufacturer to measure the density range listed on the instrument with precision and accuracy. A hydrometer consists of a sealed glass tube containing weights at the bottom. The top of the instrument is a thin, hollow tube with a scale inside that reports the density of the liquid based on the buoyancy of the hydrometer when it floats in the liquid. When a hydrometer is placed in a solution, the large part of the tube sinks below the surface of the liquid and the narrow part stays above the solution. Typically, the user fills a glass cylinder, such as a graduated cylinder, with the solution to be measured. The hydrometer is then carefully lowered into the solution and spun with the fingers, like a top. Doing this ensures that the hydrometer is free of the sides of the cylinder. Then the level of the solution is measured against the scale inside the narrow part of the hydrometer. The top of the meniscus indicates the density of the solution. Remember that when using a hydrometer, the user must read the top of the meniscus instead of the bottom (Fig. 13.2). This is the opposite of reading other scientific glassware, for which the bottom of the meniscus is used.

A hydrometer is a very cost-efficient instrument to use in the brewery, as it can provide quick and reliable results and requires minimal training to use. The value reported from the hydrometer is the density, so a calculation to provide the specific gravity is required. That calculation can be done by measuring the temperature of the solution and then looking up the density of water at that temperature (Table 13.1). Dividing the density of the solution by the density of water reveals the specific gravity of the solution.

For example, let's assume that we collect a sample of wort at 32°C (89.6°F). A hydrometer is placed in the solution, and careful examination of the scale shows the density to be 1.052 g/mL (1,052 kg/m^3). The density of water at that temperature is 995 kg/m^3 (see Table 13.1). Calculating the specific gravity of the solution involves dividing 1,052 kg/m^3 by 995 kg/m^3. The specific gravity of the solution is 1,057 kg/m^3. Note that while the specific gravity is a

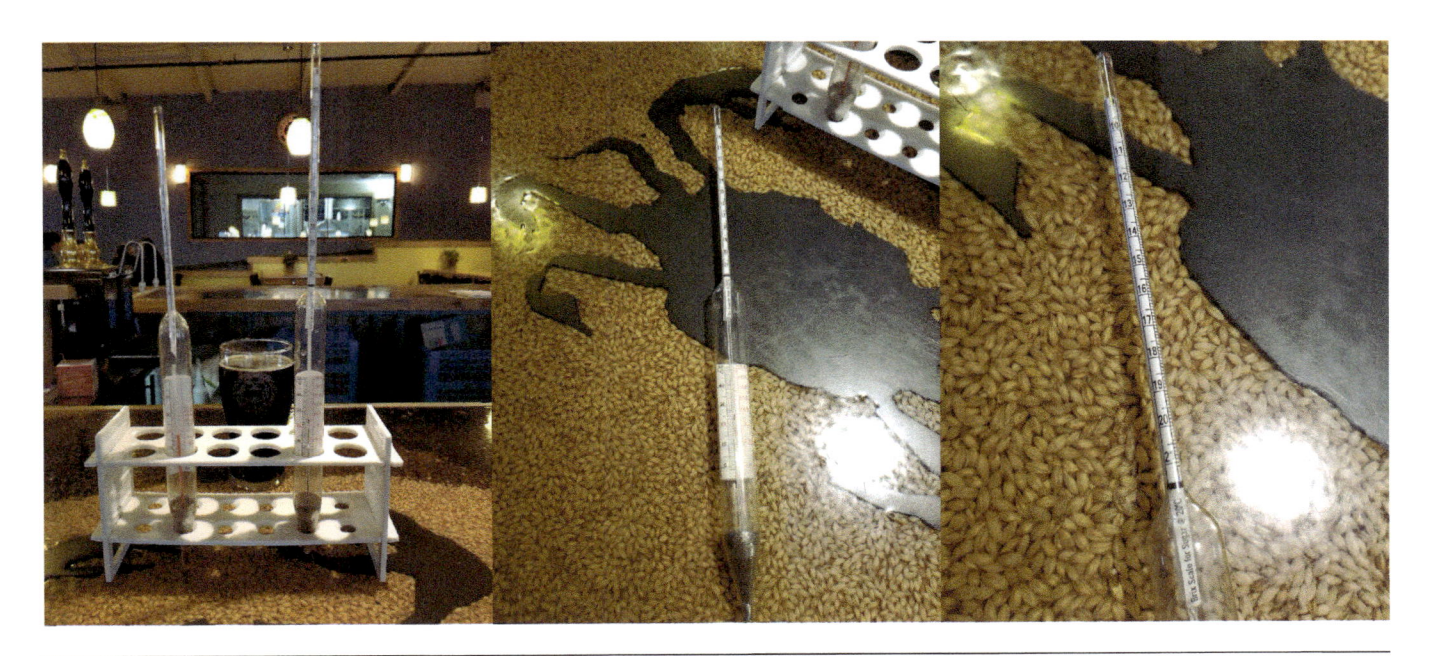

FIG. 13.1. Types of hydrometers. (Special thanks to Crabtree Brewing Company, Greeley, Colorado, for allowing us to photograph their hydrometers. Photos courtesy S. E. Johnson and M. D. Mosher—© ASBC.)

ratio of the density of the solution to the density of water, the specific gravity should not possess units (although it's often reported with units):

$$\text{specific gravity} = \frac{1{,}052 \frac{\text{kg}}{\text{m}^3}}{995 \frac{\text{kg}}{\text{m}^3}} = 1{,}057 \text{ kg/m}^3$$

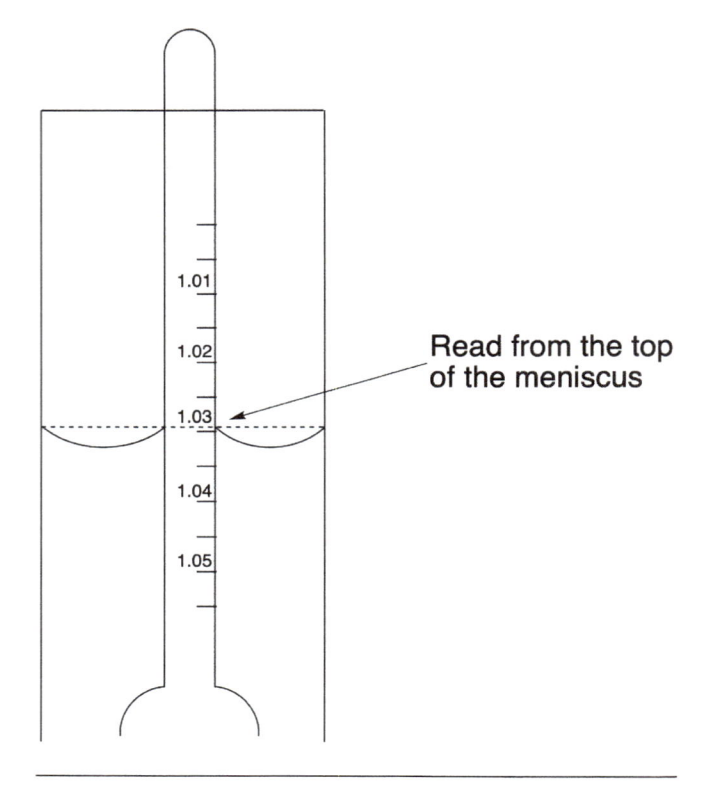

FIG. 13.2. Reading density from a hydrometer. The top of the meniscus is used to read the value of the density using a hydrometer. (Courtesy S. E. Johnson and M. D. Mosher—© ASBC)

TABLE 13.1. Density of pure water at selected temperatures[a,b]

Temperature (°C)	Density (kg/m³)	Temperature (°C)	Density (kg/m³)
4	1,000	36	994
8	1,000	38	993
12	999	42	991
16	999	46	990
20	998	50	988
24	997	54	986
28	996	58	984
32	995	62	982

[a] Courtesy S. E. Johnson and M. D. Mosher—© ASBC.
[b] Each density reported here is to the unit's place. More detailed densities can be found by searching the internet or by measuring the density in the laboratory at a specific temperature.

Gravimetric Density Meters

A **gravimetric density meter**, or densimeter, uses a hollow, flexible tube shaped like the letter "U" to measure the density of a liquid (Fig. 13.3). A beam of light is directed across the U-bend in the instrument. As the distance between the two halves of the U-bend changes, the beam is deflected by different amounts. The instrument is incredibly sensitive so that it can measure very small deflections in the beam of light. The result is that the instrument can provide highly accurate measurements of density. The sensitivity of the instrument also means that it's expensive, however.

The deflection of the beam correlates to the mass of the liquid being analyzed. This is because the mass of the liquid affects how the U-bend in the tube spreads apart. Liquids with higher masses will spread the ends of the U more. Density is a measure of the mass divided by the volume of the sample; the instrument can determine the density, because the tube contains a very well defined volume of liquid. In practice, a sample is obtained from the brewery floor and then pulled or pushed into the instrument. The deflection of the U-tube is determined and the instrument displays the density in a unit that's programmed into it. The user then converts the density to

gravimetric density meter
An instrument with a hollow, flexible, U-shaped tube that measures the density of a liquid.

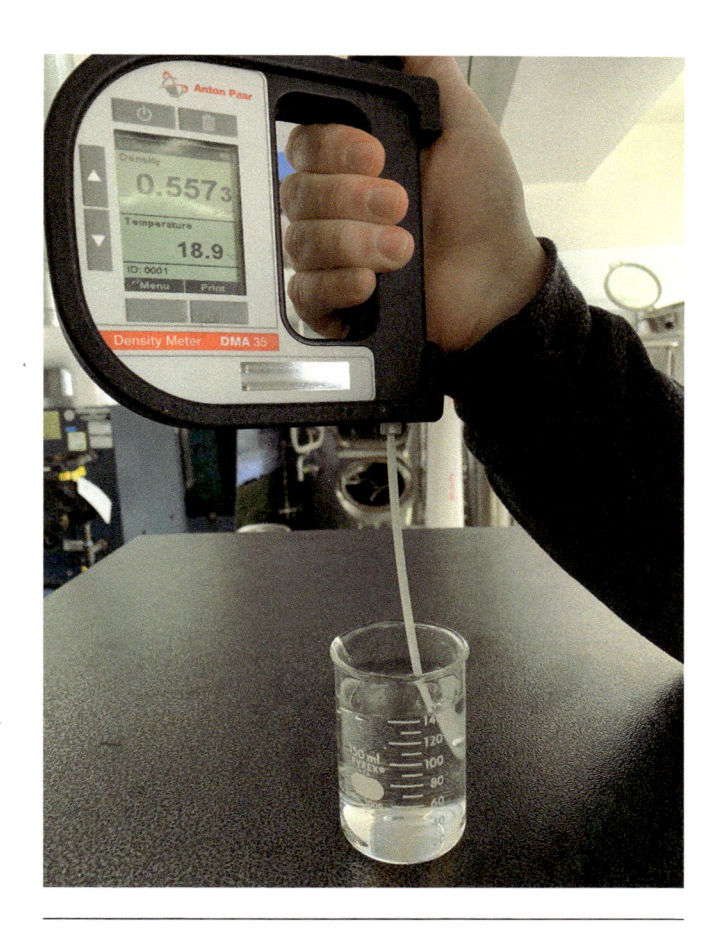

FIG. 13.3. A portable gravimetric densimeter. (Image of densimeter reproduced by permission of Anton Paar. Photo courtesy S. E. Johnson and M. D. Mosher—© ASBC.)

the specific gravity of the liquid using the temperature of the sample. Some instruments perform that calculation automatically and report the specific gravity directly.

Coriolis Density Meters

Coriolis density meters, or densimeters, are often referred to as mass-flow densimeters. These instruments are great for in-line measurement of the density of a liquid. There are two types of mass-flow densimeters:

1. In rotation mass-flow analysis, the liquid is passed through a uniquely designed curved tube. As the liquid runs through the tube, it generates a centrifugal force against the tube based on the velocity of the liquid. This causes the tubing to deflect slightly. Liquids with a greater mass have a greater centrifugal force as they move and deform the tube more. The flow rate of the liquid through the tube causes a similar effect; faster flow rates cause the tube to deform more. The amount of the deformation of the tube is measured by the instrument. When the instrument is coupled with a sensor that measures the velocity of the liquid, the density can be calculated. Alternatively, if the density of the liquid is known, the instrument can be reconfigured to calculate the velocity of the liquid through the tube.
2. Parallel flow mass-flow densimeters work in a similar manner. The liquid is directed into two parallel U-tubes separated by a small distance, and the tubes are vibrated so they move back and forth. In the absence of a moving liquid, the tubes vibrate regularly. When a liquid is moving in the tubes, it causes oscillation in their movement (Fig. 13.4). The amount of distortion is related to the flow rate and the density of the liquid. Thus, if the flow rate is known, the density can be calculated and vice versa.

Coriolis density meter An instrument made of two U-shaped tubes that oscillate; the degree of deviation in the oscillation is related to the mass and flow rate of the liquid.

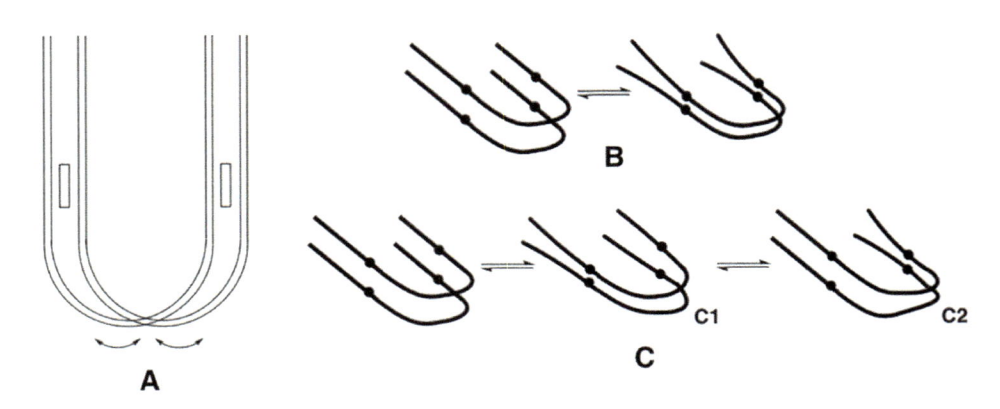

FIG. 13.4. Parallel flow mass-flow densimeter. **A,** The meter is made of two U-shaped tubes separated by a given distance. The tubes oscillate so that they move together and then separate with a regular motion. **B,** When the U-tubes are empty or the liquid inside them isn't moving, the distance between the sides of each U-tube oscillates similarly. Compare the distances between the spheres (•) on each tube to verify this fact. **C,** When the liquid inside is moving, the distance between the sides of each U-tube differs during each oscillation. The degree of deviation in the oscillation is related to the mass and flow rate of the liquid. (Courtesy S. E. Johnson and M. D. Mosher—© ASBC)

Refractometers

Another way to estimate the density of a solution is by using a **refractometer**: an instrument that measures the refraction of light as it passes through a sample of the solution (Fig. 13.5). **Refraction** is essentially the change in direction of a beam of light as it passes from one medium to another. We notice refraction when we look at a stick that we have inserted into a glass of water. The end of the stick doesn't line up with our perception of where it should align. The reason for the deflection is the refraction of light. Figure 13.6 illustrates the refraction of light that comes from the stick under the water. As the light moves up toward our eyes, it leaves the water and changes direction.

The degree of refraction is related to the concentration of materials dissolved in the water sample. Often, we compare the refraction of a sample to pure water. Because of this, we can determine the sugar content in a wort sample. This is the same analysis that's done with specific gravity. In practice, instead of reading the refractive index of a solution and then comparing the result against a table that relates the percentage of sugar in the solution, the scale of a typical refractometer is often written in other units.

refractometer An instrument that measures the refraction of light as it passes through a sample of the solution.

refraction The change in direction of a beam of light as it passes from one medium to another.

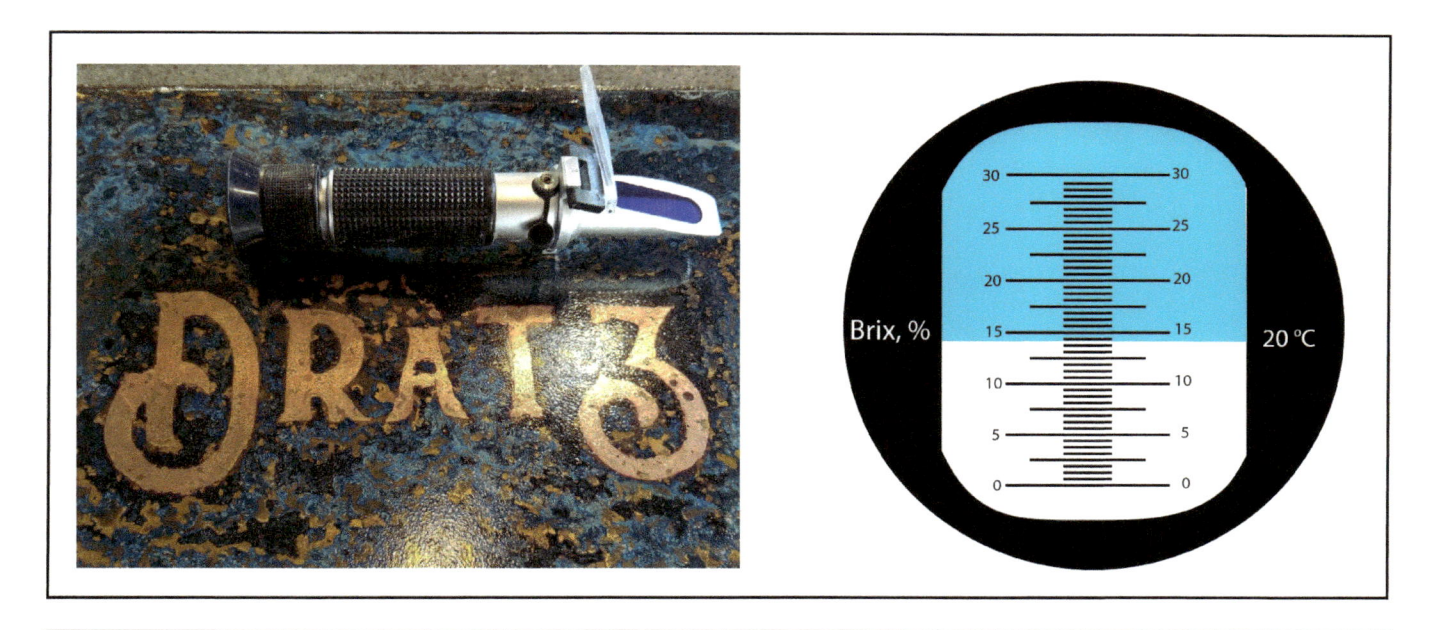

FIG. 13.5. Example of a handheld refractometer and the scale inside. (Special thanks to Dratz Brewing Company, Loveland, Colorado, for allowing us to photograph their refractometer. Photos courtesy S. E. Johnson and M. D. Mosher—© ASBC.)

A refractometer is simple to use. A drop of the sample to be measured is placed on the instrument, and the sample door is then closed and held up to the light. By looking through the eyepiece, the user can see a scale and light and dark areas. When there's water in the sample chamber, almost the entire window inside the instrument is dark. In other words, the light is refracted by the water so that it reads 0 on the scale. If the sample is wort, the light portion of the window gets bigger, because the light is refracted more than the water. The horizontal region in which the light and dark areas meet lines up along the scale and provides the measurement of the sample.

Some instruments report the scale in degrees Brix, others report the scale in degrees Plato, and still others report the index of refraction. Correlation of these units to the specific gravity of the solution is provided in Table 13.2. The **index of refraction** (n_D^{20}) is a unitless number. Its symbol is

index of refraction A unitless number that quantifies the refraction of light in a substance.

FIG. 13.6. Refraction in the real world. The straw appears to be bent because of the refraction of light. (Courtesy S. E. Johnson and M. D. Mosher—© ASBC)

TABLE 13.2. Correlation of degrees Brix, degrees Plato, and specific gravity with the index of refraction[a]

Index of refraction (n_D^{20})	Brix (°Bx)	Plato (°P)	Specific gravity (SG_{20}) (kg/m³)
1.3330	0.0	0.00	1,000.0
1.3403	5.0	4.97	1,019.5
1.3478	10.0	9.95	1,039.8
1.3557	15.0	14.90	1,060.9
1.3638	20.0	19.90	1,082.9
1.3723	25.0	24.90	1,105.8
1.3811	30.0	29.80	1,129.7

[a]Courtesy S. E. Johnson and M. D. Mosher—© ASBC.

an italic "n" with a superscript that denotes the temperature of the measurement and a subscript that denotes the wavelength of light that's used to perform the measurement. Historically, that wavelength is the "D" line found in a sodium lamp, which corresponds to 589 nanometers (nm).

Unfortunately, there is an issue associated with the use of the refractometer. Because it's a technique that measures the physical nature of the entire solution, mixtures of solutes in the solution can affect the index of refraction differently. For example, if the refractive index of a sample from a fermentation is analyzed, the presence of sugars and ethanol will contribute to the total refraction. The sugar and ethanol will both increase the refractive index. However, if we know the percentage of ethanol in the solution, then we can determine the sugar content. Similarly, if we know the sugar content, we can measure the ethanol concentration using refractometry.

If both ethanol and sugar concentrations are unknown, a laboratory experiment is needed to determine them. First, a known volume of the sample is taken and boiled gently until the sample is reduced by 25%. This boiling step removes the ethanol from the sample. Second, the sample is diluted with water back up to the original volume. The refractive index of the sample is measured and converted into the specific gravity using Table 13.2. Finally, the % ABV calculation below is completed using the prefermentation specific gravity and the percentage of alcohol in the sample determined. In addition, the refractive index of the sample relates the concentration of sugars remaining:

$$\% \text{ ABV} = (\text{OG} - \text{FG}) \times 131.25$$

where OG is the original gravity of the wort and FG is the final gravity of the beer after fermentation has been completed.

Viscometers

Viscosity is another of the physical property measurements of a liquid sample. The viscosity of a solution measures the bulk property of the solution, rather than the specific viscosity of each component of the solution. Viscosity is defined as the resistance of a liquid to flow. We can think of this as a measurement of the thickness or syrupiness of a solution. For example, if water and syrup are poured in a puddle at one end of a cutting board and then the board is tilted, the water will run faster than the syrup to the bottom of the board. Water has a lower resistance to flow and moves more quickly down the cutting board. Syrup has a higher resistance to flow and moves slowly. So, water has a lower viscosity than syrup.

Viscosity is typically measured by one of two methods, because there are two main types of viscosity. **Dynamic viscosity** was studied by Sir Isaac Newton in the 1600s. He found that the force required to move the liquid is based on the surface area of the liquid and the rate at which it can deform:

$$F = \mu \times A \times R$$

where F is the force that must be applied to cause the liquid to flow, A is the surface area of the liquid, R is the deformation rate or shear rate, and μ is

viscosity The resistance of a liquid to flow.

dynamic viscosity The measure of the internal resistance to flow.

kinematic viscosity The measure of the ratio of the internal resistance to flow versus the density.

the proportionality constant known as the dynamic viscosity. The units of μ are typically reported in mPa•s. (In the United States, they are reported in centipoise [cP], where 1 cP is equivalent to 1 mPa•s.)

Kinematic viscosity (ν) can also be measured or calculated. Kinematic viscosity is the dynamic viscosity divided by the density of the solution. It's reported in units of square meters per second, or m²/s. (In the United States, the units are in stokes [St], where 1 m²/s is equivalent to 10^4 St.) It should be noted that the variable ν is used for both the frequency of light in hertz and the kinematic viscosity in m²/s. These two meanings of ν are not related; the use of the variable to describe these two items is coincidental:

$$\frac{\mu}{\rho} = \nu$$

The viscosity of a liquid can be determined by how quickly the liquid flows through a tube or how fast a solid material falls through the liquid. Measuring viscosity by either of these methods is a lot less messy than using the cutting board described earlier. For accurate measurements, however, the liquid that moves must exhibit **laminar flow**. Laminar flow occurs when all the particles of a solution move in the same parallel direction (Fig. 13.7).

laminar flow Movement of all the particles of a substance in the same parallel direction, with no diffusion lateral to the direction of flow.

turbulent flow Movement of the particles of a substance with variations in the direction of the flow; lateral diffusion occurs.

Although we want the flow of the liquid to be laminar, **turbulent flow** can occur with moving solutions. An example can be seen when a match is extinguished. The smoke exhibits an initial laminar flow as it leaves the match head, and then it rapidly converts to chaotic turbulent flow. To see an example of this, light a match and blow it out. Watch the smoke as it leaves the surface of the match. The smoke will move upward as laminar flow but eventually become turbulent as it continues to rise.

Turbulent flow affects how long it takes for something to flow through the liquid or how fast the liquid is flowing, because turbulent flow is random and chaotic. Thus, the results are neither predictable nor reproducible. When measuring viscosity, we need to make sure that laminar flow is the only condition present in the liquid. This may seem difficult to control; however, we can predict whether laminar flow will be present using **Reynolds number (Re)**. Reynold's number is a measure of the turbulence in the flow of a liquid. As the number gets larger, the degree of turbulence increases. Typically, a Reynolds number greater than 2,300 indicates

Reynolds number (Re) A measure of the turbulence in the flow of a liquid; can be calculated to determine if a flow is turbulent or laminar.

laminar flow

turbulent flow

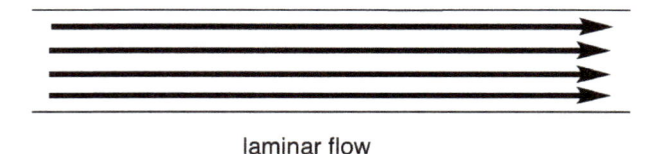

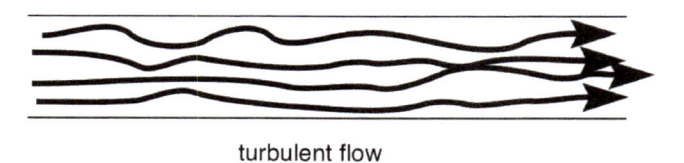

FIG. 13.7. Laminar and turbulent flow. As a solution moves along a tube, laminar flow results when all the particles in the solution move at the same rate in the same direction. Turbulent flow results when the particles move at different speeds. (Courtesy S. E. Johnson and M. D. Mosher—© ASBC)

turbulent flow and a number less than 2,300 indicates laminar flow. Here is the equation:

$$\mathrm{Re} = \frac{\rho\, u\, L}{\mu} = \frac{u\, L}{\nu}$$

where ρ is the density of the liquid, u is the velocity of the fluid (m/s), L is the diameter of the flow space (m), μ is the dynamic viscosity of the fluid (kg/m·s), and ν is the kinematic viscosity of the fluid (m²/s). Using this equation, we can increase the likelihood of laminar flow by passing the liquid through small-diameter tubes at a very slow rate.

Because viscosity must be observed under laminar flow, instruments that measure viscosity often include flow conditioners. These conditioners pool the liquid so that the flow rates are slow, and they also help to remove any flow rates prior to the location at which the measurement will be taken. The ideas of laminar flow and turbulent flow are also very important in brewhouse piping. For the movement of product, laminar flow is often best, as it's less disruptive and damaging to the product. But when the pipes need to be cleaned, turbulent flow often does a much better job. The result of turbulent flow means that the cleaners are mixed and chaotically scrubbing the sides of the tubing as they move. Since the tubing is typically the same size, the cleaners tend to be pumped at a higher velocity to increase the likelihood that they will be turbulent.

Viscosity and the other physical properties for solutions are quite temperature dependent (Table 13.3). As the temperature increases, the viscosity of a liquid decreases. Unfortunately, an increase in temperature also causes the Reynolds number to increase. A viscosity measurement at 20°C (68°F) may be very stable and easily measured, because the liquid exhibits laminar flow. But at 30°C (86°F), the same measurement might be difficult to reproduce, because the liquid exhibits turbulent flow.

TABLE 13.3. Dynamic viscosity (μ) and kinematic viscosity (ν) of water at selected temperatures[a]

Temperature (°C)	Dynamic viscosity, μ [× 10^{-3} kg/(m·s)]	Kinematic viscosity, ν (× 10^{-6} m²/s)
5	1.518	1.519
10	1.307	1.306
20	1.002	1.003
30	0.798	0.800
40	0.653	0.658
50	0.547	0.553
60	0.466	0.474
70	0.404	0.413
80	0.354	0.364
90	0.315	0.326

[a] Courtesy S. E. Johnson and M. D. Mosher—© ASBC.

Ostwald viscometer An instrument that measures the flow of a liquid by determining the amount of time that the liquid takes to traverse a specific distance.

falling ball viscometer An instrument that measures the time it takes for a ball to descend a given distance through the sample.

There are many different types of viscometers. An **Ostwald viscometer** measures the flow of a liquid by determining the amount of time that the liquid takes to traverse a specific distance. A **falling ball viscometer** measures the time it takes for a ball to descend a given distance through the sample.

The Ostwald viscometer was invented by a German chemist, Wilhelm Ostwald, in the late 1800s. It's also known as a capillary viscometer (Fig. 13.8). The device is essentially a U-shaped capillary tube between two reservoirs. The apparatus is placed in a water bath to maintain a constant temperature. The liquid to be analyzed is pulled up into the initial reservoir and then released. As the liquid flows down the capillary tube, it passes a mark where a timer is started. When the liquid passes the mark at the end of the capillary tube, the timer is stopped. The time it takes for the liquid to pass through this area is then divided by the time for water to do the same transit. The result is the kinematic viscosity.

A falling ball viscometer is a long tube with marks at the top and bottom (Fig. 13.9). The tube is filled with the liquid to analyze; a small ball that's slightly smaller in diameter than the tube is added. As the ball falls

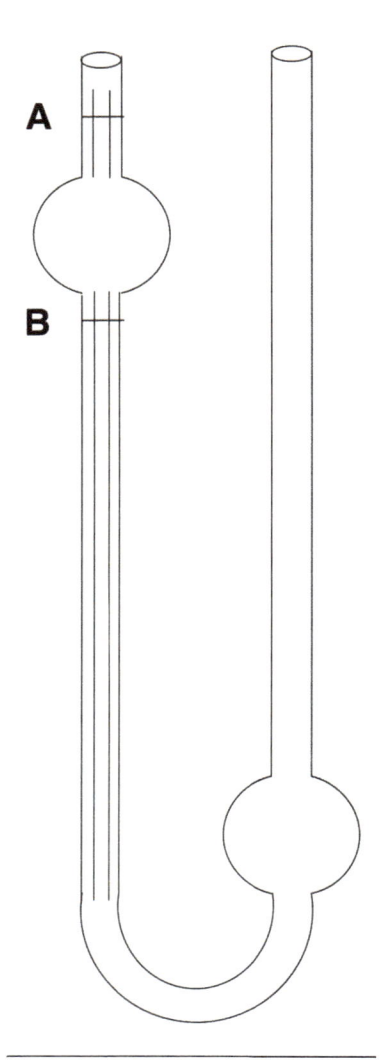

FIG. 13.8. Ostwald viscometer. The liquid is pulled above mark "A" and then released. When the level of the liquid reaches mark "A," the timer is started. When the liquid reaches mark "B," the timer is stopped. (Courtesy S. E. Johnson and M. D. Mosher—© ASBC)

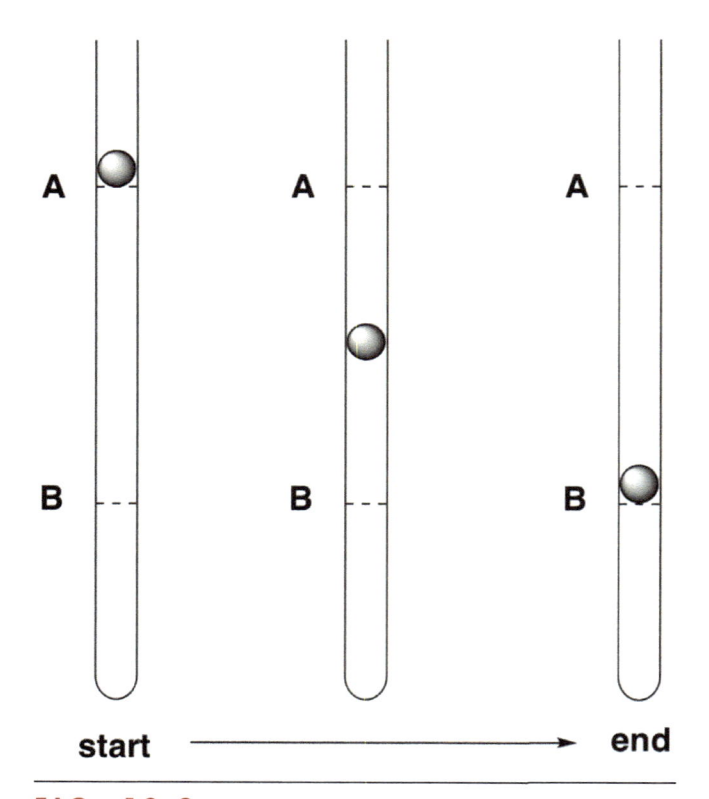

FIG. 13.9. Falling ball viscometer. The tube is filled with liquid, and a ball is added at the top of the tube. The time it takes the ball to move from the top mark, "A," to the bottom mark, "B," is related to the kinematic viscosity. (Courtesy S. E. Johnson and M. D. Mosher—© ASBC)

past the first mark, the timer is started. The ball reaches **terminal velocity** as it sinks through the solution. Terminal velocity is the fastest an object will fall through a medium based on the force of gravity. When the ball passes the bottom mark, the timer is stopped. As with a capillary tube viscometer, the time is compared with the time when water is used. The final result is the kinematic viscosity of the liquid. In some cases, falling ball instruments are automated with software that will perform the calculations before the reported value is displayed; however, using a stopwatch and a manual falling ball viscometer provides equally accurate data.

Dynamic viscosity can be measured with a device that measures the shear force directly. This instrument has a propeller-like device that's placed into a sample. The instrument then measures the force required to spin the propeller. Given the equation we used earlier, the instrument knows the value of the area of the propeller (A), the force (F), and the resistance of the liquid to that force (R). The dynamic viscosity can then be calculated.

Viscosity isn't often required in the brewing process. However, it is used in determining flow rates during sparging, measuring the efficiency of a mash-out, detecting the concentration of yeast in solution, identifying the viscosity of congress wort to gain information on filterability, and determining the concentrations of cleaners in the automated clean-in-place setup. More importantly, viscosity and the Reynolds number can be used to determine the sizes of piping and pumps that will be most efficient in the brewhouse as wort and beer are moved around.

terminal velocity The fastest an object will fall through a medium based on the force of gravity.

Olfactometers

An **olfactometer** is a unique instrument. It's basically a detector that uses the human olfactory system. Compounds are detected by a human user by noticing the smells of the compounds.

Typically, an olfactometer is attached to a chromatography system. As we discussed in Chapter 12, the chromatography system separates the individual components in a sample, and then each compound is directed into the olfactometer. This is basically a cone at the end of the chromatography column. A user sits at the instrument and smells the exhaust from the chromatography instrument periodically. The odors and aromas are noted and recorded at specific times.

In addition to the olfactometer data, the chromatography system records the presence of each compound that's separated from the sample. That data may be from a thermal conductivity detector, a flame ionization detector, a mass spectrometry detector, or any of a wide variety of detectors. The correlation of data from the two detection systems—the olfactometer and the chromatography detector—provides useful information:

olfactometer A detector that uses a human to smell and detect the individual components in a mixture; typically used with a chromatography system.

- **Identity of separated components in a sample.** While the chromatographic information can be useful in determining the identity of a particular component, the odor associated with the compound can be instructive in determining the identity of the compound. Some compounds have similar chromatographic retention times, but their aromas are likely different.

- **Relative threshold of a particular component to its actual concentration.** A chromatographic separation may reveal a large concentration of a particular compound. Such information may imply that the particular compound is important to the overall aroma of a sample. However, the olfactometer can be used to judge the intensity of the aroma. That information can be used to report the relative effect of a given concentration of a particular compound.
- **Commonality of the aroma compounds.** A sample may have a number of individual components with similar aromas. Where one compound may be thought to be the only compound of interest in the sample, the actual aroma may be composed of multiple compounds with similar odors.

In recent years, the combination of gas chromatography and mass spectrometry (GC-MS) and olfactometry has been used in research within the brewing industry. The data have provided information on wood aging, hop oil contributions, and malt analysis. Specific analyses for a product stream may not be suitable using an olfactometer. However, analysis of hop oils from different suppliers or from different crop years would be of immediate assistance in determining the quality of the particular hop.

Foam Measurements

head Beer foam.

collar Beer foam.

Also known as the **head** or **collar**, the foam is important to the consumer's perception of beer quality. Too much or too little, and the consumer is unhappy. Measurements of foam are therefore important to the brewery.

Beer foam is made when the bubbles of carbon dioxide (or nitrogen) in the beer are released. These bubbles are nonpolar, and polypeptides (small proteins from the malt that was used to make the beer), which are also nonpolar, collect around the bubbles. The polypeptides actually coat the bubbles of carbon dioxide and help to stabilize them. The iso-alpha-acids from hops are thought to cross-link between the polypeptides and strengthen the structure. As the foam persists, smaller bubbles (which have higher gas pressure) are taken up into larger bubbles (which have lower gas pressure). The larger bubbles are less stable and eventually break open and are lost.

Two important processes result in foam on top of the beer:

1. **Formation of the foam.** To create a lot of foam, the beer must be agitated when it's poured, and the temperature of the beer should be warm. These two actions will help to release as much gas as possible from the beer. Nitrogen gas, which is very insoluble in beer, immediately degasses from the beer when it's opened and poured into a glass. This results in the development of lots of very small foam bubbles, which last for a long time. The temperature has little impact on the amount of bubbles, since the gas is insoluble at all temperatures. Carbon dioxide gas, on the other hand, is more soluble in beer. When it's released, the bubbles change size based on the temperature of the beer. Even when the beer is cold, the bubbles tend to be larger than those from nitrogen. Having larger bubbles means the foam doesn't last as long.

2. **Collapse of the foam.** The other process that puts foam on the beer is the rate at which it collapses or dissipates. This rate is dependent on the size of the bubbles, as we have just explored, and on the types and quantities of the polypeptides and iso-alpha-acids in the product. Smaller bubbles couple with more polypeptides and isomerized hop acids, and the foam persists.

Laboratory methods that measure foam focus on the rate at which the foam collapses. The processes are highly dependent on the method by which the foam is produced and the temperature of the beer. This means that the laboratory analysts using these methods must be very repetitive in their operations.

The first method for measuring foam collapse is the **Rudin method**. Degassed beer is poured into a vertical glass column, and carbon dioxide gas is bubbled through the beer, creating foam. The gas is continued until the foam swells to reach the top mark on the column. Then the gas is turned off and a timer is started. The timer is stopped when the foam collapses back to the lower mark on the column. A good foam collapse rate is one that takes 90 seconds or more to traverse the two lines. The strongest advantages to this method are that it measures the potential of the beer foam and that it removes any issues with dispensed gas mixtures, because only carbon dioxide is used to create the foam. The main disadvantage is that this method measures only the potential of the beer foam—not the actual performance that a consumer might see. In addition, the size of the column is often much smaller than the pint glass a consumer might use, so it really isn't measuring what a consumer will see.

Rudin method A method of beer foam analysis.

The other method for measuring foam collapse is the **Nibem method**. It requires the use of an instrument called a foam collapse meter. A beer is poured normally into a glass jar representing a pint glass. Then a set of electrodes attached to a plate is lowered onto the foam and the instrument is turned on. An electrical signal is sent across the electrodes through the foam. As the foam collapses, the signal gets weaker and the plate moves down to restore the original signal. After the plate has moved about 30 millimeters (mm), the device calculates the foam collapse. Beer that takes more than 260 seconds to collapse is considered to have stable foam. The greatest advantage to using this method is that it measures what the consumer will observe—the actual performance of the beer, rather than its potential. The biggest issue, unfortunately, is that many factors influence foam creation. If the beer is poured from a can, a bottle, a keg, or a growler, the foam created can be quite different. Even the variability in duplicate runs from the same source tends to be large.

Nibem method A method of beer foam analysis.

CHAPTER 14

Microbes

The greatest concern of the quality control program in any brewery should be microbiology. When microbes escape from where they should be in the brewery, they wreak havoc on product quality.

The term **microbes** is used to describe all types of microscopic organisms. Note that many microbial fermentation biochemical pathways don't involve the production of ethanol but rather acetic acid, lactic acid, or another of the many other organic acids. Some microbes are needed in the brewing process. Yeast is one of these microbes. In some cases, microbes are extremely valuable to the finished product, such as the species of *Lactobacillus* and *Pediococcus* bacteria used for a kettle sour. The brewery laboratory expects to find microbes living in the product stream at certain locations during the making of a product. However, if microbes are found outside their normal places in production, the product could be affected and the quality could be compromised.

microbes Organisms so small that they can be viewed only with a microscope.

Early Research and Inventions

Microbes are organisms that are too small to be seen with the naked eye. In fact, their existence was only postulated before invention of the microscope in the early 1600s. By then, the glass lenses used to make spectacles had been around for about 300 years. Three of the main prospects for inventing the microscope came from the community of spectacle (glasses) makers. At the end of the 1500s, Middelburg, Holland, was home to many highly skilled lens manufacturers.

A microscope works by using how light travels through a convex lens to magnify an image. Glass manufacturing and polishing had been underway long before the microscope was invented. However, it wasn't until the discovery of how to utilize the way light refracts through objects (and how to precisely tune the instrument to be reliable down to microscopic measurements) that microscopic studies truly blossomed (Fig. 14.1).

There's much debate about who invented the microscope. The first claim came from Johannes Janssen, who stated that his father, Zacharias, had invented the device as early as 1590. That would have been an amazing accomplishment, because Zacharias would have been less than 10 years old at the time. In addition, it wasn't until 1615 that a set of lens-making tools was bequeathed to Zacharias. In fact, Zacharias was an interesting character who moved a few times during his life to avoid being sentenced to death for being a counterfeiter.

More likely, the microscope was invented by Zacharias's neighbor in Middelburg, Hans Lippershey. Hans was a respected spectacle maker and had applied for the first telescope patent in 1608. Although a telescope is much different than a microscope, both devices use a similar layout of lenses within a tube. Another contender for inventing the microscope was Cornelis Drebbel, a Dutch inventor who built a public fountain in Middelburg in 1600. His claim to the invention stemmed from his using a microscope in London in 1619. Regardless of whether he invented the microscope, Drebbel was appropriately credited for many other inventions, including the first navigable submarine.

Each of the contenders for inventor of the microscope was able to examine objects with a magnification of 3–7 times the normal view. However, none of them described or could view microbes using his microscope. In the latter half of the 1600s, Anton van Leeuwenhoek, a Dutch merchant with an interest in science, redesigned the microscope to have only one lens. With that simple change, he was able to greatly increase the magnification of the microscope and observe yeast, bacteria, and a wide range of other microbes. He referred to these microscopic creatures as "animalcules," because they appeared to be small versions of organisms that existed in the world. Even though van Leeuwenhoek thought that yeast were not living creatures, his work started the field of microbiology and redefined our understanding of the world around us.

Unfortunately, the world of microbiology didn't advance very quickly after van Leeuwenhoek invented the microscope. To be sure, many scientists explored microbes that lived in water and soil and even the structures of

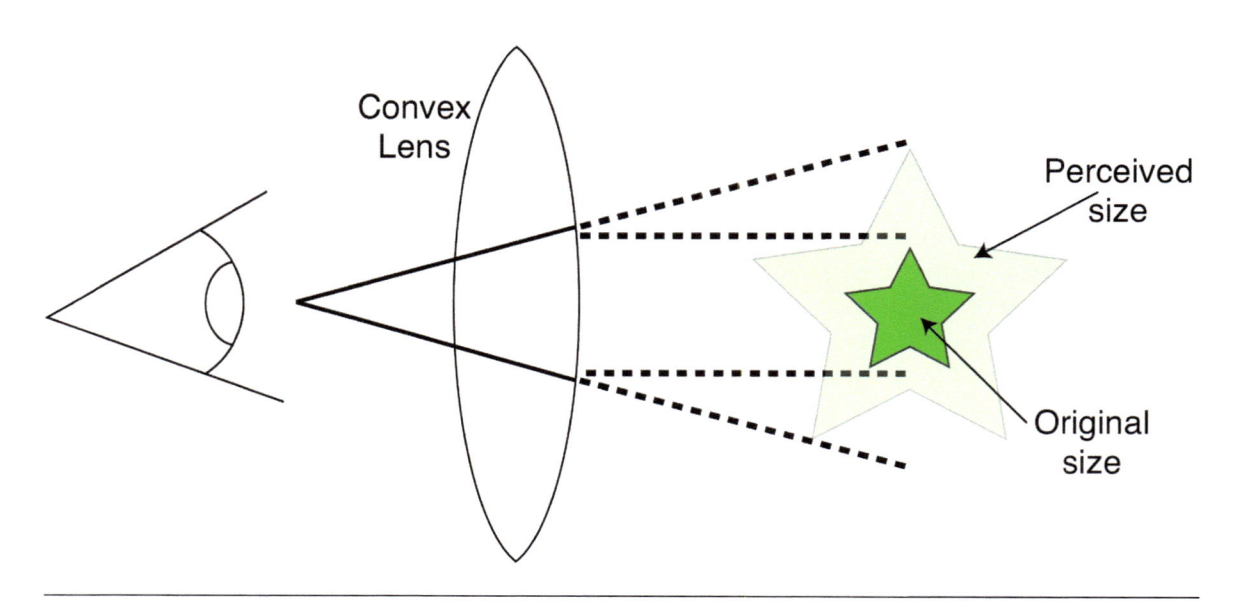

FIG. 14.1. Basic concepts of magnification. (Courtesy S. E. Johnson and M. D. Mosher—© ASBC)

plants and animals up close. But it wasn't until 1866 that the publication of a significant discovery changed people's understanding.

Louis Pasteur, a French scientist, had been trying to understand the fermentation process in wine. At the time, most people believed that fermentation was simply the decomposition of chemicals into ethanol and carbon dioxide. Pasteur performed experiments that confirmed that yeast caused the formation of alcohol and carbon dioxide in beer. In additional experiments, he proved that other microbes resulted in the formation of lactic acid, acetic acid, and other compounds. Pasteur used this information to explore what were believed to be diseases of wine fermentation, in which the wine would sour or develop other unpleasant flavors. His initial results regarding wine fermentation were published in 1866. By 1876, he had published significant research about fermentation in beer, determining that when the fermentation process was exposed to air, the amount of sugar consumed during the process dropped. His work in eliminating microbes by applying heat gave rise to the process known today as pasteurization.

With all of his work in studying microbes, Pasteur helped advance the idea that microbes likely resulted from disease—whether a soured fermentation or a human illness. The so-called germ theory of disease was opposed by many at the time but was later proven to be correct. Microbes live everywhere in our world. Because they are living creatures, it doesn't take a large number of microbes to ruin a large amount of product. And if they inhabit places that they shouldn't be, they can result in an infection.

Sterilization Revisited

Pasteur's work with microbes included studies on how to reduce spoilage in wine. His work was also directly applicable to the brewing and dairy industries, although neither industry immediately applied his theory. However, the process of sterilizing foods to make them free of microbes had been explored and put into practice long before Pasteur.

In the 1790s, Nicolas Appert, a French chef, had perfected a way to store foods free of microorganisms. He placed vegetables and meat into containers and sealed them, and then he placed the containers in boiling water and heated the food to cook it. In the process, any bacteria living on the food were killed, as well. Appert's method significantly extended the life of the food products, but it also reduced the flavor of the food. In 1800, Napoleon Bonaparte offered a prize to anyone who could demonstrate a new way to preserve food. Appert eventually won the prize in 1810 for his method, which is used today to make canned food both commercially and in the home.

Pasteur showed that it was possible to kill microbes without significantly impacting the flavor of the food. Later work on his discovery yielded this mathematical expression:

$$PU = t \times 10^{\frac{(T-60)}{Z}}$$

where T is the temperature to which the beer has been heated (in degrees Celsius), t is the time of the exposure (in minutes), Z is the relationship

pasteurization unit (PU) A measure of the effect of heat and time on the survivability of microbes.

decimal reduction time (D-value) The time required to kill 90% (a 10-fold reduction) of the microbes in a sample.

between lethality and the temperature, and PU is the number of pasteurization units. For beer, Z is typically set to 7, which implies that for every 7°C increase in temperature, the lethality of the process increases 10-fold. The value of Z can be greater when the microorganism is more resistant to heat and lower when the microorganism is less resistant. Many U.S. breweries assume that Z is 6.94; therefore, the 10-fold increase correlates to an increase of 12.5°F.

A **pasteurization unit (PU)** is the measure of the amount of lethality applied to a sample. The equation reveals that applying 60°C (140°F) of heat for 1 minute gives a value of 1 PU. For most applications, just a few PUs result in sufficient heating to kill most of the microbes in beer. Yet most breweries aim for between 15 and 30 PUs to ensure that the majority of microbes have been killed and that the beer will be stable for at least 1 year. The equation implies that using even lower temperatures will result in a significant number of PUs if the time is long enough. This isn't true, however, because the effect of heat on the lethality of microbes becomes negligible below 50°C (122°F).

When examining the effect of pasteurization on a sample, we typically use a unit known as the **decimal reduction time**, or **D-value**. This is the time required to kill 90% of the microbes. This value also corresponds to 1 logarithm reduction in the number of microbes. The amount of reduction for a product depends on the level of product stability sought by the brewery. Table 14.1 lists the D-values and z-values for some common microbes found in beer. A much more detailed set of values for a wide variety of beer spoilers can be found in the Lemgo D-value and z-value Database for Food (www.th-owl.de/fb4/ldzbase/). Note that lager yeast is more susceptible to heat treatment than ale yeast and that *Lactobacillus* species tend to be more robust than other beer spoilers.

What amount of heat should be applied? In other words, what level of microbial reduction will produce the desired increase in product stability? For beer, we often try to get a minimum of 6 D worth of reduction (6 logarithms)

TABLE 14.1. D-values and z-values for selected microbes in beer[a]

Microbe	D-values at 60°C	z-values
Aspergillus niger	0.04	3.7
Lactobacillus	0.02–0.44	6.5–15.0
Pediococcus	0.00073	4.0
Saccharomyces cerevisiae	0.01	4.6
Saccharomyces cerevisiae var. *diastaticus*	0.06	7.8
Saccharomyces pastoranus	0.004	4.4

[a] Data from Hill, A., ed. 2015. Brewing Microbiology: Managing Microbes, Ensuring Quality and Valorising Waste. Woodhead Publishing, Cambridge, UK. Table courtesy S. E. Johnson and M. D. Mosher—© ASBC.

for the specific microbe. This corresponds to reducing the number of organisms to 99.9999% of their original concentration. If the suspected microbial infestation is from a species of *Lactobacillus*, at least 2.64 minutes of exposure (6 × 0.44) at 60°C (140°F) will be required. This corresponds to 3.07 PU. To be conservative and ensure that we hit this 6 D reduction, we might be sure to have at least 3 minutes of exposure at 60°C (140°F). However, if we are trying to reach a 12 D reduction (corresponding to 99.9999999999% of the original concentration), we will need 5.28 minutes of exposure at 60°C (140°F) (6.16 PU). This time of exposure may impact the flavor profile for the product, making a shorter exposure to a hotter temperature a better choice. The same number of PUs can be obtained by heating the beer to 70°C (158°F) for 0.88 minutes.

Is the beer sterile when a 6 D reduction is implemented? No. What about when a 12 D reduction is implemented? The answer is still no. Very few microbes will survive the 12 D reduction, but some will. For example, a 6 D reduction will reduce the number of microbes to 99.9999% of the original amount. Said differently, for every 1,000,000 microbes, 1 will survive. This means that the sample isn't free of living microbes. For even a mildly infected sample of beer, the remaining few microbes might grow and multiply over time. How long that takes depends on the temperature of the beer after the reduction.

Given the information here, we might assume that the best result for pasteurization is to apply a high temperature over a long period to ensure that all the beer spoilers are removed from the beer. This is the best approach. The hotter we make the beer and the longer we hold it at that temperature, the greater the reduction in the number of living microbes. Unfortunately, this isn't the best way to preserve the beer. High temperatures significantly impact the flavor of the product. In fact, temperatures greater than 20°C (68°F) can change the flavor profile over time. A **thermal degradation unit (TDU)** has been proposed to measure the relative impact of temperature on the flavor of a beer:

$$\text{TDU} = t \times 2^{\frac{(T-20)}{10}}$$

thermal degradation unit (TDU) A measure of the impact of heat and time on the quality of the beer.

where T is the temperature in degrees Celsius and t is the time in minutes. Note that as the number of TDUs increases, the effect on the flavor also increases. Also note that temperatures lower than 20°C (68°F) don't have any effect on the flavor, but when the temperature rises above 20°C (68°F), the flavor is affected. That effect can be negative or positive, depending on the beer, but most of the time, the effect is harmful to the flavor profile.

In our example of pasteurizing at 60°C (140°F) for 5.28 minutes, we will end up with 84.5 TDUs. The same number of PUs will be obtained at 70°C (158°F) for 0.88 minutes and will give us only 28.2 TDUs. This is only one-third of the impact of the lower temperature on the flavor of the product. However, with shorter times, small errors in the amount of time that the beer spends at the given temperature can result in very significant changes in the number of PUs.

Microbes and the Brewing Process

Fortunately, microbes that tend to be harmful to human health can't survive in most craft beers on the market, with the exception of very low alcohol-by-volume (ABV) and no-hop beers. So why do we hear about recalls of beer due to microbial contamination? These recalls are done because a given beer didn't live up to the brand's expectations, not because of a risk to human health. In fact, most beer recalls result from the beer's quality not being up to par. Most of the human health hazards that force recalls involve broken glass or metal shavings from failure of brewing equipment.

The reason that beer isn't recalled like beef and other food products is that the brewing process eliminates microbes that are hazardous to human health. Wort is boiled during the brewing process for at least 1 hour. While the boiling process is used to isomerize alpha acids in the beer, it also sterilizes the wort. In addition, hops function as antibacterial compounds, acting on their own as natural preservatives of beer. Why don't any contamination risks to human health arise after the boil? Risks can arise when the wort temperature drops below 55°C (130°F) quite quickly. However, as soon as the wort is cooled, it's inoculated with a large batch of yeast. The yeast rapidly consumes the available nutrients and produces acid, ethanol, and carbon dioxide. These byproducts are also antimicrobial compounds, especially to human pathogens.

Given all of the challenges to microbes surviving in, say, a moderate-ABV, hopped beer, it's incredibly unlikely that microbes hazardous to human health will get through the brewing process. And if beer were contaminated by microbes, wouldn't it likely be noticeable to the sensory panel testing the beer? Such a product would never be distributed to the public. In reality, most contaminations are noted only after the beer has been released and the microbes have had additional time to grow. That's why having a microbiological program at the brewery is so important.

The product may contain microbial growth, but that would only make the beer less appealing. At the worst, it would smell bad and be undrinkable, but it wouldn't be harmful to your health. Because there's little risk to human health from microbial contamination, the U.S. Food and Drug Administration (FDA) hasn't been heavily involved in monitoring the products that breweries produce. However, as the industry has continued to grow, the FDA has begun to implement some regulations for breweries, as noted in Chapter 1. Namely, breweries are expected to have in place a hazard analysis and critical control points (HACCP) program, just like any food production facility must have under the law. Moving forward, all breweries will likely have to show that they have implemented some form of HACCP to avoid fines and interventions by the FDA.

It's important to track the microbes present in the brewing process in a HACCP or quality control program. The focus of this recordkeeping is typically to monitor the yeasts that are present in the brewhouse. Often, the largest contamination issue is cross-contamination of yeast from one batch to another. For example, a Belgian-style strain that gets into a blonde ale will implement a lot of undesirable flavors in the finished beer.

Yeast Cell, 3-4 μm

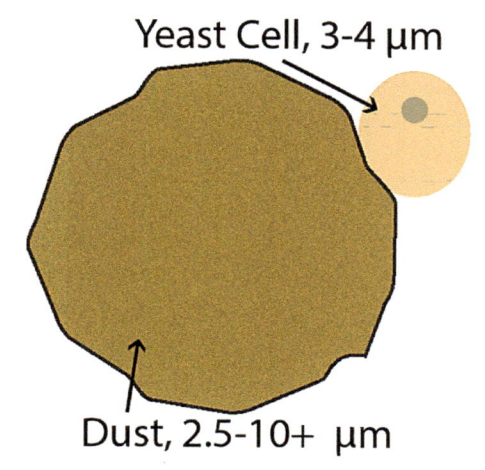

Dust, 2.5-10+ μm

*The smallest object that the naked eye can see is about 50 μm

FIG. 14.2. Dust can transport microorganisms. (Courtesy S. E. Johnson and M. D. Mosher—© ASBC)

Another potential problem is dust, which can be generated from milling, blown in from outdoors, tracked in on shoes, or created during almost any activity. Microscopic particles that break off any substance during regular use will result in the formation of dust. Most dust particles can't be observed by the naked eye, and unfortunately, dust is a great substance for microbes to cling to (Fig. 14.2). Then when dust is blown around the brewery, it carries those microbes with it. This is yet another reason that cleanliness is so important in the brewery. Dust can't be completely prevented or eliminated, but under good brewery practices (GBPs), the chance of cross-contamination is significantly reduced. Fortunately, in the rare case that contamination occurs, many different analyses are available for identifying various bacteria and yeast strains. These analyses have been developed by the brewing industry and are essential to the brewery micro lab.

Making Media

Setting up a laboratory within the brewery to analyze microbes involves making some very specific adjustments:

1. The area must be relatively clean and free of dirt and debris that may harbor microbes.
2. The benches on which the analysis will be performed must be wiped down on a regular basis with a disinfectant, such as isopropyl alcohol.
3. The laboratory must be equipped with specific equipment and instruments that aren't used for other laboratory experiments.

A laminar flow hood pulls air from the room into and through a filter, so that microbes are ejected from the area in which work requires sterile conditions. The hood can be omitted when setting up of the lab, but contamination of samples will be much more likely. The incubator in the micro lab doesn't need to be a commercial microbiology incubator. Any small oven will do, as long as the temperature can be maintained at 20–40°C (68–104°F) on a regular basis. A shaker that swirls large volumes of broth over multiple hours will reduce the workload (and sore arms for lab workers), and a refrigerator and freezer are useful for storing yeast samples.

The first step in working with microbes is to create a place for them to live, which is known as the **media**. The media is the base material that provides microbes with food and nutrients and an atmosphere that's conducive to their survival. Food and nutrients can be provided as either a liquid medium (a broth) or a solid medium (an agar plate).

Broths, Plates, and Slants

A **broth** is a liquid solution used to grow microbes. More specifically, a broth is a water solution containing the appropriate amounts of food and nutrients that microbes need to survive. There are many different types of broths, but they differ only in the kinds and amounts of chemical compounds added to the water:

- A **culture broth** contains all the nutrients (such as sugars, amino acids, and minerals) that microbes need to grow. This type of broth will grow almost any microbe and do it well.
- A **minimal broth** contains only minimal amounts of the nutrients needed for microbes to grow. Typically, a minimal broth lacks some of the amino acids needed, which causes the microbes to be stressed. Minimal broths tend to be used more in research-type settings.
- A **selective broth** contains a compound that's toxic, such as an antibiotic. Alternatively, a selective broth can be made to lack some amino acid or compound that a particular microbe needs to survive. The result is that certain microbes will grow in a selective broth while others will not. For example, in the brewery laboratory, a sample of fermenting wort might be analyzed to find out if any microbes are growing in the product. Brewery lab personnel would add to the broth a small amount of cycloheximide—a compound that's toxic to yeast. Then if any microbes were found growing in the broth, they would signal that the fermenter is contaminated with nonyeast microbes.
- A **differential broth** contains an indicator that changes color when a particular microbe grows in it. For example, a lactose-rich broth containing a compound known as neutral red will change color if microbes can use the lactose to grow. When they do, they produce acid as a waste product. The acid changes the color of the broth from yellow to red. The color change signals that microbes in the broth can utilize the lactose.

media The base material used to grow microbes.

broth A liquid material used to grow microbes.

culture broth A liquid containing nutrients to support the growth of microbes.

minimal broth A liquid that lacks sufficient amounts of a single nutrient or multiple nutrients required to support the growth of microbes.

selective broth A liquid that contains a chemical or chemicals that allow only certain microbes to grow.

differential broth A liquid that contains chemicals that signal the presence of certain microbes.

While using broths is a great way to propagate large quantities of microbes, the brewery laboratory often wants to know only if a particular microbe exists in the sample. To do that, only a small amount of media is needed, and a dish is the best place to do the analysis. An **agar plate** is a petri dish containing an **agar** (also known as agar-agar). Agar is a jelly-like substance that comes in powder form but solidifies when dissolved in water. Agar likely was discovered in the mid-1600s in Japan, and it became popular in cooking to make jams, jellies, custards, and other gelatinous foods. It's still used in cooking across many different cultures as a thickener and in making desserts. By the late 1800s, agar began to replace gelatin as the medium used to grow microbes, because agar has a higher melting point and doesn't turn back into a liquid at elevated media growth temperatures. Agar also isn't easily digestible by many organisms, so it maintains its shape as microbes grow on or in it. Gelatin, which is still used in cooking today, is made from processing collagen from animals. (Collagen is the main component of the connective tissue found in animals.)

Agar comes from red algae and is a mixture of two main components: agarose (~70%) and agaropectin (~30%). Agaropectin is a polymer that consists mainly of glucuronic acid monomers with small amounts of pyruvic acid monomers. Agarose is also a polymer, but it's made up of approximately 800 galactose monomers. These polymers are structurally similar to cellulose and starch.

When polymers are dissolved in water, the water interacts with them. Intermolecular associations are formed between the hydrogen atoms on the water and the oxygen atoms on the polymer. This hydrogen bonding, as it's known, is very strong and further links all the polymer molecules and water. The network becomes solid. If the appropriate nutrients are added to the water as the powdered agar is added, the final solidified mixture will then contain those nutrients.

In practice, either a broth or an agar plate is made by following the directions on the bottle of dry reagent. The powder is added to the water in an Erlenmeyer flask, and the flask is heated to at least 71°C (160°F) and held there until the reagent is dissolved. To prevent heating the reagent to a boil (which will destroy it), the best method is generally to use an autoclave. That way, high pressures and temperatures can be used to sterilize the broth and the container at the same time. If a broth is being prepared, it's removed from the heat and capped and then cooled to room temperature before being used. If agar plates are being prepared, the agar solution is poured into sterilized glass dishes and the tops placed on them. The dishes are cooled to room temperature, which will make them set up. Typically, agar plates are stored upside down inside their lids to minimize the chance of a stray microbe finding the agar.

In some cases, the lab may want to grow microbes on a **slant**. A slant is a smaller version of the agar plate in which the agar is placed in a sterile test tube. The test tube is then tilted so that the agar gelatinizes on a slant. The test tube is capped with a sterile cap and then stored at an angle until the agar hardens. At room temperature, the test tubes can be stored vertically to save space.

agar plate A petri dish containing an agar used to grow microbes.

agar A gelatin-like substance that's used to support the growth of microbes.

slant A gelatin-like substance that's poured into a test tube and solidified at an angle.

Growing Microbes

Now that we have made the media, we can grow microbes. The growing process starts in the laminar flow hood. A sample of the microbes to be grown is collected and brought to the hood. The brewery personnel then sit at the hood and transfer the microbes into the media. How the sample is collected and how it's transferred depend on what the brewery laboratory wants to do.

Liquid Media

Typically, broths are used to grow large quantities of yeast for use in the brewing process. These broths are created from slants of yeast; however, a plate or broth can be used as the starting point. The broth used for growing is usually made on the brewery floor.

A container known as a Carlsberg flask is taken to the brewery floor, where it's filled with wort from the brewing process. The wort can be obtained before the addition of hops or hop oils; however, the bacteriostatic properties of the hops can be beneficial during the growth of the yeast. If a hopped wort is used, it should be only lightly hopped, because high concentrations of hop oils can inhibit the adequate growth of yeast. The wort should have a gravity in the range of 1,048–1,050 kilograms per cubic meter (kg/m^3). This corresponds to 12.0–12.5 degrees Plato (°P). Studies have shown that yeast grown in wort within this gravity range have good viability after growth and don't get overly stressed when pitched on high-gravity worts (>1,072 kg/m^3, or 17.5°P) on the brewery floor.

The Carlsberg flask is then heated to sterilize the wort. The wort is boiled and the interior of the flask is also heated until everything is sterilized. At this point, a small sample of the wort can be withdrawn into a 100 milliliter (mL) sterile flask containing a stir bar. The brewery laboratory and the yeast requirements may require a smaller initial volume of wort. Then, in the hood, a lab worker heats up the inoculation loop in a Bunsen burner until it glows red. With his or her other hand, the worker removes the cap from a slant of yeast and passes the opening of the test tube through the flame a few times to kill any microbes near the opening. Next, the loop is used to reach down inside the test tube and "grab" a colony of yeast cells by scraping them off the surface of the agar. The loop is transferred into the 100 mL sterile flask containing the wort and swirled to remove the yeast from the loop. The opening of the flask is passed through the flame a few times, and the cap is put back on the flask.

The flask is then placed on the shaker and shaken for 24 hours to encourage rapid growth of the yeast. The result should be approximately 1 million cells per milliliter for every degree Plato of the wort. A sample can be removed and checked on the microscope to count the yeast in the starter. The yeast can be continued in the growth process until the volume needed by the brewery is obtained. Typically, the volume of the wort is increased by a factor of 10. The smaller sample of yeast is then poured completely into the new volume; after 24 hours, the new larger sample has the same number of cells per milliliter that the smaller sample contained.

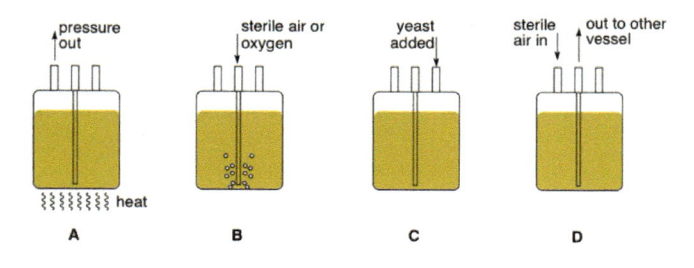

FIG. 14.3. Use of a Carlsberg flask to grow yeast. **A,** The flask is heated to sterilize the wort and the flask. **B,** Sterile air or oxygen is added to the cooled flask to aerate the wort. **C,** A syringe is used to add a sample of yeast to the flask. **D,** When the yeast has completed its growth, the flask is attached to another larger vessel and the yeast is pushed out using sterile air or oxygen. (Courtesy S. E. Johnson and M. D. Mosher—© ASBC)

Alternatively, the growth can be accomplished in the Carlsberg flask by inoculating the flask with a small yeast sample (Fig. 14.3). Then after the required time, the entire solution can be transferred to a larger flask or small fermentation vessel, allowing the volume of yeast to be built up. When the next 10-fold increase in volume will equal the volume of the fermenter on the brewery floor, the yeast sample is handed over to the brewery. That sample is pitched into the large fermenter, along with the wort from the brewing process, to create the final product.

Solid Media

Slants and plates are used for two main purposes. The first is to grow and save yeast for making broths for the brewery floor. A plate is typically used first to find a single healthy colony of pure yeast within a sample. Then that colony is placed on a slant to grow more of the same colony. After growth, the slants of the various types of yeast are stored in the refrigerator to minimize their growth rates.

The other main reason for using gelatinous media is to evaluate samples to find out if any nonyeast organisms are growing within them. This method focuses on the use of agar plates. Slants tend not to be used in this method, although a laboratory may keep examples of beer spoilers on a slant for future reference.

An agar plate is inoculated similarly to a slant. A sample is obtained and brought to the hood. The worker heats up the loop in the Bunsen burner. With his or her other hand, the worker passes the sample container through the flame to kill microbes living on the surface of the container. Then the loop is dipped into the sample and swirled back and forth. The loop is withdrawn from the sample and moved across the agar plate. The first pass should be similar to "coloring in" a quadrant of the plate. Before the second pass, the loop is reheated in the Bunsen burner. Then the loop is placed just inside the first quadrant and dragged into the second quadrant. The worker moves the loop across the second quadrant and "colors it in," as with the first quadrant. The worker must be careful not to go back into the first quadrant other than for that first stroke. The loop is reheated again, and the third quadrant is completed like the second one. Finally, the loop is reheated again and the fourth quadrant is completed (Fig. 14.4). The lid is replaced onto the agar plate, and the plate is turned upside down. Based on the particular requirements, the plate may be stored at room temperature or placed in an incubator at an elevated temperature.

Plates can be evaluated after the requisite amount of time. They are often incubated for 7–11 days before they are examined. A differential agar plate will report a color change if particular microbes are growing on the agar. Often, that color change can be observed without having to open the plate or even turn it over. Selective agar plates can show if any microbes are growing at all.

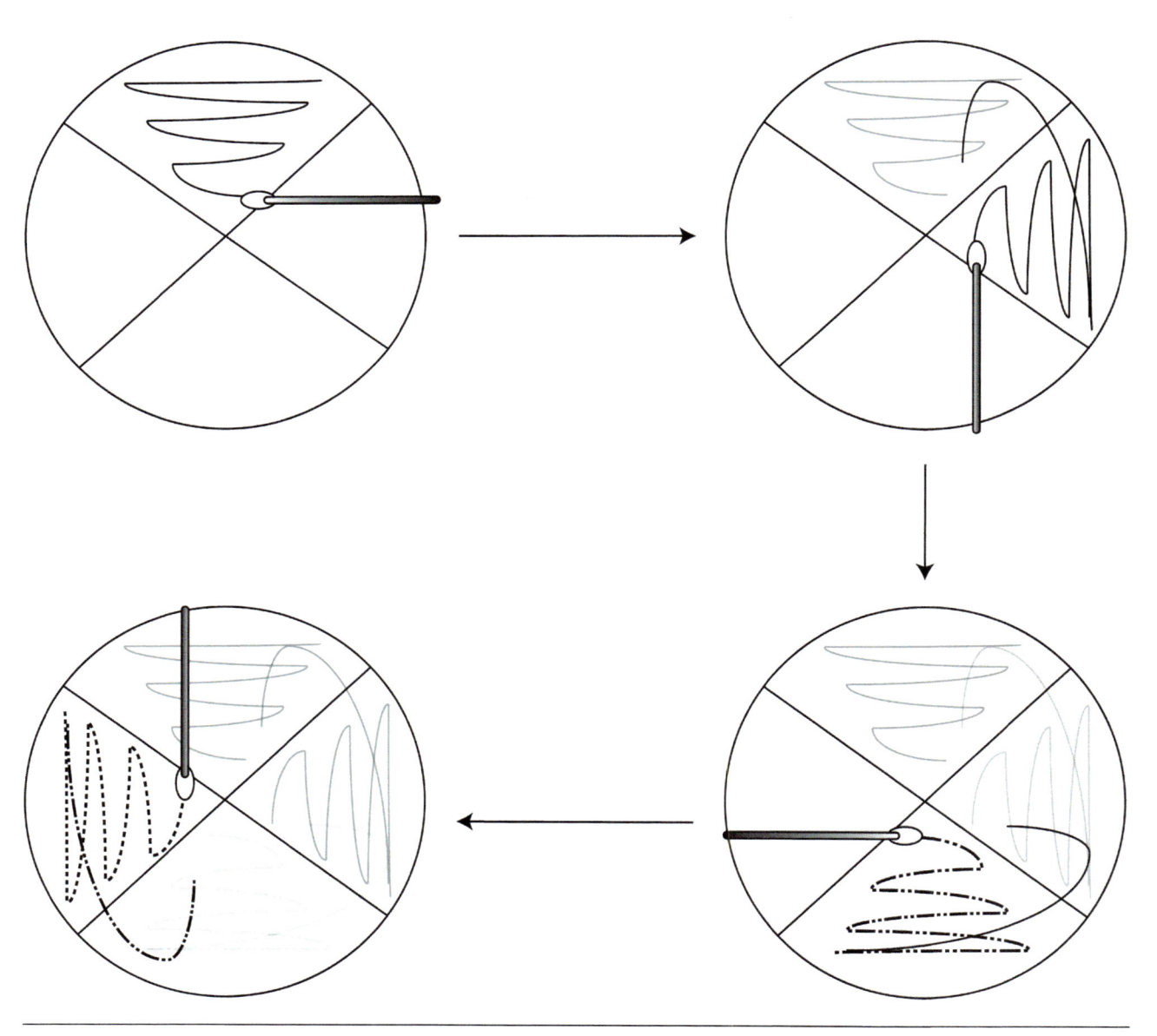

FIG. 14.4. Proper inoculation of an agar plate (beginning with the upper-left plate and moving clockwise). (Courtesy S. E. Johnson and M. D. Mosher—© ASBC)

Aseptic Sampling

When a sample needs to be obtained to evaluate the microbes in it, we have to be careful not to contaminate the sample. Just as important, we need to make sure we don't contaminate the product stream that we are sampling. Collecting a sample using **aseptic sampling** is exactly the process we need to follow. Not only will doing so ensure that the sample we collect is representative of the product stream, but it will also ensure that our results can be trusted.

Aseptic sampling from fermenters illustrates the techniques that we need to follow. Typically, the fermenter contains a sample port. The port looks similar to a tap in the taproom, but instead of a tap handle, it has a knob that's turned to open the valve. Some sample ports have valves that are similar to the tap-handle design, but the knob is better for ensuring that the valve is closed when not in use. This sample port is where liquid from the product stream will be collected.

First, the valve is opened and a few seconds' worth of product is drained into a bucket for waste. Then after the valve is closed, the entire sample port is sprayed with 70% ethanol from a spray bottle. The spray is even directed up into the open end of the sample port. Next, a sterile swab or brush is rubbed on every surface of the valve that will come into contact with the sample. A hand-held burner is then directed to the sample port; the flame is directed to every surface that will touch the sample. The flame is held on the valve until all the ethanol has evaporated or burned away. Because the valve is now hot, it's opened and the first portion of the product that runs out is thrown away. Then the sterile sample container is placed under the valve and the product is added to the container. The container is then sealed.

Aseptic sampling at other locations in the brewery follows similar procedures. Most importantly, the sample valves must be sterilized with alcohol and then flamed to burn off the alcohol. Doing this ensures the elimination of any microbes on the sample valve.

aseptic sampling The process of withdrawing a sample for analysis so as not to contaminate or alter the sample or the source of the sample.

Membrane Filtration

One method of verifying that microbes don't exist in a sample is to concentrate the sample. Unfortunately, evaporating liquid from a wort sample will likely destroy any microbes and produce inconsistent data on their presence in the wort. In addition, if we try to swab a petri dish from a wort sample that has a very minor contamination, we might not even find the microbes.

The way around these issues is to conduct a **membrane filtration**. This process involves filtering the liquid containing the microbes across a membrane that has pores small enough not to allow the microbes to cross (pores about 0.22 mm in diameter). To start, an absorbent pad is placed in a petri dish and broth is added until the pad is soaked. The broth is typically a nutrient-rich culture broth that contains **cycloheximide**, which inhibits the growth of yeast but doesn't inhibit the growth of other microbes. Next, the liquid to be analyzed is added to the filter apparatus and a vacuum is applied to pull the liquid through the membrane. Sterile tweezers are used to transfer the membrane to the absorbent pad in the petri dish. The lid is placed on the dish and the sample is incubated for at least 24 hours.

membrane filtration The process of separating and concentrating microbes from a liquid using filtration.

cycloheximide A chemical that inhibits the growth of brewer's yeast but not other microbes.

The growth of any nonyeast microbes can be easily observed by examining the petri dish. If some are found, they can be removed and plated on differential or selective agar media to identify them.

Identifying Bacteria

Microbes can be difficult to identify. Examination of a sample under a microscope may not even reveal their presence if the sample is very dilute. This section will examine some of the ways that the brewery laboratory can do the identification. If the identification is difficult in the brewery laboratory, a sample of the microbes can be sent to an external company that specializes in microbial identification. This process can be expensive, however, so the brewery laboratory should do what it can to avoid sending a lot of samples off site.

Reading Plates

One of the simplest ways to attempt identification is to plate the unknown microbes onto various media that can select or differentiate between microbes. Many sources are available to help identify the microbes that might be growing on a particular plate, including the American Society of Brewing Chemists (ASBC) identification guide titled *Common Brewery-Related Microorganisms*. This guide provides color photos of specific microbes growing on plates and filters and what the microbes look like under the microscope. The online guide can be very helpful in identifying the microbes in a product stream (http://methods.asbcnet.org/summaries/microrefguide.aspx).

Table 14.2 lists the specific types of agar plates that can be used for determining microbial growth in a sample. This list is provided to give readers an idea of what sources exist. Many other sources on the internet or in your library provide additional specific information and can be quite helpful.

Gram Stainings

A lab experiment can be performed on an isolated microbe to determine to which of two classes the microbe belongs. While doing so often helps in identifying a troublesome microbe, it doesn't provide a definitive classification. The lab experiment that provides a definitive result relies on the fact that some bacteria have thick cell walls and other bacteria have thin cell walls. The microbes with thick cell walls are referred to as **gram positive**, because they absorb a stain first used by Hans Christian Gram in the late 1800s. These microbes appear violet after the experiment has been completed. The microbes with thin cell walls are **gram negative**, because they don't absorb the stain. These microbes appear red after the experiment has been completed.

The experiment involves placing the microbes to be analyzed on a microscope slide. The slide is heated briefly in a Bunsen burner to fix the microbes to the slide. A few drops of crystal violet are added to the slide, and after 1 minute, the slide is rinsed off with distilled water. An iodine solution is then

gram positive A microbe that turns purple when treated with Gram's stain.

gram negative A microbe that turns red when treated with Gram's stain.

added to the slide, and after 1 minute, the slide is rinsed off with ethanol and then quickly again with distilled water. Then the secondary stain, safranin, is added to the slide; after 1 minute, the safranin is rinsed off with distilled water. Next, the slide is observed under a microscope. If the microbes appear violet, they are gram positive. If they appear red, they are gram negative.

Lactobacillus and *Pediococcus* are examples of gram-positive microbes. *Pectinatus* and *Megasphaera* are examples of gram-negative microbes. Other sources can also be used to help classify common beer spoilers into their appropriate categories.

Polymerase Chain Reaction

By far the most interesting and useful tool for identifying microbes in the brewery is **polymerase chain reaction**, or **PCR**. In this technique, the DNA of the microbe is mixed with a small fragment of DNA known as a **primer**

polymerase chain reaction (PCR) A technique used to make copies of DNA using primers specific for a particular organism; can be used to detect the presence of an organism in a mixture.

primer A short, single-stranded version of DNA specific for a particular organism.

TABLE 14.2. Some differential and selective agars[a]

Agar	Description
Barney–Miller Brewery Medium (BMBM)	Selects for lactic acid-producing bacteria. May still take up to 7 days to identify all microbes.
DeMan–Rogosa–Sharpe (MRS) agar	Selects for *Lactobacillus* and *Pediococcus* bacteria.
Hsu's Lactobacillus and Pediococcus (HLP) medium	Typically used in test tubes instead of plates. Differential for *Lactobacillus* and *Pediococcus* bacteria.
Lee's Multi-Differential Agar (LMDA)	A differential agar that identifies *Lactobacillus*, *Pediococcus*, and *Enterobacter* bacteria.
Lin's Cupric Sulfate Medium (LCSM)	Selective for non-*Saccharomyces* yeasts.
MacConkey agar	A differential agar that is specific for *Enterobacter* bacteria.
Raka-Ray Agar (Raka-Ray Lactic Acid Bacteria Medium [RRLM])	Selects for lactic acid-producing bacteria. Takes 3–7 days to fully identify *Pediococcus* bacteria.
Selective Medium Megasphaera and Pectinatus (SMMP)	Selects for *Megasphera* and *Pectinatus* bacteria. Suppresses all yeast and other bacteria.
Universal Beer Agar (UBA)	A good general-purpose media for growing all the microbes in the brewery. Not differential, but with the addition of cycloheximide, can be used to focus on nonyeast microbes.
Wallerstein Laboratory Differential (WLD) medium	Contains cycloheximide, which inhibits yeast growth. Good for growing most nonyeast microbes. *Obesumbacterium* bacteria don't grow well.
Wallerstein Laboratory Nutrient (WLN) medium	Good for growing all the microbes common in the brewery. Takes 1–2 weeks for full development of plates. *Obesumbacterium* bacteria don't grow well.

[a] Courtesy S. E. Johnson and M. D. Mosher—© ASBC.

from a known microbe. If the DNA of the microbe is the same as that of the known microbe, the primer binds to the DNA. Then enzymes are added and the DNA is copied multiple times. Once enough copies have been made to allow detection, the experiment can determine that enzymes were present.

Recent advances in PCR technology allow an instrument called a thermal cycler to grow up multiple copies of the DNA (if the primer binds to the DNA of the unknown microbe). The instrument then reads the copies and determines whether the result is positive. The process has become so automated that results on the identification of any of a set of beer-spoilage microbes can be obtained within 3 hours. Remember that growing microbes on a petri dish can take up to 7 days. The PCR process is considerably faster.

The advantage to using PCR is obvious: rapid identification of a set of microbes with minimal sample preparation, performed according to the manufacturer's specifications. The disadvantages include the cost of the PCR instrument (although it may pay for itself quickly, considering the time savings) and the fact that only microbes supported by the PCR technique can be detected. If the unknown microbe isn't one of those supported, it can't be identified.

CHAPTER 15

Data Sheets for *ASBC Methods of Analysis*

This chapter includes data sheets for some of the *ASBC Methods of Analysis*. The methods are organized by the matrixes with which they are associated. For instance, methods that involve determining analytes in wort are grouped in the Wort category. Each of the methods in this chapter has been reformatted from its original version to increase ease of use in the laboratory. To view the original version of the method and its associated information, visit the *ASBC Methods of Analysis* online.*

Table 15.1 lists the methods in the same order as they appear in this chapter and identifies the types of instruments used with specific methods. The instrument recommended for a specific measurement is also indicated. In those cases in which an alternative instrument can be used, that is also indicated. Note that if an alternative instrument is suggested, a significant amount of research and experimentation may be needed to implement use of that instrument.

Any associated tables required by these data sheets are found in the online *ASBC Methods of Analysis*. Additional information, training videos, and online calculators can also be found on the *ASBC Methods of Analysis* website (methods.asbcnet.org). Readers should become familiar with this website to access additional information not included in this book.

---

*Access to the *ASBC Methods of Analysis, 14th Edition*, is provided to ASBC members and made available through corporate and academic subscriptions (methods.asbcnet.org).

TABLE 15.1. Correlation of instruments with *ASBC Methods of Analysis*[a,b]

Method of Analysis	Colorimetry	UV-Vis Spectrometry	Chromatography	Other Instrument	Chemical
Malt 4 (Extract)				hydrometer	
Malt 6A (Diastatic Power)					titration
Malt 6B (Diastatic Power)	alternative	UV-vis (415 nm)			
Malt 15 (Grist)				sieves	
Hops 6A (α- and β-Acids)		UV-vis (275, 325, 355 nm)			
Hops 13 (Total Essential Oils by Steam Distillation)					distillation
Hops 15 (Iso-α-Acids)			HPLC (UV)		
Hops 17 (Hop Essential Oils)			GC (FID)		
Wort 5 (Yeast Fermentable Extract)				hydrometer	
Wort 12 (Free Amino Nitrogen)	alternative	UV-vis (570 nm)			
Wort 22 (Fermentable and Total Carbohydrates)			HPLC (ELSD)		
Wort 23A (Bitterness, Spectrophotometry)		UV-vis (275 nm)			
Wort 23C (Bitterness, Iso-α-Acids)			HPLC (UV)		
Beer 4A (Alcohol, Volumetric)				hydrometer	
Beer 4C (Alcohol, Refraction)				refractometer	
Beer 4D (Alcohol, Chromatography)			GC (FID or TCD)		
Beer 10A (Color)	alternative	UV-vis (430, 700 nm)			
BEER 11C (Protein)		UV-vis (215, 225 nm)			
BEER 13C (Dissolved CO_2)					syringe
Beer 22A (Foam Collapse Rate)					foam tube
Beer 23A (Bitterness by UV-vis)		UV-vis (275 nm)			
Beer 23C (Bitterness by HPLC)			HPLC (UV)		
Beer 25D (Diacetyl)	alternative	UV-vis (530 nm)			
Beer 26 (Turbidity)	alternative	UV-vis (580 nm)		nephelometer	
Beer 38 (Magnesium and Calcium)					titration
Yeast 4 (Cell Counting)				microscope	
Yeast 6 (Viability)				microscope	
Yeast 9 (Giant Colony Morphology)				microscope	
Yeast 11B (Flocculation)	alternative	UV-vis (600 nm)			
Sensory Analysis 7 (Triangle Test)					people
Sensory Analysis 18 (Tetrad Test)					people

[a] Courtesy S. E. Johnson and M. D. Mosher—© ASBC.
[b] ELSD = evaporative light-scattering detector; FID = flame ionization detector; GC = gas chromatography; HPLC = high-performance liquid chromatography; TCD = thermal conductivity detector; UV = ultraviolet; UV-vis = ultraviolet visible.

ASBC Malt 4: Extract

Reagents Needed for Analysis

Malt as used for analysis standards

Use malt that confirms to the following specifications:

1. Variety: Malt made from six-row malting variety of barley recommended by the American Malting Barley Association. Available from ASBC, St. Paul, MN.
2. Moisture, % = 4.0
3. Color of laboratory wort = not over 2.0°L
4. Ratio of soluble protein to total protein in % = 41.0
5. Fine-coarse difference, not greater than 2.0% extract, dry basis
6. Viscosity, not over 1.50 cP
7. Assortment: From a B grade barley malt meeting above specifications, take the portion passing through a 7/64-inch screen and remaining on a 6/64-inch screen for actual standardizing operation.

Malt sample of interest

Distilled or deionized (DI) water, 46–48°C and 70–71°C

Materials Needed

Analytical balance
Filter paper (509 fluted paper recommended)
20 cm short-stem glass funnels
Mashing apparatus*
20 cm watch glasses

Densitometer
500 mL Erlenmeyer flasks
Mash beakers
Mill
Thermometer

Procedure

To determine if mill is correctly adjusted for milled malt size, perform analysis: Malt 15.

Prepare samples for analysis:
1. Weigh 50 g of milled malt into a preweighed mash beaker.
2. Add 200 mL of 46–48°C DI water so that the resulting solution is 45°C.
3. Stir solution with a glass rod to prevent clumps.
4. Rinse the glass rod and beaker with a small amount of DI water.
5. Report the odor as aromatic/slightly aromatic.
6. Report unpleasant odors as musty/green/stale/other.

---

*A circulating water bath that will firmly hold a number of mash beakers. Heating must be able to raise the temperature 1°C/minute up to 70°C. Each mashing beaker has a stirrer operating at a uniform speed (80–100 rpm).

Perform analysis:
1. Allow the samples to sit at 45°C for 30 minutes.
2. After 30 minutes at 45°C, raise the mash temperature 1°C/minute until solution is 70°C.
3. Add 100 mL DI water previously heated to 70–71°C.
4. Hold the mash at 70°C for 60 minutes. Temperature deviations during the mashing procedure must not exceed 0.5°C.
5. After 60 minutes at 70°C, cool mash to room temperature in 10–15 minutes by gradual addition of cold water.
6. Dry the outside of each beaker and place the beaker on the balance.
7. Add DI water to the beaker to adjust the weight of the contents inside the beaker to 450 g.
8. Stir the mixture thoroughly, without splashing, and then pour the entire contents of the beaker into a funnel fitted with fluted filter paper and discharging into a 500 mL Erlenmeyer flask.
9. Collect 200 mL for analysis. This should take less than 1 hour.
10. Measure the specific gravity of the 200 mL sample using a hydrometer. Be sure to compare with the density of water at the same temperature. Convert specific gravity to degrees Plato.

Calculations

Obtain the moisture (M) and degrees Plato (°P) of the wort obtained, and use the values in the calculation below to determine extract %. Moisture should be ~5% and listed on the malt specifications you receive when the malt arrives.

Extract of malt, E (as is), %:

$$E \text{ (as is)} = \frac{°P \times (M + 800)}{100 - °P}$$

Extract of malt, E (dry basis), %:

$$E \text{ (dry basis)} = \frac{100 \times E \text{ (as is)}}{100 - M}$$

Data Sheet ASBC Malt 4

Sample	Moisture (*M*)	Specific gravity	Degrees Plato (°P)	*E* (as is)	*E* (dry basis)
Malt Standard 1					
Malt Standard 2					
Malt Standard 3					

Notes:

Data sheets may be reproduced and are available online: asbcnet.org/BSL

ASBC Malt 6A: Diastatic Power

Reagents Needed for Analysis

Acetate buffer solution (Reagent A)

In a 1,000 mL volumetric flask:

1. Add 500 mL of 1 N acetic acid.
2. Dissolve 68 g of sodium acetate.
3. Fill to 1,000 mL with distilled or deionized (DI) water.

Store in glass bottle. Mixture is stable for 1 month.

2% special starch solution (Reagent B)

In an appropriately sized volumetric flask:

1. Weigh the equivalent of 2 g of starch for each 100 mL of solution required.
2. Dissolve the starch in a small amount of cold water until it makes a paste.
3. Add paste to boiling water that accounts for 75% of solution's final volume.
4. Rinse remaining starch into solution and boil for 2 minutes.
5. Add 10% of final volume of cold DI water to solution; mix and cool to 20°C.
6. Once cooled, add 2 mL of acetate buffer for each 100 mL of final volume.
7. Fill to final volume with DI water and invert to mix.

Store in glass bottle. Mixture is stable for 48 hours.

Acetic acid–salt solution (Reagent C)

In a 1,000 mL volumetric flask:

1. Add 300 mL of DI water.
2. Dissolve 70 g of potassium chloride and 20 g of zinc sulfate.
3. Add 200 mL of glacial acetic acid.
4. Fill to volume with DI water.

Store in glass bottle. Mixture is stable for 1 month.

Alkaline ferricyanide solution 0.05 N (Reagent D)

In a 1,000 mL volumetric flask:

1. Add 500 mL of DI water.
2. Dissolve 16.5 g of potassium ferricyanide and 22 g of sodium carbonate.
3. Fill to volume with DI water.

Store in glass bottle. Mixture is stable for 1 month.

Potassium iodide solution (Reagent E)

In a 100 mL volumetric flask:

1. Add 50 mL of DI water.
2. Dissolve 50 g of potassium iodide.
3. Add 1–2 drops of concentrated sodium hydroxide.
4. Fill to volume with DI water.

Store in glass bottle. Mixture is stable for 1 month.

Sodium thiosulfate solution (Reagent F)

In a 1,000 mL volumetric flask:

1. Add 200 mL of DI water.
2. Dissolve 12.41 g of sodium thiosulfate and 3.8 g of sodium borate.
3. Fill to volume with DI water.

Store in glass bottle. Mixture is stable for 1 month.

Potassium dichromate (Reagent G)

Concentrated hydrochloric acid (Reagent H)

Sodium chloride solution (Reagent I)

In a 1,000 mL volumetric flask:

1. Add 500 mL of DI water.
2. Dissolve 5 g of sodium chloride.
3. Fill to volume with DI water.

Store in glass bottle. Mixture is stable for 1 month.

Materials Needed

Analytical balance	Hot plate
Filter paper, 32 cm	Funnels, glass, short-stem, 20 cm
Mash beakers	Erlenmeyer flasks, 125 mL, 250 mL, 1,000 mL
Mill, fine grind	Burettes, 10 mL semimicro, 50 mL
Stopwatch	Water bath, 20°C, boiling

Procedure

To determine if mill is correctly adjusted for milled malt size, perform analysis: Malt 15.

Prepare blank for analysis:

1. Perform the same steps as for samples. However, be sure to add the 20 mL of 0.5 N sodium hydroxide before adding the 200 mL of special starch solution (Reagent B).

Prepare samples (digested starch sample) for analysis:
1. Weigh 25 g of milled malt into a preweighed mash beaker.
2. Transfer malt to a 1,000 mL Erlenmeyer flask.
3. Add 500 mL of Reagent I and swirl every 20 minutes for 150 minutes at 20°C.
4. Filter the solution into a 1,000 mL Erlenmeyer flask.
5. Refilter the first 50 mL of filtrate by pouring it back over the filter.
6. After filtration, transfer 20 mL of filtrate into a 100 mL volumetric flask.
7. Dilute to volume with Reagent I.
8. Transfer 10 mL of solution to a 250 mL volumetric flask.
9. Add 200 mL of Reagent B and mix for 30 minutes at 20°C.
10. Add 20 mL of 0.5 N sodium hydroxide and quickly mix by inverting.
11. Fill flask to volume with DI water and invert to mix.

Perform analysis:
1. Pipette 10 mL of alkaline ferricyanide solution into a 125 mL Erlenmeyer flask.
2. Pipette 5 mL of sample (blank or digested starch solution) into the 125 mL flask.
3. Mix solution for 20 minutes in a boiling water bath.
4. After 20 minutes, cool the flask under cool, running water to room temperature.
5. Add 25 mL acetic acid–salt solution (Reagent C).
6. Add 1 mL of potassium iodide solution (Reagent E) and mix well.
7. Titrate with 0.05 M sodium thiosulfate solution (Reagent F).
8. Titrate until blue color disappears. Report this value to 0.01 mL.

Calculations

For the calculations, the volume (in milliliters) of Reagent F needed to titrate the sample is the variable A; the volume (in milliliters) of Reagent F needed to titrate the blank is the variable B; and the % moisture of the malt is the variable M.

Diastatic power, as is:

$$DP° \text{ (as is)} = (B - A) \times 23$$

Diastatic power, dry basis:

$$DP° \text{ (dry basis)} = \frac{DP° \text{ (as is)} \times 100}{100 - M}$$

Data Sheet   ASBC Malt 6A

Sample	Moisture (*M*)	mL of Reagent F needed for blank (*B*)	mL of Reagent F needed for sample (*A*)	DP° (as is)	DP° (dry basis)
Malt Standard 1					
Malt Standard 2					
Malt Standard 3					

Notes:

ASBC Malt 6B: Diastatic Power (Rapid Method)

Reagents Needed for Analysis

Dextrose

Acetate buffer solution (Reagent A)

In a 1,000 mL volumetric flask:

1. Add 500 mL of 1 N acetic acid.
2. Add 68 g of sodium acetate slowly and swirl to dissolve.
3. Fill to volume with distilled or deionized (DI) water.

Store in glass bottle. Mixture is stable for 1 month.

2% special starch solution (Reagent B)

In an appropriately sized volumetric flask:

1. Weigh the equivalent of 2 g of starch for each 100 mL of solution required.
2. Muddle the starch in a small amount of cold water until it makes a paste.
3. Add paste to boiling water that accounts for 75% of solution's final volume.
4. Rinse remaining starch into solution and boil for 2 minutes.
5. Add 10% of final volume of cold DI water to solution; mix and cool to 20°C.
6. Once mixture has cooled, add 2 mL of acetate buffer for each 100 mL of final volume.
7. Fill to final volume with DI water and invert to mix.

Store in glass bottle. Mixture is stable for 48 hours.

Starch manufactured specifically for diastatic power determinations is available from ASBC.

Sodium hydroxide solution, 0.5 N (Reagent C)

Alkaline diluent solution (Reagent D)

In separate 250 mL beakers:

1. Dissolve 14.7 g of trisodium citrate in one beaker.
2. Dissolve 1.47 g of calcium chloride in the other beaker.
3. Combine the solutions in a 1,000 mL volumetric flask.
4. Dissolve 20 g of sodium hydroxide.
5. Mix and dilute to volume with DI water.

Store in glass bottle. Mixture is stable for 1 month.

PAHBAH working solution (Reagent E)
In a 1,000 mL volumetric flask:

1. Add 500 mL of alkaline diluent solution (Reagent D).
2. Dissolve 5 g of solid PAHBAH (para-hydroxybenzoic acid hydrazide).
3. Mix and dilute to volume with Reagent D.

Store in amber glass bottle. Mixture is stable for 24 hours. Do not use if reagent yellows.

Ammonium hydroxide solution, 6.0 mM (Reagent F)
In a 1 L volumetric flask:

1. Pipette 65 mL of ammonium hydroxide.
2. Dilute to volume with DI water. This is now a 1.0 M solution.
3. Pipette 6.0 mL of the 1.0 M solution into a different 1 L volumetric flask.
4. Fill to volume with DI water.

Store in glass bottle. Mixture is stable for 1 month.

Materials Needed

Spectrophotometer (415 nm)	Quartz cuvettes
Analytical balance	Volumetric pipettes, 5, 10, and 20 mL
200 mL fast pipette	Volumetric flasks, 1 L
Micropipette 0.2, 1.0 mL	Erlenmeyer flasks, 50 mL, 500 mL
Test tubes, 25 × 150 mm	Filter paper, fluted
Mash beaker*	Mill, fine grind
Stopwatch	Water bath, 20°C, boiling

Procedure

To determine if mill is correctly adjusted, perform analysis: Malt 15.

Prepare blank for analysis:
1. Perform the same steps as for samples. However, be sure to add the 1.2 mL of 0.5 N sodium hydroxide before adding the 20 mL of special starch solution (Reagent B).

Prepare samples (digested starch sample) for analysis:
1. Weigh 10 g of fine milled malt into a preweighed mash beaker.
2. Transfer malt to a 500 mL Erlenmeyer flask.
3. Add 200 mL of 6.0 mM ammonium hydroxide solution (Reagent F).
4. Stopper the flask, swirl, and let solution stand for 10 minutes at 20°C. Swirl solution every 2 minutes. Do not invert.
5. After 10 minutes, filter the solution into a 500 mL Erlenmeyer flask.

---

*Mash beakers and stirrers of brass, pure nickel, or stainless steel may be used; copper mash beakers must not be used.

6. Refilter the first 50 mL of filtrate by pouring it back over the filter.
7. After filtration, pipette 0.2 mL of filtrate into a 50 mL Erlenmeyer flask at 20°C.
8. Add 20 mL of special starch solution (Reagent B) slowly, mixing as you add.
9. Keep solution at 20°C for exactly 10 minutes from the start of Reagent B addition.
10. After 10 minutes, add 1.2 mL of 0.5 N sodium hydroxide and mix.

Standardize samples:
1. Perform the same steps as for samples, but prepare starch solution from dextrose that has been dried at 103°C for 4 hours.
2. Weigh accurately 0.2, 0.4, 0.6, 0.8, and 1.0 g of anhydrous dextrose and add to five separate 1,000 mL volumetric flasks.
3. Fill the flasks to volume, and then treat them as if they are samples. DI water can be used as a blank for standardization. Plot each absorbance for the standard curve. From the line of best fit, the slope will be referred to as "*b*" and the *y*-intercept *f* will be referred to as "*a*" in the calculations below.

Perform analysis:
1. Add 5 mL of PAHBAH to a test tube.
2. Pipette 0.2 mL of sample (blank or digested starch solution) into the test tube.
3. Mix solution and submerge for 4 minutes in a boiling water bath.
4. After 4 minutes, cool the flask to 20°C.
5. Add 10 mL of DI water, mix, and read the absorbance at 415 nm. This value will be referred to as "A_{415}" in the calculations below.

Calculations

The absorbance is the variable A_{415}, and the % moisture is the variable M. The slope of the standard curve is b, and the *y*-intercept is the variable a.

Reducing sugars (dextrose, g/L):

$$\text{sugars} = a + (b \times A_{415})$$

Diastatic power, as is:

$$\text{DP}° \text{ (as is)} = a + (b \times \text{sugars})$$

Diastatic power, dry basis:

$$\text{DP}° \text{ (dry basis)} = \frac{\text{DP}° \text{ (as is)} \times 100}{100 - M}$$

Data Sheet ASBC Malt 6B

Sample	Moisture (*M*)	Absorbance at 415 nm	Reducing sugars (dextrose g/L)	DP° (as is)	DP° (dry basis)
Dextrose Standard 1					
Dextrose Standard 2					
Dextrose Standard 3					
Dextrose Standard 4					
Dextrose Standard 5					

Notes:

ASBC Malt 15B: Grist—By Manual Sieve

Reagents Needed for Analysis
Malt

Materials Needed
US standard test sieves, 8-inch diameter (nos. 10, 30, 60, 100), a sieve cover or lid, and a bottom pan

Brush for cleaning

Rubber balls, 5/8-inch diameter

Analytical balance

Dark-colored paper

Timer

Procedure

Perform analysis:
1. Place three rubber balls on each US standard test sieve.
2. Stack the four sieves in order by size, with 10 on the top and 100 on the bottom. Place the stack onto the sieve bottom pan.
3. Weigh 100–130 g of milled malt and place it on top of the no. 10 sieve.
4. Cover with the lid and sieve for 3 minutes by sliding sieves vigorously over a smooth surface. Ideally, move sieve back and forth 18 inches each direction per second.
5. Tap the sieve stack sharply against a flat surface every 15 seconds.
6. After 3 minutes, empty and use the brush to clean each sieve, as well as what is left in the bottom of the pan, onto different weighing dishes.
7. Weigh each fraction to the nearest 0.1 g. Each fraction will be referred to a "W_x" in the calculations, where x is the sieve size. "W_p" will refer to the weight of flour in the bottom of the pan.
8. The total weight combined will be referred to as "W_t" in the calculations. W_t should be within 0.5 g of the original mass of malt placed into the sieves. If W_t is not within 0.5 of the original mass, do not use the collected data. Instead, repeat the entire analysis using a fresh sample.

Calculations

The masses (in grams) of the contents of the sieves are the variables W_{10}, W_{30}, W_{60}, and W_{100}; W_p is the mass of the contents landing in the bottom pan. The x values in W_x and $\%_x$ correspond to the sieve size being used.

Weight, total:

$$W_t = W_{10} + W_{30} + W_{60} + W_{100} + W_p$$

Percent composition:

$$\%_x = \frac{W_x}{W_t}$$

Data Sheet ASBC Malt 15B

Sample	W_t	$\%_{10}$	$\%_{30}$	$\%_{60}$	$\%_{100}$	$\%_P$
Malt Standard 1						
Malt Standard 2						
Malt Standard 3						

Notes:

Data sheets may be reproduced and are available online: asbcnet.org/BSL

ASBC Hops 6A: α- and β-Acids in Hops and Hop Pellets— α- and β-Acids by Spectrophotometry

Reagents Needed for Analysis

Toluene, high-performance liquid chromatography (HPLC) grade

Alkaline methanol

In a 500 mL volumetric flask:

1. Add 200 mL of methanol.
2. Add 1 mL of 6.0 N sodium hydroxide solution.
3. Fill to volume with distilled or deionized (DI) water and mix well.

Store in glass bottle with plastic cap. Mixture is stable for 1 month.

Materials Needed

Spectrophotometer (275, 325, and 355 nm)
Analytical balance
Erlenmeyer flasks, 250 mL
Pipette, 5 mL
Quartz cuvettes
Shaker, mechanical, rotary
Volumetric flasks, 100 mL

Procedure

Prepare blank for analysis:

1. Perform the same steps as for sample preparation, but use pure toluene in place of the extract solution.

Prepare samples (digested starch sample) for analysis:

1. Weigh 5.000 ± 0.001 g of freshly ground hops into a 250 mL Erlenmeyer flask.
2. Add 100 mL of toluene.
3. Stopper solution and shake for 30 minutes, preferably on a mechanical or rotary shaker at 200 rpm.
4. Let solution stand until solids have settled. The time to settle can be reduced by using a centrifuge.

Perform analysis:

1. Dilute 5 mL of extract to 100 mL with methanol.
2. Pipette 3 mL of this solution into a 50 mL volumetric flask and dilute to the mark with alkaline methanol.
3. Record the absorbances at 355 nm, 325 nm, and 275 nm. The absorbances at 355 nm and 325 nm should be below 1.0. If not, repeat step 2 but dilute to 100 mL with alkaline methanol. The dilution factor (see next page) will be 0.333 if this is done.
4. Samples must be analyzed quickly to avoid decomposition of alpha and beta acids.

Calculations

The dilution factor shown below is based on the quantities used in this procedure as written. Alternatively, in step 2 of the analysis, 1.5 mL of the solution can be diluted to 25 mL with alkaline methanol to obtain a dilution factor of 0.667, or 1.5 mL can be diluted to 50 mL to obtain a dilution factor of 0.333. The variables A_{355}, A_{325}, and A_{275} refer to the absorbances at those wavelengths.

Dilution factor, d:

$$d = 0.667$$

α-acids, %:

$$a = d \times (-51.56 A_{355} + 73.79 A_{325} - 19.07 A_{275})$$

β-acids, %:

$$\beta = d \times (55.57 A_{355} - 47.59 A_{325} + 5.10 A_{275})$$

Data Sheet   ASBC Hops 6A

Sample	mL toluene used for extraction	mL methanol used for first dilution	mL first dilution pipetted	mL alkaline methanol solution added to first dilution	Absorbance 355 nm	Absorbance 325 nm	Absorbance 275 nm	% α-acids	% β-acids

Notes:

Data sheets may be reproduced and are available online: asbcnet.org/BSL

ASBC Hops 13: Total Essential Oils in Hops and Hop Pellets by Steam Distillation

Reagents Needed for Analysis
Hops

Materials Needed

Distillation setup	Heating mantle
Analytical balance	5,000 mL round-bottom flask

Procedure

Prepare samples (digested starch sample) for analysis:
1. Weigh the appropriate quantity of hops from the list below (to the nearest 0.5 g) into the 5,000 mL round-bottom flask:
   - 100–120 g of coarsely ground hops or unground hop pellets
   - 300–400 g of dried, frozen, unground whole-hop compressed minibales
2. Add 3,000 mL of water to the round-bottom flask.

Perform analysis:
1. Bring the solution to a boil. Using Teflon chips may assist in providing an even boil.
2. Regulate the heat so that 25–35 drops are collected per minute. Be careful to limit the heat so that foam does not proceed into the distillation head.
3. Boil for 4–7 hours. Add water to the round-bottom flask periodically to maintain a minimum of 1,500 mL of water.
4. After boiling, separate the oil from the water in the receiver using either a separatory funnel or a disposable pipette. Record the volume of oil to the nearest 0.05 mL. If using a graduated cylinder, the oil will be the top layer in the cylinder.

Calculations

The volume of oil is reported in milliliters, and the mass of the hops added is reported in grams.

Total essential oils, ml/100 g of hops:

$$\text{total essential oil} = \frac{\text{volume of oil} \times 100}{\text{mass of hops added}}$$

Data Sheet ASBC Hops 13

Sample	Grams of hops added	Boil time	Volume of essential oil	Total essential oil (mL/100 g)

Notes:

Data sheets may be reproduced and are available online: asbcnet.org/BSL

ASBC Hops 15: Iso-α-Acids in Isomerized Hop Pellets by High-Performance Liquid Chromatography 1993

Reagents Needed for Analysis

n-Butyl acetate, reagent grade

Phosphoric acid, 3M

In a 50 mL volumetric flask:

1. Add 10.2 mL of 85% phosphoric acid, reagent grade.
2. Fill to volume with distilled or deionized (DI) water.

Store in glass bottle. Mixture is stable for 1 month.

Water, high-performance liquid chromatography (HPLC) grade

Ethylenediaminetetraacetic acid (EDTA), 0.1M in acidic methanol

In a 1,000 mL volumetric flask:

1. Add 0.5 mL of 85% phosphoric acid, reagent grade.
2. Dilute to 1,000 mL with methanol, HPLC grade, to make an acidic methanol solution.
3. In a second 1,000 mL volumetric flask, add 29.229 g of dry EDTA salt.
4. Add 500 mL of the acidic methanol solution.
5. Stir until dissolved and then fill to volume with the acidic methanol solution.

Store in glass bottle. Mixture is stable for 1 month.

Iso-α-acid standard

The international calibration standard with known concentration of iso-α-acids obtained from ASBC or equivalent

Materials Needed

HPLC with 270 nm ultraviolet detector, 10 µL injector valve

Quaternary pump, degasser, and column heater

Chromatographic column, 250 × 4.6 mm, 5 µm Nucleosil C18; XDB-C18 Rapid Resolution, 4.6 × 12.5 mm, 5 µ; or equivalent

Guard column and assembly, Eclipse XDB-C18, 4.6 × 12.5 mm, 5 µ or equivalent

Data acquisition software	Erlenmeyer flasks, 250 mL
Analytical balance	Volumetric pipettes, 3, 5, and 100 mL
Mechanical shaker	Volumetric flasks, 50, 100, and 1,000 mL
Aluminum foil	Ultrasonic bath

Procedure

HPLC operating conditions:

Mobile phase: Methanol:water:3M phosphoric acid: EDTA solution 81.2:18.4:0.25:0.10

Temperature: 20–40°C

Flow rate: 1.0–1.4 mL/minute

Prepare standard for analysis:
1. Weigh ~50 g (record to 0.1 g) of the iso-α-acids standard into a 100 mL volumetric flask.
2. Add 40 mL of methanol, HPLC grade, and dissolve.
3. Sonicate in ultrasonic bath for 5 minutes to degas the sample.
4. Bring to volume with methanol, HPLC grade.
5. Wrap flask in aluminum foil and store at −10°C.

Solution will be stable for 1 month.

Prepare samples for analysis:
1. Weigh 5 g (record to 0.001 g) of freshly ground hop pellets into a 250 mL Erlenmeyer flask.
2. Pipette 3 mL of 3 M phosphoric acid solution.
3. Pipette 100 mL of n-butyl acetate into the Erlenmeyer flask.
4. Stopper tightly and shake for 30 minutes.
5. Clarify solution by centrifuge, or let solution stand for 15 minutes.
6. Pipette 5.0 mL of the solution into a 50 mL volumetric flask.
7. Fill to volume with methanol, HPLC grade.
8. Mix contents and then filter or centrifuge.

Perform analysis:
1. Inject 10 µL of standard into the HPLC twice.
2. Inject 10 µL of samples, one at a time, into the HPLC.
3. Inject 10 µL of standard into the HPLC twice.

Calculations

The areas under the peaks in each HPLC chromatogram are determined using the HPLC software. The areas of all the standard peaks are averaged before being used in the calculations.

Response factor, RF:

$$RF = \frac{\text{average combined area of the three iso-}\alpha\text{-acid standard peaks} \times 100}{\text{mass of standard used} \times \%\text{ iso-}\alpha\text{-acids reported in the standard}}$$

% Iso-α-acids, IAA, in pellets:

$$\%IAA = \frac{\text{combined area of the three peaks} \times 100}{RF \times \text{mass of hops used}}$$

Data Sheet ASBC Hops 15

Sample	Average peak area of standard	Mass of standard used	% IAA reported in standard	RF value	Combined area of sample peaks	Mass of pellets used	% IAA in pellets

Notes:

Data sheets may be reproduced and are available online: asbcnet.org/BSL

ASBC Hops 17: Hop Essential Oils by Capillary Gas Chromatography–Flame Ionization Detection

Reagents Needed for Analysis

Pentane or hexane, HPLC grade

2-Octanol, reagent grade

Materials Needed

Gas chromatograph (GC) with flame ionization detector (FID)

GC column: 30 m × 0.25 mm, 25 µm film thickness Supelcowax 10, or equivalent column

Sample vials, 1.8 mL

Crimp seals for sample vials

Procedure

GC operating conditions:

Carrier gas:	Helium
Split ratio:	1:50
Injection port:	200°C
FID:	260°C
Oven temperature:	50°C for 1 minute. Increase temperature from 50 to 260°C at 3°C per minute. Remain at 260°C for 15 minutes

Prepare sample with internal standard for analysis:
1. Pipette 1 mL of 2-octanol, reagent grade, into a 100 mL volumetric flask.
2. Fill to volume with pentane or hexane, HPLC grade.
3. Pipette 100 µL of hop oil into a 1.8 mL sample vial.
4. Pipette 900 µL of 2-octanol solution into the sample vial.
5. Mix the solution until dissolved.

Perform analysis:
1. Inject 1 µL of solution into the GC.
2. Analysis run time will be 86 minutes (Fig. 15.1).

Calculations

The individual components in the sample are determined by comparison to the known standard sample. The HPLC software will provide the area under each peak in the chromatogram.

Mass of internal standard in prepared sample:

Using the procedure as written, the mass of the internal standard in the sample should be 0.00737 g.

Component in oil, g/mL:

$$\text{component} = \frac{\text{weight of internal standard}}{\text{area of internal standard peak}} \times \frac{\text{area of component}}{\text{volume of essential oil in sample}}$$

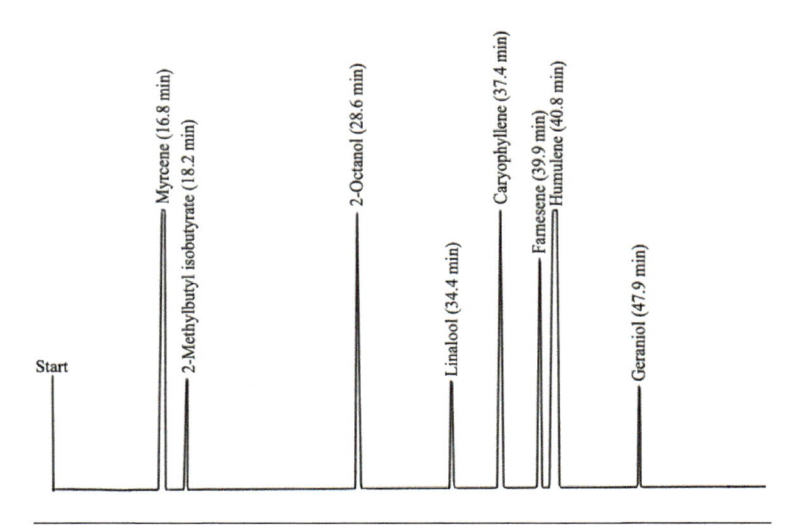

FIG. 15.1. Sample chromatogram, with selected components shown. (Adapted from ASBC Hops 17. Courtesy S. E. Johnson and M. D. Mosher—© ASBC.)

Data Sheet ASBC Hops 17

Sample	Component	Mass of internal standard (0.00737 if prepared as described)	Area for internal standard peak	Area for component peak	Volume of essential oil in sample (0.1 if prepared as described)	Component amount (g/mL)

Notes:

Data sheets may be reproduced and are available online: asbcnet.org/BSL

ASBC Wort 5: Yeast Fermentable Extract

Reagents Needed for Analysis
Brewer's yeast

Materials Needed

Stir plate and stir bar
Funnel
Water bath
Erlenmeyer flask, 500 mL
Filter paper
Fermentation lock

Procedure

Perform regular fermentation analysis:
1. Determine specific gravity of wort (original gravity, OG) by hydrometer or pycnometer, and calculate degrees Plato as shown below (initial °Plato).
2. Add 250 mL of wort sample to be analyzed to a 500 mL Erlenmeyer flask fitted with a fermentation lock filled with water.
3. Add 5 g of Brewer's yeast, and hold solution at 15–25°C during fermentation.
4. Swirl flask several times each day for 48 hours.
5. Filter wort through filter paper, and determine specific gravity of fermented wort (final gravity, FG) by hydrometer or pycnometer. Calculate degrees Plato using the equation below (end °Plato).

Perform rapid fermentation analysis:
1. Determine original gravity (OG) of wort by hydrometer or pycnometer, and calculate degrees Plato using the equation below (initial °Plato).
2. Add 32 g of active Brewer's yeast (from fermenter) to 200 mL of wort in a 500 mL Erlenmeyer flask fitted with a fermentation lock.
3. Hold solution at 20°C in water bath.
4. Stir solution at 120 rpm for 5 hours.
5. Filter wort through filter paper, and determine specific gravity of FG by hydrometer or pycnometer. Calculate degrees Plato using the equation below (end °Plato).

Calculations

°Plato:

$$°Plato = 135.997(SG)^3 - 630.272(SG)^2 + 1111.14(SG) - 616.868$$

where SG is specific gravity.

Real Degree of fermentation, %:

$$\text{fermentation \%} = \frac{100(\text{inital °Plato} - \text{end °Plato})}{\text{initial °Plato}} \times \frac{1}{1 - (0.005151 \times \text{end °Plato})}$$

Data Sheet ASBC Wort 5

Sample	Original specific gravity (OG)	Initial °Plato	Final specific gravity (FG)	End °Plato	Fermentation (%)

Notes:

Data sheets may be reproduced and are available online: asbcnet.org/BSL

ASBC Wort 12: Free Amino Nitrogen (International Method)

Reagents Needed

Ninhydrin solution (Reagent A)

In a 100 mL volumetric flask:

1. Add ~75 mL of distilled or deionized (DI) water.
2. Add 10.0 g of $Na_2HPO_4 \cdot 12\ H_2O$, 6.0 g of KH_2PO_4, and 0.5 g of ninhydrin. Swirl to dissolve; gently heat if needed.
3. Add 0.3 g of fructose and swirl to dissolve.
4. Dilute to 100 mL.

Store in an amber bottle and keep cold. Solution should be remade every 2 weeks.

Dilution solution (Reagent B)

In a 1 L amber reagent bottle:

1. Add 600 mL of DI water.
2. Add 2 g of KIO_3. Swirl to dissolve.
3. Measure out 400 mL of 95% ethanol and add it to the solution.

Store in an amber bottle and keep cold. Solution should be remade every month.

Glycine stock solution (Reagent C)

In a 100 mL volumetric flask:

1. Add ~50 mL of DI water.
2. Add 0.1072 g of glycine. Swirl to dissolve.
3. Dilute to 100 mL.

Store in a bottle and keep cold. Solution should be remade every month.

Glycine standard solution (Reagent D)

In a 100 mL volumetric flask:

1. Add 1.0 mL of Reagent C.
2. Dilute to 100 mL with DI water.

Store in a bottle. Solution should be remade daily.

Materials Needed

16 × 150 mm test tubes (3 per sample)
Volumetric flask (100 mL wort; 50 mL beer)
500 mL beaker (for waste)
Glass marbles to place on tops of test tubes
Volumetric pipette (2 mL)
Vortex mixer

Procedure

Prepare samples:
1. Standard: Transfer 2 mL of glycine standard (Reagent D) to test tubes in triplicate.
2. Wort sample: Dilute 1 mL of wort to 100 mL with DI water in a volumetric flask.
3. Beer sample: Dilute 1 mL of degassed beer to 50 mL with DI water in a volumetric flask.
4. Blank sample: Transfer 2 mL of DI water to test tubes in triplicate.

Perform analysis:
1. Add 1 mL of ninhydrin reagent (Reagent A) to each test tube. Stopper all tubes with glass marbles. Briefly vortex each tube.
2. Place test tubes in a boiling water bath for 16 minutes.
3. Cool tubes for 20 minutes in a room-temperature water bath.
4. Add 5 mL of dilution solution (Reagent B) to each test tube using a volumetric pipette.
5. Mix contents of each test tube thoroughly using a vortex mixer.
6. Measure absorbance of each solution at 570 nm within 30 minutes.

Calculations

Free amino nitrogen (FAN) is determined by measuring the absorbance at 570 nm of the sample (A_{sample}), the average of the blank (avg A_{blank}), and the average of the standard (avg $A_{standard}$). Wort samples have a dilution factor of 100. Beer samples have a dilution factor of 50. FAN should be reported as a whole number.

$$\text{FAN (ppm)} = \frac{(A_{sample} - \text{avg } A_{blank})}{(\text{avg } A_{standard} - \text{avg } A_{blank})} \times 2 \times \text{dilution}$$

Data Sheets ASBC Wort 12

Trial	A_{blank}	Trial	$A_{standard}$
Blank 1		Standard 1	
Blank 2		Standard 2	
Blank 3		Standard 3	
Average blank		Average standard	

Sample	Dilution (100 for wort; 50 for beer)	A_{sample}	A_{sample} – avg A_{blank}	FAN (ppm)	Average FAN for sample (ppm)

Notes:

ASBC Wort 22: Wort and Beer Fermentable and Total Carbohydrates by High-Performance Liquid Chromatography (HPLC)

Reagents Needed for Analysis

Reagent water, 18MΩ resistivity or greater
D-(–)-Fructose
D-(+)-Glucose
Maltose monohydrate
Maltotriose
Acetonitrile, high-performance liquid chromatography (HPLC) grade

Materials Needed

HPLC with 270 nm ultraviolet (UV) detector, quaternary pump, vacuum degasser, thermostatted column compartment, evaporative light scattering detector (ELSD) detector, N_2 generator

Chromatographic column: Prevail carbohydrate ES column, 250 × 4.6 mm, 5 µm

Guard column: Prevail carbohydrates ES guard column 7.5 × 4.6 mm, 5 µm

Pipettes, 1.0, 5.0, and 20.0 mL

Volumetric flask, 25 mL

Analytical balance

Procedure

ELDS settings:

Tube temperature:	110°C
Gas flow:	3.0 L/minute
Gain:	1

HPLC operating conditions:

Mobile phase:	A = acetonitrile; B = Reagent water
Temperature:	30°C
Flow rate:	1.0 mL/minute
Upper pressure:	300 bar
Flow ramp:	100.0 mL/minute
Draw speed:	100 µL/minute
Eject speed:	100 µL/minute

Flow timetable:

45-minute run time	%A	%B
0	80	20
22	70	30
30	50	50
40	80	20
45	80	20

Prepare standard for analysis:
1. Add ~10 mL of reagent-grade water to a 25 mL volumetric flask.
2. Add 0.1010 g of dry fructose to flask.
3. Add 0.2020 g of dry glucose to flask.
4. Add 0.1010 g of dry maltose to flask; the monohydrate must be dried.
5. Add 0.0354 g of dry maltotriose to flask.
6. Swirl to dissolve. Then dilute flask to volume with reagent-grade water.

Prepare samples for analysis:
1. Degas all beer samples.
2. Dilute samples according to the method guidelines provided by the ELSD manufacturer.

Perform analysis:
1. Inject 10 µL of standard and samples, one at a time (or the volume recommended by the ELSD manufacturer).
2. Establish a second-order fit calibration curve for each carbohydrate. This can be done using the HPLC software.

Calculations

The area of each carbohydrate in the standard is determined by the HPLC software. Using a second-order curve fit equation (see below), the values of the intercept, A, and B are determined from the concentration of a specific carbohydrate in the standard (as %w/w) and the area of that peak in the chromatogram. Then the concentration of the same carbohydrate in the sample (as %w/w) can be determined from the area and the second-order curve.

Assuming that the standards are 99% pure when they are prepared according to the procedure, the standard concentration of fructose is 0.400 %w/w, of glucose is 0.800 %w/w, of maltose is 0.400 %w/w, and of maltotriose is 0.142 %w/w.

Peak area:

area = intercept + A(concentration of Std carb) + B(Concentration of Std carb)^2

Data Sheets ASBC Wort 22

Standard	Concentration (%w/w)	Area (from HPLC)	Intercept (from equation)	A (from equation)	B (from equation)
Fructose	0.400				
Glucose	0.800				
Maltose	0.400				
Maltotriose	0.142				

Sample	Fructose area	Glucose area	Maltose area	Maltotriose area	Fructose concentration (%w/w)	Glucose concentration (%w/w)	Maltose concentration (%w/w)	Maltotriose concentration (%w/w)

Notes:

Data sheets may be reproduced and are available online: asbcnet.org/BSL

ASBC Wort 23A: Wort Bitterness—Bitterness Units by Spectrophotometry

Reagents Needed for Analysis

Reagent water, 18MΩ resistivity or greater

3 N Hydrochloric acid
In a 100 mL volumetric flask:

1. Add 29.4 mL of concentrated HCl.
2. Fill to volume with reagent water.
3. Transfer to 250 mL glass container for safety.

Store in glass bottle. Mixture is stable for 1 month.

Isooctane

Materials Needed

Spectrophotometer, 270 nm
Centrifuge
Plastic stoppers
Graduated cylinders, 30 mL, 100 mL
Volumetric flask, 100 mL
Analytical balance

Quartz cuvette
50 mL centrifuge tubes
Pipettes, 1.0 mL, 5.0 mL, 20.0 mL
Glass container, 250 mL
Mechanical shaker

Procedure

Prepare blank for analysis:
1. Use isooctane as the blank for the instrument.

Prepare samples for analysis:
1. Pipette 5.0 mL of wort or degassed beer into a 50 mL centrifuge tube.
2. Pipette 5.0 mL of reagent water into the centrifuge tube.
3. Pipette 1 mL of 3N HCl and 20 mL of isooctane into the centrifuge tube.
4. Shake samples using a mechanical shaker for 15 minutes
5. To remove emulsion, centrifuge samples for 3 minutes at $1{,}164 \times g$.
6. Transfer the isooctane layer to a quartz cuvette.

Perform analysis:
1. Read the absorbance of the isooctane layer in a quartz cuvette at 275 nm.

Calculations

If the absorbance of the isooctane layer is greater than 1.0, 2.0 mL of the isooctane layer should be diluted with 2.0 mL pure isooctane and the absorbance remeasured.

International bitterness units (IBU):

$$\text{IBU} = \text{absorbance at 275 nm} \times 100$$

Data Sheet ASBC Wort 23A

Sample	Absorbance at 275 nm	IBU	Sample	Absorbance at 275 nm	IBU

Notes:

Data sheets may be reproduced and are available online: asbcnet.org/BSL

ASBC Wort 23C: Wort Bitterness—Iso-α-Acids in Wort by High-Performance Liquid Chromatography

Reagents Needed for Analysis

Reagent water, 18MΩ resistivity or greater

Acidic methanol

In a 1,000 mL volumetric flask:

1. Pipette 0.5 mL of 85% phosphoric acid.
2. Fill to volume with high-performance liquid chromatography- (HPLC-) grade methanol.

Store in glass bottle. Mixture is stable for 1 month.

Mobile phase solution

In a 1,000 mL volumetric flask:

1. Add 750 mL of HPLC-grade methanol.
2. Pipette 10 mL of 85% phosphoric acid.
3. Pipette 1 mL of 0.1 M ethylenediaminetetraacetic acid (EDTA).
4. Fill to volume with reagent water and mix.

Store in glass bottle. Mixture is stable for 1 month.

ASBC hop standard solutions

Standard stock solution

In a 100 mL volumetric flask:

1. Add 30–33 mg (record to 0.1 mg) of iso-α-acid standard.
2. Add 40 mL of acidic methanol and sonicate until dissolved.
3. Fill to volume with acidic methanol.
4. The standard stock concentration (in ppm) is equal to the mass of the iso-α-acid (in mg) times 100 (i.e., 30.5mg of iso-α-standard will result in a 305 ppm standard stock solution).

Store in freezer in light-shielded, clear glass bottle. Mixture is stable for 1 month.

10% aqueous solution of antifoam FG-10

Materials Needed

HPLC: with column heater and ultraviolet (UV) detector

Column: C18, 250 × 4 mm (Nucleosil 5 or equivalent), or 125 × 4 mm (Nucleodur 100-5 C18 ec or equivalent)

Ultrasonic bath

Volumetric flasks, 100 mL, 1,000 mL

Analytical balance

Precolumn (optional)

Procedure

HPLC operating conditions:

Mobile phase: Acidic EDTA solution
Temperature: 35°C
Flow rate: 1.0–1.4 mL/minute
Detector setting: 270 nm

Prepare standard curve for analysis:
1. Prepare four standard solutions by pipetting 1.0 mL, 5.0 mL, 10.0 mL, and 20.0 mL of the standard stock solution into four separate 100 mL volumetric flasks.
2. Dilute each standard with acidic methanol.
3. The four solutions have concentrations of 1/100th, 1/20th, 1/10th, and 1/5th of the concentration of the standard stock solution. For example, a 300 ppm stock solution will result in standards of 3 ppm, 15 ppm, 30 ppm, and 60 ppm.

Prepare samples for analysis:
1. Obtain wort or beer that has been degassed with 1 drop of 10% antifoam.
2. Filter the sample to remove any dissolved gas.
3. Dilute sample 1:1 with acidic methanol.
   - Highly hopped samples may need additional dilution to be under 30 international bitterness units (IBUs).

Perform analysis:
1. Inject 20 µL of each standard to prepare a standard curve, and then inject 20 µL of sample.

Calculations

A standard curve of the iso-α-acid (IAA) is prepared by plotting the sum of the peak areas for each IAA (y-axis) versus the standard ppm (x-axis). The equation of the best-fit line, in the form $y = mx + b$, provides the slope of the line used in the calculations below.

Response factor (RF):

$$RF = \frac{1}{\text{slope of standard curve}}$$

Iso-α-acid concentration (mg/L):

$$IAA = \text{sum of IAA peak areas} \times RF$$

Data Sheet ASBC Wort 23C

Sample	Standard curve response factor	Sum of IAA peak areas	IAA concentration

Notes:

ASBC Beer 4A: Alcohol—Beer and Distillate Measured Volumetrically

Reagents Needed for Analysis
None

Materials Needed
Distillation setup with entrainment trap and vertical condenser (Fig. 15.2)
Volumetric flask, 100 mL
Digital density meter (or hydrometer)
Pipette, 100 mL (optional)

Procedure
Prepare samples for analysis:
1. Pipette or measure via a volumetric flask 100 mL of decarbonated beer at 20°C into distilling flask.
2. Add 50 mL of distilled or deionized (DI) water into distilling flask. Use 50 mL of DI water to rinse the volumetric flask if sample is added that way.
3. Place collection flask in ice bath; do not use a funnel to collect distillate.

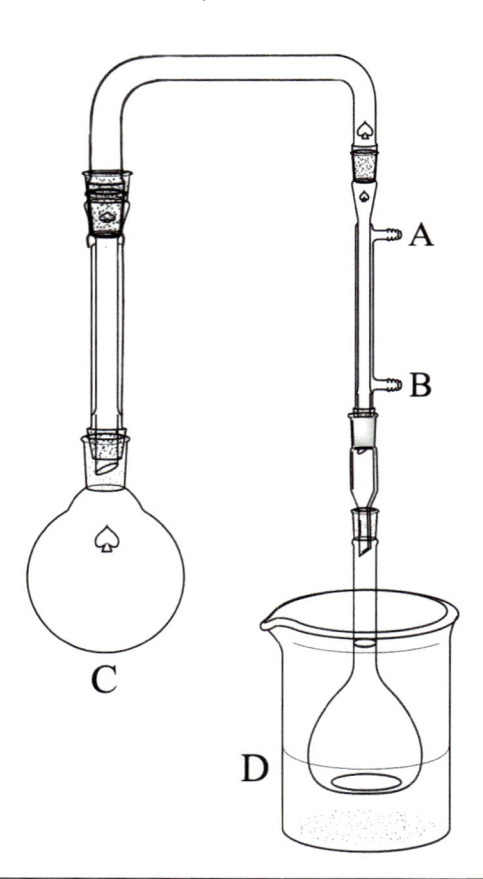

FIG. 15.2. Distillation setup with entrainment trap and vertical condenser. (Courtesy S. E. Johnson and M. D. Mosher—© ASBC)

Perform analysis:
1. Distill the sample at a rate in which ~96 mL is collected in 30–60 minutes.
2. Mix distillate well and dilute to 100 mL with DI water.
3. Read specific gravity of beer using digital density meter (or hydrometer for less accurate results).

Calculations

Alcohol % by volume and weight for the distillate can be determined by using the specific gravity (SG) of distillate when referencing the ASBC publication *Determinations on Wort, Beer, and Brewing Sugars and Syrups* (see the appendix in that book, Table 2). Alternatively, a similar table may be found online. If no table is available, the calculations below can be used to determine the alcohol content.

Alcohol, % by weight in distillate:

$$\%ABW_{distillate} = 6{,}282.3 \times (SG\ of\ distillate)^2 - 13{,}072 \times (SG\ of\ distillate) + 6{,}789.7$$

Alcohol, % by weight in beer:

$$\%ABW_{beer} = \frac{\%ABW_{distillate}}{specific\ gravity\ of\ beer}$$

Data Sheet ASBC Beer 4A

Sample	Distillate specific gravity	Alcohol, % by weight in distillate	Beer specific gravity	Alcohol, % by weight in beer

Notes:

Data sheets may be reproduced and are available online: asbcnet.org/BSL

ASBC Beer 4C: Alcohol—Alcohol Determined Refractometrically

Reagents Needed for Analysis
Distilled or deionized (DI) water

Materials Needed
Immersion refractometer with prisms to provide a range of 1.32–1.37
Water bath, 20°C
Cuvettes for refractometer
Digital density meter (or hydrometer)

Procedure

Prepare standard curve:
1. Analyze several beers of the same style that cover the alcohol range of interest using the steps below.
2. Plot the *y*-axis as R-N (see variables below).
3. Plot the *x*-axis as alcohol, % by weight.
4. Add a trendline to determine accuracy. Record the slope of this line as F and the intercept of this line as C.

Perform analysis:
1. Determine the specific gravity of the sample at 20°C.
2. Prepare and zero, if needed, the refractometer.
3. Adjust the temperature of the DI water and sample to 20°C in water bath.
4. Once at temperature, fill refractometer cuvette with DI water samples and record values over a 10-minute span. The average of these readings should be close to 14.50. Record this average value as R_{water}.
5. Rinse and dry prism for refractometer.
6. Fill with sample, wait 1 minute, and then read refractometer. Record this value as R_{beer}.

Calculations

The equation of the standard curve trendline provides the slope (F) and *y*-intercept (C), which are used in the calculation of percent alcohol by weight (%ABW).

Refractometer value, R:
$$R = R_{beer} - R_{water}$$

N value:
$$N = 1{,}000 \times (\text{beer specific gravity} - 1.00000)$$

Alcohol, % by weight in beer:
$$\%ABW = \frac{(R - N) - C}{F}$$

Data Sheet ASBC Beer 4C

Sample	R_{beer}	R_{water}	N value	F value	C value	Alcohol, % by weight

Notes:

Data sheets may be reproduced and are available online: asbcnet.org/BSL

ASBC Beer 4D: Alcohol—Ethanol Determined by Gas Chromatography

Reagents Needed for Analysis

n-Propanol internal standard

In a 100 mL volumetric flask:

1. Pipette 5 mL of n-propanol.
2. Fill to volume with distilled or deionized (DI) water and mix.

Store solution in glass bottle. Solution is stable for 1 month.

5% ethanol stock solution(s)

In a 100 mL volumetric flask:

1. Pipette 5 mL of ethanol.
2. Fill to volume with DI water and mix.

Store solution in glass bottle. Solution is stable for 1 month.

Materials Needed

Gas chromatograph with flame ionization detector (FID)

GC column: Chromosorb 103, 80–100 mesh, 6 feet × 1/8 inch, stainless steel or glass

Syringe, 1 µL

Volumetric flasks, 100 mL

Erlenmeyer flasks, 25 mL, 50 mL

Variable pipette

Procedure

GC operating conditions:

Carrier gas:	Helium
Flow rate:	20 mL/minute
Injection port:	175°C
FID:	250°C
Oven temperature:	185°C
Fuel:	Hydrogen and air

Prepare standard curve with internal standard:

1. Prepare aqueous solution of ethanol containing 3%, 4%, 6%, 7%, and 8% ethanol in the same manner that the 5% ethanol solution was prepared above.
2. Pipette 5 mL of each standard curve solution into individual 25 mL Erlenmeyer flasks.
3. Pipette 5 mL of n-propanol standard curve solution into each 25 mL Erlenmeyer flask.
4. Stopper each flask and mix well.

Prepare sample with internal standard for analysis:
1. Degas the sample through appropriate filter paper.
2. Pipette 5 mL of each sample into individual 25 mL Erlenmeyer flasks.
3. Pipette 5 mL of n-propanol standard curve solution into each 25 mL Erlenmeyer flask.
4. Swirl to mix solutions; inject solutions shortly after they are prepared.

Perform analysis:
1. Inject 0.2 µL of each standard and samples into the GC.
2. Analysis run time will be 2 minutes; allow 3.5 minutes between injections.

Calculations

The calculations can be performed using peak height or peak area. These are shown below in the calculations below as "signal."

F value:

$$F = \frac{\text{signal of 1\% ethanol standard}}{\text{signal of n-propanol internal standard}}$$

Ethanol, %v/v:

$$\% \text{ ABV} = \frac{\text{signal of ethanol in sample}}{F \times (\text{signal of n-propanol})}$$

Data Sheet ASBC Beer 4D

Sample	Signal of 1% ethanol standard	Signal of n-propanol standard	F value	Signal of ethanol in sample	Ethanol, (%v/v)

Notes:

Data sheets may be reproduced and are available online: asbcnet.org/BSL

ASBC Beer 10A: Color—Spectrophotometric Color Method

Reagents Needed for Analysis
Reagent water, 18MΩ resistivity or greater

Materials Needed
Spectrophotometer, 430 nm, 700 nm
Quartz cuvette, 10 mm

Procedure
Prepare blank for analysis:
1. Use reagent water as the blank for the instrument.

Prepare samples for analysis:
1. Degas sample.

Perform analysis:
1. Place sample into cuvette and record absorbance at 430 and 700 nm. Record these values as A_{430} and A_{700}.
2. If the value of $A_{700} \leq 0.039$ times that of A_{430}, the beer is considered free of turbidity.
   - If beer is turbid, clarify by centrifuge, filtration, or time.
3. If the absorbance reading is higher than 1.5, dilute the sample. Record this dilution factor as F.
   - If no dilution is done, F = 1.

Calculations
The values of the absorbance at 430 nm are used in the calculation of beer color. Multiply by the dilution factor (F), as determined while performing the analysis.

Beer color, standard reference method (SRM):

$$\text{SRM color} = 10 \times 1.27 \times A_{430} \times F$$

Beer color, European Brewery Convention (EBC):

$$\text{EBC color} = A_{430} \times 25 \times F$$

Data Sheet ASBC Beer 10A

Sample	A_{700}	A_{430}	F	Turbid?	SRM/EBC

Notes:

ASBC Beer 11C: Protein—By Spectrophotometer

Reagents Needed for Analysis

Reagent water, 18MΩ resistivity or greater

Materials Needed

Spectrophotometer, 215 nm, 225 nm
Pipette, 6 mL, Pasteur
Top-loading balance, 1 kg capacity

Quartz cuvette, 10 mm
Erlenmeyer flasks, 500 mL, 1,000 mL

Procedure

Prepare blank for analysis:

1. Use Reagent water as the blank for the instrument.

Prepare samples for analysis:

1. Degas 10 mL of sample.
2. Pipette 6 mL of sample into a tared 1,000 mL Erlenmeyer flask, and record the weight to the nearest 0.1 g. Record this value as W_1.
3. Add 500 mL of Reagent water to flask and mix. Record this weight value as W_2.

Note: Steps 1–3 are for a double-beam ultraviolet-visible (UV-vis) spectrophotometer. For a single-beam UV-vis spectrophotometer, the sample should be twice as dilute.

Perform analysis:

1. Place sample into cuvette and record absorbances at 215 and 225 nm. Record these values as a_{215} and a_{225}.

Note: If beer has been stabilized (no longer able to form colloids), total phenol content must be determined by ASBC Beer 35.

Calculations

Percent protein is determined as the percent mass versus the total mass (%m/m). The three equations allow for determination of the protein in stabilized, unstabilized, and dark (SRM > 15) beers. If the beer has been stabilized, the total polyphenol (TP) in ppm must first be determined.

Weighted absorbance values:

$$A_{215} = W_2/W_1\ (a_{215}) \qquad A_{225} = W_2/W_1\ (a_{225})$$

Change in weighted absorbance, ΔA:

$$\Delta A = A_{215} - A_{225}$$

Unstabilized beer protein, %m/m:

$$\%\ \text{protein} = (0.00982 \times \Delta A) + (0.000029 \times \Delta A^2)$$

Stabilized beer protein, %m/m:

$$\%\ \text{protein} = (0.110 \times \Delta A) + (0.000029 \times \Delta A^2) - (0.00029 \times TP)$$

Dark beer protein, %m/m:

$$\%\ \text{protein} = (0.0107 \times A_{215}) - (0.0115 \times A_{225}) + (0.00003 \times \Delta A^2)$$

Data Sheet ASBC Beer 11C

Sample	W_1	W_2	a_{215}	a_{225}	A_{215}	A_{225}	ΔA	Protein (%m/m)

Notes:

Data sheets may be reproduced and are available online: asbcnet.org/BSL

ASBC Beer 13C: Dissolved Carbon Dioxide—Manometric/Volumetric Method

Reagents Needed for Analysis
Octyl alcohol

Materials Needed
Syringe, graduated with cap, 60 cm^3
Barometer reads lb/in^2
Parafilm

Procedure

Prepare samples for analysis:
1. Record the temperature of the sample and the local barometric pressure. Record these values as T and P, respectively.
2. Remove plunger from syringe, and add 2 drops of octyl alcohol to the syringe.
3. Slowly draw (or pour) ~25 mL of sample into the syringe, being careful to minimize agitation.
4. Invert, remove cap, and expel any bubbles. Adjust to 20 mL. Record the exact volume as V_b. There should be no air in the syringe at this point.
5. Place a piece of parafilm that has been folded over four times to the syringe opening, and cap the syringe.

Perform analysis:
1. Pull back on the plunger until it reaches the 50 mL mark.
2. Hold the plunger firmly at this volume, and record the exact mark as V_s.
3. Shake the syringe vigorously 50 times.
4. Release the plunger, and allow contents to equilibrate for 1 minute.
5. Read where the plunger now marks the volume. Record this value as V_t.

Calculations

The temperature should be recorded in °C. The pressure should be the local barometric pressure reported in psi. All volumes are reported in milliliters.

Volume of gas, V_g:

$$V_g = V_t - V_b$$

Concentration of CO$_2$ in beer, C:

$$C = \frac{P}{14.7} \times \frac{273.15}{T + 273.15} \times \frac{V_g}{V_b} + (0.100943 - 0.00236T) \times P \times \frac{V_g}{V_s - V_b}$$

Data Sheet ASBC Beer 13C

Sample	P	T	V_t	V_b	V_g	V_s	CO_2 (vols.)

Notes:

Data sheets may be reproduced and are available online: asbcnet.org/BSL

ASBC Beer 22A: Foam Collapse Rate—Sigma Value Method (Modified Carlsberg Method)

Reagents Needed for Analysis
Isopropyl alcohol

Materials Needed

Special foam funnel marked at 800 mL
Graduated cylinders, 25 mL, 100 mL
Stopwatch
Watch glass, ≥100 mm diameter
Pipette, 2 mL
Water bath, 25°C

Procedure

Prepare samples for analysis:
1. Bring sample to 25°C in water bath.
2. Clean foam funnel with warm detergent and rinse well with warm water and then with 25°C water.
3. Clamp funnel to stand and allow to drain for 1 minute.

Perform analysis:
1. Make all foam determinations immediately after draining.
2. Pour beer directly onto center of funnel until foam reaches 800 mL mark.
3. Start stopwatch and cover funnel with watch glass.
4. After 30 seconds, open the stopcock and drain all liquid in 25–30 seconds.
5. Allow small amount of foam to drain out, and then immediately close the stopcock.
6. Reset stopwatch to 0 and begin keeping track of time.
7. After 200 seconds, open stopcock and collect liquid in a 100 mL graduated cylinder within 25–30 seconds.
8. Once the last drop of liquid has drained, close the stopcock, stop the timer, and record the time value as t.
9. Record the volume of liquid as value b.
10. At the end of this step, the time should be 225–250 seconds.
11. Wash inside the funnel with 2 mL of isopropyl alcohol to collapse any remaining foam.
12. Once the foam has been collapsed, open the stopcock and collect liquid in a 25 mL graduated cylinder. Record the amount of liquid, minus the amount of isopropyl alcohol used, as value c.

Calculations

Foam collapse rate, σ:

$$\sigma = \frac{t}{2.303\left(\log \frac{b+c}{c}\right)}$$

Data Sheet ASBC Beer 22A

Sample	Time, t	First collection, b	Second collection, c	Foam collapse rate, σ

Notes:

Data sheets may be reproduced and are available online: asbcnet.org/BSL

ASBC Beer 23A: Beer Bitterness— Bitterness Units (International Method)

Reagents Needed for Analysis

Isooctane

Hydrochloric acid, 3N

Octyl alcohol

Materials Needed

Spectrophotometer, 275 nm
Mechanical shaker
Centrifuge tubes, 50 mL
Quartz cuvette, 10 mm
Centrifuge (optional)

Procedure

Prepare blank for analysis:
1. Use 20 mL of isooctane with 1 drop of octyl alcohol as the blank for the instrument.

Prepare samples for analysis:
1. Place 1 drop of octyl alcohol into centrifuge tube.
2. Transfer 10 mL of 10°C sample to centrifuge tube.
3. Pipette 1 mL of 3N HCl into tube.
4. Pipette 20 mL of isooctane into tube and then stopper or seal tube.
5. Place sample on shaker for 15 minutes and shake vigorously.
6. Allow the mixture to stand upright for 5–10 minutes to break up the emulsion. It may be necessary to centrifuge the sample to speed up separation of the two layers.
7. Collect the upper isooctane layer.

Perform analysis:
1. Fill a cuvette with some of the isooctane layer, and measure the absorbance at 275. Record this value as A_{275}.
2. If the absorbance reading is higher than 1.5, dilute the sample. Record this dilution factor as F.
   - If no dilution is done, $F = 1$.

Calculations

The IBU values are reported to one decimal place.

International bitterness units (IBU):

$$IBU = 50 \times A_{275} \times F$$

Data Sheet ASBC Beer 23A

Sample	A_{275}	F	IBU

Notes:

Data sheets may be reproduced and are available online: asbcnet.org/BSL

ASBC Beer 23C: Beer Bitterness—Iso-α-Acids by Solid-Phase Extraction and High-Performance Liquid Chromatography

Reagents Needed for Analysis

Iso-α-acid calibration standard

Tetraethylammonium hydroxide $(CH_3CH_2)_4NOH$, 10% in water

Water, high-performance liquid chromatograph (HPLC) grade

Methanol, HPLC grade

Octanol

Desorbing solvent A

In a 250 mL Erlenmeyer flask:

1. Add 100 mL of water, HPLC grade.
2. Pipette 0.2 mL of phosphoric acid, 85%.

Stopper or seal solution in Erlenmeyer flask. Solution is stable for 1 month.

Desorbing solvent B

In a 250 mL Erlenmeyer flask:

1. Add 50 mL of water, HPLC grade.
2. Add 50 mL of methanol, HPLC grade.
3. Pipette 0.2 mL of phosphoric acid, 85%.

Stopper or seal solution in Erlenmeyer flask. Solution is stable for 1 month.

Desorbing solvent C

In a 250 mL Erlenmeyer flask:

1. Add 100 mL of methanol, HPLC grade.
2. Pipette 0.1 mL of phosphoric acid, 85%.

Stopper or seal solution in Erlenmeyer flask. Solution is stable for 1 month.

HPLC mobile phase

In a 2,000 mL beaker on a stir plate:

1. Add 725 mL of methanol, HPLC grade.
2. Add 275 mL of water, HPLC grade, and mix with stir bar.
3. Add 17 g of phosphoric acid, 85%, and mix with stir bar.
4. Add 29.5 g of $(CH_3CH_2)_4NOH$, 10% in water, and mix with stir bar.
5. Once mixed, filter through 0.45 μm filter and degas.
6. Check pH. Final pH should be 3.0–3.5.

Store solution in glass bottle to use for analysis. Make fresh solution daily.

Materials Needed

HPLC with 270 nm UV detector, 20 µL injection loop

Chromatographic column: Shimadzu, Shim-pack CLC-ODS, 25 cm × 4.6 mm or equivalent

Other equipment: C8 solid phase extraction (SPE) octyl column, 500 mg, 3 mL

Pipettes, volumetric, 20.0 mL

Pipettes, graduated, 1 mL, 5 mL, 10 mL

Volumetric flask, 2 mL, 100 mL

Erlenmeyer flask, 250 mL

Syringe, 30 mL

Analytical balance

Procedure

HPLC operating conditions:

Mobile phase: See "Reagents" section on previous page
Temperature: 20–50°C
Flow rate: 1.5 mL/minute
Detector: 280 nm

Column conditioning:

To achieve maximum resolution among the three iso-α-acid analogs, condition the column with mobile phase by pumping at 1 mL/minute overnight; recycling of the solvent is acceptable. Longer conditioning periods of up to 65 hours are acceptable.

Prepare standard for analysis:

1. Weigh 20 mg of calibration standard into a 100 mL volumetric flask. Record this mass to the nearest 0.1 mg.
2. Add ~40 mL of methanol, HPLC grade, and dissolve.
3. Fill to volume with methanol, HPLC grade.

Prepare samples for analysis:

1. Transfer 200 mL of sample to an Erlenmeyer flask and degas by transfer 20 times.
2. Add 1 drop of octanol and continue to transfer until no foam remains.
3. Adjust pH to 2.5. (Typically requires ~200 µL of phosphoric acid, 85% per 100 mL of sample.)

To prepare samples for injection, follow the steps below.

Prepare C8 SPE column equipment with adsorption/desorption sequence of the following:

- 2 mL methanol, HPLC grade, discard elute.
- 2 mL water, HPLC grade, discard elute.
- 20 mL degassed sample using volumetric pipette, discard elute.
- 6 mL desorbing solvent A, discard elute.
- 2 mL desorbing solvent B, discard elute.
- 0.6 mL desorbing solvent C, three successive aliquots.

Collect in a 2 mL volumetric flask and bring to volume with desorbing solvent C.

Perform analysis:
1. Inject 20 μL of standard twice, and then inject 20 μL of each sample.
2. Finish by injecting 20 μL of standard twice.

Calculations

The chromatogram of the standard should show three iso-α-acid (IAA) peaks. The area of these peaks is used to determine the IAA (in mg/L or ppm) using the equations below.

Response factor, RF:

$$RF = \frac{\text{total area of 3 IAA peaks in calibration}}{\text{mg of calibration standard} \times \left(\dfrac{\%\text{IAA purity in standard}}{100}\right)}$$

Iso-α-acids, IAA, mg/L:

$$IAA = \frac{\text{Total area of 3 IAA peaks in sample}}{RF}$$

Data Sheet ASBC Beer 23C

Sample	Mg of standard	% IAA purity in standard	Total area of 3 IAA peaks in spectra	RF	Total area of 3 IAA peaks in sample chromatogram	IAA (mg/L)

Notes:

Data sheets may be reproduced and are available online: asbcnet.org/BSL

ASBC Beer 25D: Diacetyl— Ultraviolet Spectrophotometer Method

Reagents Needed for Analysis

α-Naphthol solution (Reagent A)

In a 100 mL volumetric flask:

1. Add 40 mL of isopropanol.
2. Dissolve 4.0 g of α-naphthol.
3. Fill to 100 mL with isopropanol.
4. Transfer solution to 250 mL Erlenmeyer flask and add 0.5 g of activated charcoal. Swirl for 30 minutes.
5. Filter the solution.

Store in dark in amber glass bottle. Mixture is stable for 1 month.

KOH-creatine solution (Reagent B)

In a 250 Erlenmeyer flask:

1. Add 80 mL of 40% KOH solution.
2. Dissolve 0.3 g creatine.
3. Filter solution.

Store cold in a polyethylene bottle. Mixture is stable for 1 month.

Diacetyl stock solution (Reagent C)

In a 1 L volumetric flask:

1. Pipette 0.506 mL of diacetyl.
2. Fill to 1 L with distilled or deionized (DI) water.

Store in amber glass bottle. Mixture is stable for 1 month.

Diacetyl working solution (Reagent D)

In a 100 mL volumetric flask:

1. Pipette 1 mL of Reagent C.
2. Dilute to 100 mL with DI water.

Store in amber glass bottle. Mixture is stable for 1 month.

Materials Needed

Spectrophotometer
Volumetric flasks, 10 mL
Graduated cylinder, 50 mL
Distillation equipment, 500 mL
Heating mantle

Procedure

Prepare standard curve:
1. Pipette 0.5, 1.0, 1.5, 3.0, and 4.0 mL of Reagent D into individual 10 mL volumetric flasks.
2. Fill each flask approximately halfway to volume with DI water.

Prepare samples for analysis:
1. Distill 100 mL of degassed beer into 50 mL graduated cylinder.
2. When 15 mL of distillate has been collected, stop distillation.
3. Add 10 mL of DI water to bring volume to 25 mL.
4. Pipette 5 mL of solution into a 10 mL volumetric flask.

Perform analysis:
1. Pipette 1 mL of Reagent A into standards and sample flasks.
2. Pipette 0.5 mL of Reagent B into up to four or five flasks at a time. This compound is time sensitive. This step must be done shortly before photometric analysis.
3. Fill flasks to volume with DI water, and shake vigorously for exactly 60 seconds.
4. Measure absorbance at 530 nm between 5 and 6 minutes after shaking has been completed.

Calculations

Plot absorbance values of standards to produce a standard curve. This can be done using Figure 15.3 but is more efficiently done using Microsoft Excel or a related program.

Values for standards are 0.025, 0.050, 0.075, 0.150, and 0.200 mg/L respectively.

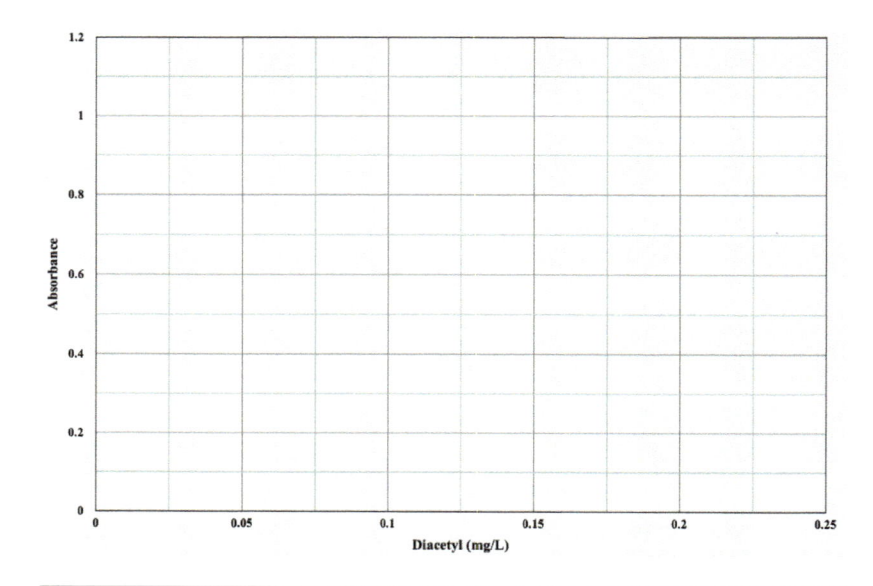

FIG. 15.3. Graph for plotting a diacetyl standard curve by hand. (Courtesy S. E. Johnson and M. D. Mosher—© ASBC)

Data Sheet ASBC Beer 25D

Sample	Absorbance	mg/L diacetyl
Standard 1		
Standard 2		
Standard 3		
Standard 4		
Standard 5		

Notes:

Data sheets may be reproduced and are available online: asbcnet.org/BSL

ASBC Beer 38: Magnesium and Calcium

Reagents Needed for Analysis

Calcium indicator (Reagent A)

In a ceramic pestle:

1. Blend 0.1 g Cal-Red and 10 g reagent grade potassium sulfate powder.
2. Stop when dye is evenly distributed.

Store in amber glass bottle. Mixture is stable for 1 year.

Magnesium indicator (Reagent B)

In a 100 mL volumetric flask:

1. Add 40 mL absolute methanol.
2. Dissolve 0.5 g Eriochrome Black T and 4.5 g hydroxylamine hydrochloride.
3. Fill to 100 mL with absolute methanol.

Store in amber glass bottle. Mixture is stable for 2 months.

Ethylenediaminetetraacetic acid (EDTA) 0.01M (Reagent C)

In a 1 L volumetric flask:

1. Add 600 mL of distilled or deionized (DI) water.
2. Dissolve 3.723 g of disodium ethylenediamine tetraacetate dehydrate powder.
3. Fill to 1 L with DI water.

Store in amber glass bottle. Mixture is stable for 1 year.

Calcium indicator (Reagent D)

In a 500 mL Erlenmeyer flask:

1. Add 1 g of anhydrous calcium carbonate powder.
2. In small increments, add 6N HCl until all the solid is dissolved.
3. Add 200 mL DI water and boil for 5 minutes. Let cool.
4. Cool to room temperature and add 3 drops methyl red indicator.
5. Adjust color to orange by adding 3N NH_4OH or 6N HCl as required.
6. Transfer to 1 L volumetric flask and fill to volume with DI water.

Store in amber glass bottle. Mixture is stable for 1 month.

Buffer solution (Reagent E)

In a 1 L volumetric flask:

1. Add 570 mL of concentrated ammonium hydroxide.
2. Dissolve 70 g of ammonium chloride.
3. Fill to volume with DI water.

Store in amber glass bottle. Mixture is stable for 1 month.

Acid and base solutions (Reagents F, G)

F: 8N Potassium hydroxide
G: 6N Hydrochloric acid

Materials Needed

Volumetric pipette, 10 mL
Microburet, 10 mL (graduated to 0.5 mL)
Beakers, 2 150 mL
Magnetic stirrer
Stir bar
pH meter

Procedure

For calcium titration, prepare and analyze samples:
1. Degas beer samples by pouring back and forth between two clean beakers.
2. Pipette 10 mL of degassed beer into 150 mL beaker containing 100 mL of DI water.
3. Add 3 mL of 8N Reagent F and mix well. Then add 0.05 g of Reagent A.
4. Titrate with Reagent 5 until pink color disappears. The final color is a gray–blue.

For magnesium titration, prepare and analyze samples:
1. Neutralize the solution from the calcium titration to pH 6–8 with Reagent G.
2. Adjust pH to 10 with Reagent E. Then add 3 drops of Reagent B.
3. Titrate the purple solution with Reagent C until purple color changes to blue.

Calculations

Best results for this analysis occur when at least three separate samples are titrated. The average and standard deviation of the values can then be determined.

A = mL of ethylenediaminetetraacetic acid (EDTA) used for calcium titration

B = mL total of EDTA used in both titrations

$$Ca^{2+}(ppm) = 40.08 \times A$$

$$Mg^{2+}(ppm) = 24.32 \times (B - A)$$

Data Sheet ASBC Beer 38

Sample	mL EDTA used for first titration, A	mL EDTA used total, B	Ca^{2+} (ppm)	Mg^{2+} (ppm)

Notes:

ASBC Yeast 4: Microscopic Yeast Cell Counting

Reagents Needed for Analysis

Diluent

0.5% sulfuric acid or 0.5 M disodium ethylenediaminetetraacetic acid (EDTA) solution adjusted to pH 7.0 with NaOH

Materials Needed

Microscope, 400–500× magnification
Pipettes, 1.0 mL, 10.0 mL
Magnetic stirrer and stir bar
Beaker, 250 mL
Hemocytometer*
Counting device, such as hand tally
Volumetric flask, 100 mL

Procedure

Prepare samples for analysis:

1. Degas 100 mL sample for analysis and place in 250 mL beaker.
2. Stir for at least 5 minutes using magnetic stirrer and stir bar.
3. If highly flocculated, 0.5% sulfuric acid or 0.5M EDTA, pH 7, can be used to deflocculate cells.
4. If diluting, record dilution amount as the "dilution factor."
5. Final sample volume should be 100 mL for accurate results.

Perform analysis:

1. Add clean, dry coverslip to hemocytometer.
2. Use pipette to add a drop of well-mixed yeast sample to gap between coverslip and hemocytometer.
3. Let slide stand for 3 minutes to allow yeast to settle.
4. Count number of yeast cells, being careful not to double count.
5. Each square should have no more than 48 cells. If there are more, dilute sample.
6. It's good practice to break up the plate into 10 counting areas. These areas should be averaged to ensure statistical validity for even coverage of yeast cells.

Calculations

The total cells in the ruled area is determined by multiplying the average cell count in a single square by 25.

Yeast cells/mL:

$$\text{yeast} = \text{total cells in ruled area} \times \text{dilution factor} \times 10^4$$

*Recommend use of the Neubauer Improved Hemocytometer containing squares that are 0.0025 mm².

Data Sheet ASBC Yeast 4

Sample	Cells in ruled area	Dilution factor	Yeast cells/mL

Notes:

ASBC Yeast 6: Yeast Viability by Slide Culture

Reagents Needed for Analysis

Petri dish MYGP (malt extract–yeast extract–glucose–peptone) medium

If MYGP medium is not available, it can be made by combining the following ingredients and then autoclaving the solution at 1 bar (15 psi) and 121°C for 20 minutes. The mixture is used while it is still liquid.

- Malt extract (Difco or equivalent), 0.3 g
- Yeast extract (Difco or equivalent), 0.3 g
- Glucose (Fisher Scientific)
- Dextrose anhydrous (or equivalent), 1.0 g
- Peptone (Difco or equivalent), 0.5 g
- Maltose (Sigma Grade II or equivalent), 6.0 g
- Agar (BBL granulated or equivalent), 1.5 g
- Zinc sulfate (1.5 g $ZnSO_4 \cdot 7H_2O$ dissolved in 100 mL of water), 1.0 mL
- Water, 100 mL

Autoclave at one bar (15 lb/in^2), 121°C, for 20 minutes

Materials Needed

Bunsen burner	Forceps
Slides, glass, 230 × 20 mm	Cover slips, glass
Pipette, 1 mL, 100 µL	Petri dish, sterile, 100 mm × 15 mm
Microscope, 200–250× magnification	Incubator, 25°C

Procedure

Prepare petri dishes for analysis:

1. Flame a glass slide over a Bunsen burner to sterilize.
2. Place slide into sterile petri dish.
3. Pipette 1 mL of molten MYGP medium evenly over the slide.
4. Let agar cool.

Perform analysis:

1. Place 2 drops of suspended yeast solution of ~1×10^6 cells/mL onto agar surface.
2. Place drops at end of slide so a cover slip can be easily placed.
3. Don't apply pressure to cover slips.
4. Check slide on microscope to ensure cells are not crowded.
5. Cover with petri dish lid and incubate at 25°C for 12–16 hours (and not longer than 18 hours).
6. Examine cultures on microscope. Cells that gave rise to microcolonies are considered viable. Single cells are considered dead.
7. There should be at least 500 microcolonies and cells recorded for calculation.

Calculations

Viable cells, %:

$$\text{viable cells, \%} = \frac{\text{number of microcolonies}}{\text{total number of cells and microcolonies}} \times 100$$

Data Sheet   ASBC Yeast 6

Sample	Number of microcolonies	Total number of cells/microcolonies	Viable cells (%)

Notes:

Data sheets may be reproduced and are available online: asbcnet.org/BSL

ASBC Yeast 9: Morphology of Giant Yeast Colonies (International Method)

Reagents Needed for Analysis

Gelatin, Difco Bacto Gelatin, cat. no. 0143-01 or equivalent

12 °P wort, hopped or unhopped, always same brand

Distilled or deionized (DI) water

Materials Needed

Petri dishes, 100 × 15 mm
Erlenmeyer flask, 2,000 mL
Incubator, 25°C

Procedure

Prepare samples for analysis:
1. Collect 1,000 mL of 12 °P wort in sterilized 2,000 mL Erlenmeyer flask.
2. If wort is greater than 12 °P, dilute to 12 °P with DI water.
3. Slowly add 125 g of gelatin to wort while stirring.
4. Heat solution in oven to 100°C for 40 minutes once a day for 3 days.
5. After the third heating, pour 75 mL of hot wort gelatin onto petri dish.
6. Allow 24 hours for gelatin to harden, and then refrigerate plates.

Perform analysis:
1. Dilute a sample of suspended yeast slurry so that 0.5 mL of solution will result in about 50 colonies.
2. If diluting, be sure to vortex solution to ensure solution is homogeneous.
3. Pipette 0.5 mL of yeast slurry solution into five petri dishes.
4. Incubate at 18–20°C for 3–4 weeks.
5. Evaluate shapes of colonies and compare with known samples.

Important things to consider:
- Keep the temperature between 18 and 20°C. Temperatures greater than 21°C can result in changes to the morphology of the colonies.
- The thickness of the medium will affect the morphology. Always use exactly 75 mL of the hot wort gelatin to ensure uniform thickness.
- The morphology can change depending on the wort composition. Always use the same type of wort when making the plates.
- Avoid extremes in humidity.
- Some brewing yeasts can produce distinct variants in colony morphology, even when the colonies have come from the same pure culture.
- Adequately vortex the cultures before plating. Failing to do so may cause morphology variations.

- Overcrowding colonies on a plate must be avoided because this will inhibit complete morphological development.
- Although many yeasts can be differentiated based on their giant colony morphology, some strains can produce colonies that are of similar size, shape, and coloration. Care should be taken when comparing or analyzing strains known to be similar or closely related, such as lager-type strains.
- For assistance with identifying colonies, refer to this source: Richards, M. (1967). The use of giant colony morphology for the differentiation of brewing yeasts. J. Inst. Brew. 73(2):162–166.

ASBC Yeast 11B: Flocculation— Absorbance Method

Reagents Needed for Analysis

Calcium sulfate washing solution

In a 1,000 mL volumetric flask:

1. Add 400 mL of distilled or deionized (DI) water.
2. Dissolve 0.51 g of calcium sulfate.
3. Fill to volume with DI water.

Store in glass bottle. Mixture is stable for 1 month.

Buffered calcium sulfate suspension solution

In a 1,000 mL volumetric flask:

1. Add 400 mL of DI water.
2. Dissolve 0.51 g of calcium sulfate.
3. Dissolve 6.8 g of sodium acetate.
4. Add 4.05 g of glacial acetic acid.
5. Adjust pH to 4.5, if necessary.
6. Fill to volume with DI water.

Store in glass bottle. Mixture is stable for 1 month.

Disodium ethylenediaminetetraacetic acid (EDTA) solution, 0.5 M

In a 1,000 mL volumetric flask:

1. In 400 mL of DI water, dissolve 168.1 g of disodium EDTA dihydrate.
2. Adjust pH to 7.0 with NaOH, if needed.
3. Fill to volume with DI water.

Store in glass bottle. Mixture is stable for 1 month.

Hopped wort (adjusted to 10–12 °P)

Hopped wort agar

In a 250 mL Erlenmeyer flask:

1. Add 100 mL of hopped brewery wort and 2 g of agar.
2. Warm and dissolve at 121°C for 20 minutes.

Pour agar onto petri dishes. Store cool. Plates are stable for 1 month.

Materials Needed

Spectrophotometer or colorimeter, 600 nm	Hemocytometer
Centrifuge tubes, 15 mL with lids	Centrifuge
Water bath, at 5°C, accurate to ±1°C	pH meter
Culture flasks, sterile, 250 mL or 300 mL	Analytical balance
Petri dishes, 100 mm × 15 mm	Graduated cylinder, 100 mL
Shaker incubator, 25°C, accurate to ±1°C	Vortex
Graduated pipettes or equivalent, 1 mL, 10 mL	Incubator, 25°C

Procedure

Blank the spectrophotometer with water.

Prepare samples for analysis:
1. Grow yeast on wort agar plates for 4 days at 25°C.
2. Inoculate yeast into culture flask at pitching rate of 15×10^6 cells/mL into 100 mL of hopped wort.
   - Pitch rate can be checked using Yeast 4.
3. Incubate with shaking for 2 days at 25°C.

Perform analysis:
1. Transfer 10 mL of the freshly grown cultures into two 15 mL centrifuge tubes marked A and B.
2. Centrifuge tube A at 2,500 rpm for 2.5 minutes.
3. Decant liquid and suspend yeast in 9.9 mL of DI water and 0.1 mL of 0.5 M EDTA.
4. Vortex for 15 seconds to homogenize.
5. Dilute 1 mL of suspension with 9 mL of DI water, vortex, and read absorbance at 600 nm. Record this value as A.
6. Centrifuge tube B at 2,600 rpm for 2.5 minutes.
7. Decant the liquid and suspend the yeast in 10 mL of washing solution.
8. Vortex for 15 seconds; centrifuge at 2,500 rpm for 2.5 minutes.
9. Decant liquid and suspend yeast in 10 mL of suspension solution.
10. Vortex for 15 seconds to homogenize.
11. Invert slowly five times in 15 seconds, and leave standing upright for 6 minutes.
12. Without disturbing yeast, dilute 1 mL from top of solution with 9 mL of DI water, vortex, and read absorbance at 600 nm. Record this value as B.

Calculations

The outcome of this analysis improves when three separate samples are analyzed. The results can then be averaged and the standard deviation reported.

Flocculence, %:

$$\text{flocculence, \%} = \frac{(A - B) \times 100}{A}$$

Data Sheet ASBC Yeast 11B

Sample	Absorbance of *A* at 600 nm	Absorbance of *B* at 600 nm	Flocculence (%)

Notes:

ASBC Sensory Analysis 7: Triangle Test (International Method)

Eighteen to 36 assessors preferred for difference testing (minimum of 7).

Purpose

Use this test to determine whether:

1. two samples are significantly different or
2. two samples are significantly similar.

Procedure

To perform the triangle test in a traditional fashion:

1. Define the desired objective of the test:
   - Similarity testing should be the objective when a new production process is evaluated to ensure the beer has not changed flavor.
   - Difference testing should be the objective when a new formula or recipe is being evaluated.
2. Out of the sight of the assessors, two beer samples are randomly distributed among three sample containers.
3. The analyst randomly codes each sample container and records the contents.
4. Each assessor is given a set of three coded samples, two of which are identical.
5. Each assessor evaluates each sample and determines which two are the same.
6. The analyst collects this information and determines the required level of sensitivity for the test, including acceptable levels of α, β, and P_d. (See the "Terminology" section for explanations of terms.) The minimum number of assessors required is determined from Table 15.2.

To perform the triangle test for a large group or class:
See Figure 15.4, a fill-in-the-blank guide for a self-run triangle test analysis, at the end of this section.

1. The teacher or group leader assists with the test administration. Each assessor receives three sample cups, three pieces of paper, a marker, and a coin. Each piece of paper and each cup should be marked with a circle, a square, or a triangle.
2. Each assessor is given samples of the two beers to be evaluated. One sample is coded Beer 1 and the other Beer 2. These are uniform for the entire group.
3. Each member of the group or class will do the following:
   a. Set up a triangle test for another member to perform.
   b. Perform a triangle test set up by another member.

4. Use the blue section of Figure 15.4 to help keep track of the test as each group member follows these steps:
   a. Flip a coin and circle heads or tails. For heads, circle Beer 1; for tails, circle Beer 2.
   b. Put the three labeled slips of paper in a hat or pocket. Pull out two of the slips.
   c. Pour the beer selected by the coin flip into the two cups selected by the slip pulls.
   d. Pour the beer not selected by the coin flip into the remaining cup.
   e. Order the cups from left to right as circle, square, and triangle.
5. Each group member changes seats with another member.
6. Each group member completes the green section of Figure 15.4 as he or she performs the triangle test.
7. Each group member returns to his or her original seat and fills out the orange section with the answers of the analyst who performed the test on that set-up.
8. The teacher or leader collects the sheets and counts the number of correct responses. Table 15.3 is consulted to analyze the data. If the number of correct responses is greater than or equal to the number given in the table corresponding to the number of assessments (n), then a significant difference exists between the samples at a specified α risk.
9. Conducting repeated evaluations by the same assessor is not recommended.

Terminology

α (alpha) risk: The risk of concluding that a significant difference exists when in reality there is not one (also known as a risk of false positive or a type I error).

β (beta) risk: The risk of concluding that no significant difference exists when in reality there is one (also known as a risk of false negative or a type II error).

P_d: The allowable proportion of discriminators in the population; those who can detect the difference between the products.

TABLE 15.2. Minimum number of assessors needed for a triangle test[a,b]

Value of α	Value of β				
	0.2	0.1	0.05	0.01	0.001
P_d = 50%					
0.2	7	12	16	25	36
0.1	12	15	20	30	43
0.05	16	20	23	35	48
0.01	25	30	35	47	62
0.001	36	43	48	62	81
P_d = 40%					
0.2	12	17	25	36	55
0.1	17	25	30	46	67
0.05	23	30	40	57	79
0.01	35	47	56	76	102
0.001	55	68	76	102	130
P_d = 30%					
0.2	20	28	39	64	97
0.1	30	43	54	81	119
0.05	40	53	66	98	136
0.01	62	82	97	131	181
0.001	93	120	138	181	233
P_d = 20%					
0.2	39	64	86	140	212
0.1	62	89	119	178	260
0.05	87	117	147	213	305
0.01	136	176	211	292	397
0.001	207	257	302	396	513
P_d = 10%					
0.2	149	238	325	539	819
0.1	240	348	457	683	1,011
0.05	325	447	572	828	1,181
0.01	525	680	824	1,132	1,539
0.001	803	996	1,165	1,530	1,992

[a]From ASBC Sensory Analysis 7. Courtesy American Society of Brewing Chemists—Reproduced by permission.
[b]Enter the table in the section corresponding to the chosen value of P_d and the column corresponding to the chosen value of β. Read the number of assessors from the row corresponding to the chosen value of α.

TABLE 15.3. Minimum number of correct responses for samples to be considered significantly different at the stated α-level[a–d]

Assessments, n	Significance Level of α				
	0.20	0.10	0.05	0.01	0.001
18	9	10	10	12	13
19	9	10	11	12	14
20	9	10	11	13	14
21	10	11	12	13	15
22	10	11	12	14	15
23	11	12	12	14	16
24	11	12	13	15	16
25	11	12	13	15	17
26	12	13	14	15	17
27	12	13	14	16	18
28	12	14	15	16	18
29	13	14	15	17	19
30	13	14	15	17	19
31	14	15	16	18	20
32	14	15	16	18	20
33	14	15	17	18	21
34	15	16	17	19	21
35	15	16	17	19	22
36	15	17	18	20	22
48	20	21	22	25	27
54	22	23	25	27	30
60	24	26	27	30	33
66	26	28	29	32	35
72	28	30	32	34	38
78	30	32	34	37	40
84	33	35	36	39	43
90	35	37	38	42	45
96	37	39	41	44	48
102	39	41	43	46	50

[a] From ASBC Sensory Analysis 7. Courtesy American Society of Brewing Chemists—Reproduced by permission.
[b] Adapted from Meilgaard, M., Civille, G. V., and Carr, B. T. (2007.) Sensory Evaluation Techniques, 4th ed. CRC Press, Boca Raton, FL.
[c] n = number of assessments (panelists or replicates × panelists); α = risk of concluding that a significant difference exists when there is none.
[d] For values not in the table, calculate approximate values according to the formula below or refer to more complete tables:

$$x = (n/3) + z\sqrt{(2n/9)},$$

where x is the minimum number of correct responses and z = 0.84 for α = 0.20, 1.28 for α = 0.10, 1.64 for α = 0.05, 2.33 for α = 0.01, and 3.09 for α = 0.001.

THE BLUE SECTION SHOULD BE COMPLETED BY **YOU**, THE PERSON PREPARING THE TRIANGLE TEST	
Beer 1 name:	
Beer 2 name:	
Coin flip **(Select one)**	Beer 1 (Heads) Beer 2 (Tails)
First two paper strips pulled **(Select two)** This is the beer selected by the coin flip	Circle Square Triangle
Which cup is the other beer in? **(Select one)** This is the beer **NOT** selected by the coin flip	Circle Square Triangle

THE ORANGE SECTION SHOULD BE FILLED OUT USING THE ANSWERS THAT **A DIFFERENT ANALYST** HAS GIVEN AFTER HE OR SHE PERFORMED THE TEST THAT YOU SET UP IN THE BLUE SECTION	
Which two cups did the analyst say were the same? **(Select two)**	Circle Square Triangle

THE GREEN SECTION SHOULD BE COMPLETED BY **YOU** WHEN YOU ARE PERFORMING THE TRIANGLE TEST	
Which two cups are the same? **(Select two)**	Circle Square Triangle
Note the sensory differences below	
Appearance	
Aroma	
Flavor	
Mouthfeel	

FIG. 15.4. Guide for performing a self-run triangle test. (Courtesy S. E. Johnson and M. D. Mosher—© ASBC)

ASBC Sensory Analysis 18: Tetrad Test

Eighteen to 36 or more assessors preferred for sensory testing (minimum of 15).

Purpose

Use this test to determine whether:

1. two samples are significantly different or
2. two samples are significantly similar.

Procedure

To perform the tetrad test in a traditional fashion:

1. Determine the number of assessors required for the desired level of sensitivity for the test by selecting the number needed to achieve acceptable levels of α, β, and P_d from Table 15.4. The minimum number of assessors required is determined from Table 15.4.
2. Out of the sight of the assessors, two beer samples are randomly distributed among four sample containers.
3. The analyst randomly codes each sample container and records the contents.
4. Each assessor is given a set of four coded samples, two of each beer.
5. Each assessor evaluates the four samples and groups them into two pairs.
6. The analyst collects and records this information and compares the results against Table 15.5 to determine if the samples are significantly different.

TABLE 15.4. Minimum number of assessors needed for a tetrad test[a]

Value of α		Value of β			
		0.2	0.15	0.1	0.05
0.05	$P_d = 50\%$	15	18	22	27
0.01		25	27	32	37
0.05	$P_d = 40\%$	26	29	33	41
0.01		38	44	50	59
0.05	$P_d = 30\%$	43	49	59	73
0.01		66	73	85	104
0.05	$P_d = 20\%$	94	110	124	154
0.01		143	164	183	223
0.05	$P_d = 10\%$	351	401	478	597
0.01		553	621	711	858

[a] From ASBC Sensory Analysis 18. Courtesy American Society of Brewing Chemists—Reproduced by permission.

TABLE 15.5. Minimum number of correct responses required for the samples to be considered significantly different[a]

Assessments, n	Significance level of α	
	0.05	0.01
15	9	10
16	9	11
17	10	11
18	10	12
19	11	12
20	11	13
21	12	13
22	12	14
23	12	14
24	13	15
25	13	15
26	14	15
27	14	16
28	15	16
29	15	17
30	15	17
31	16	18
32	16	18
33	17	18
34	17	13
35	17	13
36	18	20
48	22	25
54	25	27
60	27	30
66	29	32
72	32	34
78	34	37
84	36	39
90	38	42
96	41	44
102	43	46

[a] From ASBC Sensory Analysis 18. Courtesy American Society of Brewing Chemists—Reproduced by permission.

To perform the tetrad test for a large group or class:

See Figure 15.5, a fill-in-the-blank guide for a self-run tetrad test analysis, at the end of this section.

1. The teacher or group leader assists with the test administration. Each group member receives four sample cups, four pieces of paper, a marker, and a coin. Each piece of paper and cup should be marked with AA, AB, BA, or BB.
2. Each assessor is given samples of the two beers to be evaluated. One sample is coded Beer 1 and the other Beer 2. These are uniform for the entire group.
3. Each member of the group or class will do the following:
    a. Set up the tetrad test for another member of the group.
    b. Perform a tetrad test set up by another group member.
4. Use the blue section of Figure 15.5 to help keep track of the test as each group member follows these steps:
    a. Flip a coin and circle heads or tails. For heads, circle Beer 1, and for tails, circle Beer 2.
    b. Put the four labeled slips of paper in a hat or pocket. Pull out two of the slips.
    c. Pour the beer selected by the coin flip into the two cups selected by the slip pulls.
    d. Pour the beer not selected by the coin flip into the remaining two cups.
    e. Order the cups from left to right as AA, AB, BA, and BB.
5. Each group member changes seats with another member of the group.
6. Each group member completes the green section of Figure 15.5 as he or she performs the tetrad test, selecting the two samples that match to place in each pair.
7. Each group member returns to his or her original seat after pairing the beers into two sets of two.
8. The teacher or leader collects the sheets and counts the number of correct responses. Table 15.5 is consulted to analyze the data. If the number of correct responses is greater than or equal to the number given in the table corresponding to the number of assessments (n), then a significant difference exists between the samples at a specified α risk.

Terminology

α (alpha) risk: The risk of concluding that a significant difference exists when in reality there is not one (also known as a risk of false positive or a type I error).

β (beta) risk: The risk of concluding that no significant difference exists when in reality there is one (also known as a risk of false negative or a type II error).

P_d: The allowable proportion of discriminators in the population; those who can detect the difference between the products.

Sensitivity: A general term used to summarize the performance characteristics of the test.

Data Sheets for *ASBC Methods of Analysis*   **331**

THE BLUE SECTION SHOULD BE COMPLETED BY **YOU**, THE PERSON PREPARING THE TETRAD TEST		THE GREEN SECTION SHOULD BE COMPLETED BY **YOU** WHEN YOU ARE PERFORMING THE TETRAD TEST SET UP BY ANOTHER PERSON	
Name:		Name:	
Beer 1 name:		Appearance	
Beer 2 name:			
Coin flip **(Select one)**	Beer 1 (Heads) Beer 2 (Tails)		
		Aroma	
Which cups contain Beer 1? **(Select two)** This should be the beer that was selected by the coin flip and paper slips	AA AB BA BB		
		Flavor	
		Mouthfeel	
Which cups contain Beer 2? **(Select two)** This should be the beer that was **NOT** selected by the coin flip and paper slips	AA AB BA BB	Indicate the pairs that are the same	Pair 1 Pair 2

FIG. 15.5. Guide for performing a self-run tetrad test. (Courtesy S. E. Johnson and M. D. Mosher—© ASBC)

Glossary

absorbance A measurement of the amount of light that is absorbed by the sample.

accuracy The closeness of a measurement to the actual value.

acid-base reaction The reaction between an acid and a base; the products of the reaction are an ionic compound (salt) and a small molecule (usually, water).

acidic A solution with a pH less than 7.

acrospire The part of the seed that becomes the stem of the barley plant during germination.

action boundary A value for a specification that indicates the process needs to be adjusted.

adenosine triphosphate (ATP) An organic molecule used to store energy in living organisms; the low-energy form is ADP.

adjunct A nonbarley source of starch or sugar used in brewing.

aerobic conditions An environment characterized by the presence of oxygen.

agar A gelatin-like substance that's used to support the growth of microbes.

agar plate A petri dish containing an agar used to grow microbes.

albumins Smaller proteins that tend to be fairly soluble during the mash and get extracted into the wort.

ale Beer fermented with *Saccharomyces cerevisiae*, a top-fermenting yeast prone to producing more ester flavors.

alkaline A solution with a pH greater than 7.

alpha acids Compounds in hop oils that isomerize to iso-alpha acid during the boil process.

American National Standards Institute (ANSI) A private organization that sets standards for products, processes, and people.

American Society of Brewing Chemists (ASBC) An international organization of scientists and technical professionals that focuses on applying scientific principles to the brewing industry.

anabolism The making of new compounds via metabolism.

anaerobic conditions An environment characterized by the absence of oxygen.

analyte A compound being analyzed within a solution.

analytical lab A lab that handles analysis of everything from raw materials to finished product.

anion A negatively charged atom or collection of atoms.

appearance A measure of the look of a beer.

aroma A measure of the smell of a beer.

ASBC fishbone diagram A cause-and-effect diagram that aids in evaluating issues and determining solutions in the brewing process.

ASBC Methods of Analysis An online collection available to ASBC members that provides numerous analyses that can be performed in the brewing laboratory.

aseptic sampling The process of withdrawing a sample for analysis so as not to contaminate or alter the sample or the source of the sample.

atom The smallest indivisible unit of matter.

atomic number The number of protons in an atom.

autoclave A device (or the use of the device) that sterilizes objects by applying heat, moisture, and pressure.

autopipette A mechanical pipette operated to deliver a specific volume of liquid.

average The mean of a set of measurements; determined by adding all the measurements and dividing by the number of measurements.

balance A piece of equipment used to measure the mass of an item.

bar graph A plot of data with specific values on the *y*-axis that are broken into different categories on the *x*-axis.

barley The most common grain used in brewing; a member of the grasses family of organisms.

beaker Glass shaped into a cylinder that looks like a cup.

Beer Judge Certification Program (BJCP) An organization that develops and maintains a set of specification ranges for all beer styles; members are certified to serve as judges during competitions.

Beer 10A An ASBC method used to determine the color of beer.

Beer 25D An ASBC method used to determine the presence of diacetyl.

Beer 26 An ASBC method used to determine the turbidity of beer.

Beer-Lambert law Provides the direct relationship of molar absorptivity, path length, and concentration to the absorbance of a solution.

Beer-Lambert law plot A standard curve prepared using samples with known concentrations of analyte that have been experimentally measured.

beta acids Compounds in hop oils that are not very water soluble but are prone to oxidation, which can result in a variety of flavors.

beta-glucans Pieces of plant cell walls that encompass starch granules.

bine An annual vine-like structure that dies off at the end of each season; wraps clockwise around objects as it grows toward light and touch.

bitter wort Wort that has been flavored with hops; typically, the liquid obtained after the boiling step.

blank A sample that lacks the analyte in question.

boiling Heating the sweet wort to boil in a separate vessel free of any grain or grain particles.

borosilicate glass Relatively fragile glass that can't withstand large, rapid temperature changes.

bracteoles Small leaves in the hop cone that protect the lupulin glands.

bracts Leaves of the hop cone.

broth A liquid material used to grow microbes.

Büchner funnel A ceramic or plastic funnel with a flat, perforated bottom.

Bunsen burner A gas-powered device that produces a single open flame.

burette A tube that's similar in appearance to a graduated pipette but contains a stopcock at the bottom.

Carlsberg flask A sample vessel used to hold quantities of wort that can be sterilized.

carrier gas The mobile phase in a gas chromatography analysis.

catabolism The breaking down of compounds via metabolism.

cation A positively charged atom or collection of atoms.

Centers for Disease Control and Prevention (CDC) A U.S. federal organization that conducts and supports health promotion, prevention, and preparedness with the goal of improving public health.

chelating agent A chemical that binds to metal ions.

chemistry The science that studies matter, its changes, and the energy associated with those changes.

chilling Cooling the beer, postboil; can be done to reach the desired fermentation temperature, to reach the cold storage temperature, or to adjust the temperature for filtration or packaging.

chitin A hard, rigid, structural compound; makes up a bud scar on the yeast cell.

chromatography The process used to separate a mixture of compounds.

clean Free of soil, grime, dirt, and residue.

collar Beer foam.

column oven A heater used to warm the gas chromatography column.

concentration The amount of a particular analyte per a given volume.

condenser A tube surrounded by a jacket that can be filled with cold water or another fluid.

Conditional Formatting A set of rules in a computer program that adjust how a cell is displayed in a spreadsheet.

continuous wave spectrometer A spectrometer that produces a spectrum across a wide range of wavelengths by rotating a monochromator.

control point A specific location along the process stream at which the product specifications can be measured and modified.

Coriolis density meter An instrument made of two U-shaped tubes that oscillate; the degree of deviation in the oscillation is related to the mass and flow rate of the liquid.

counterion An ion with the opposite charge; an anion is the counterion to a cation.

cover slip A thin square of glass that's placed on top of a sample on a microscope slide.

critical control point (CCP) A location in the production line where evaluation of a hazard that risks the safety of the consumer can be reduced or eliminated.

cultivar A variety of hops, barley, or yeast that is in the same genus but has different properties.

culture broth A liquid containing nutrients to support the growth of microbes.

culture of safety The core values and behaviors that underlie a commitment to safety within the brewery by workers and leaders.

CUSUM chart A chart that plots the cumulated sum of differences between measurements and the target value; shows trending in a measurement.

cuvette A sample holder with a known path length.

cycloheximide A chemical that inhibits the growth of brewer's yeast but not other microbes.

decimal reduction time (D-value) The time required to kill 90% (a 10-fold reduction) of the microbes in a sample.

degrees of freedom The ability of a molecule to move in three dimensions along the x-axis, the y-axis, and the z-axis.

densitometer An instrument that measures the optical density of a sample but doesn't physically measure the density of a solution.

density meter An instrument that measures the density of a liquid.

deoxynivalenol (DON) The chemical name of vomitoxin, a toxic compound.

detector The part of an instrument that converts analyte information into a computer signal.

detergent A chemical that solubilizes oils and fats, particularly during cleaning.

diastatic power The enzymatic activity as it relates to converting starch into sugar.

differential broth A liquid that contains chemicals that signal the presence of certain microbes.

diffraction grating A monochromator with carefully cut and closely spaced grooves; acts like a prism.

dilution equation Used to calculate how to dilute a solution to obtain a particular concentration.

diode array A detector made up of a series of photodiodes aligned in an array.

dioecious A plant species that has male and female versions.

dipole moment The field that results from the separation of charges in a molecule because of an uneven distribution of electrons.

dipole–dipole interaction A medium-strength intermolecular force between molecules with a dipole moment.

Duboscq colorimeter A microscope-like device that compares the intensity of light in the sample to a standard by adjusting the path length of each sample.

dynamic viscosity The measure of the internal resistance to flow.

electromagnetic spectrum The range of energy that can be transmitted through space.

electronegativity The measure of how electrons are polarized in a bond within the molecule.

electrons Negatively charged particles that surround the nucleus of an atom.

element The smallest collection of atoms of the same type.

eluent The liquid used as the mobile phase in TLC.

elute The process of moving analytes through a chromatography system.

energy The power within a system that can be transferred.

Environmental Protection Agency (EPA) A U.S. federal organization that provides regulations for the protection of the environment.

enzymes Molecules that help reactions occur; enzymes are temperature and pH sensitive.

equipment Simple laboratory tools, such as spatulas, balances, hot plates, mixers, and so on.

equivalence A mathematical construct in which the numerator and the denominator are equivalent.

Erlenmeyer flask A conical-shaped beaker.

European Brewery Convention (EBC) A numerical rating that describes the color of a beer based on the absorbance of light at 430 nanometers; approximately equal to twice the SRM value.

excited state Any energy state of a molecule that's higher than the ground state.

falling ball viscometer An instrument that measures the time it takes for a ball to descend a given distance through the sample.

ferment Converting sugars into ethanol and carbon dioxide under anaerobic conditions.

fermentation The metabolism pathway that occurs under anaerobic conditions to convert pyruvate to ethanol.

fixed wavelength spectrometer A spectrometer that can measure only the absorbance of a single wavelength at a time.

flavor A measure of the taste of a beer.

flavor threshold The minimum concentration at which a given flavor can be perceived.

Florence flask A spherical-shaped beaker with a flat bottom.

Food Safety Modernization Act (FSMA) A law enacted in 2011 that requires the Food and Drug Administration (FDA) to regulate food production and processing.

formazin turbidity units (FTUs) Used to measure the clarity of beer based on the chemical formazin.

formula A list of the elements and the number of each found in a compound.

free amino nitrogen (FAN) The number of amino acids that the yeast will ultimately have access to during fermentation; depends on the type of malt and the mashing process.

frequency The number of times an electromagnetic wave passes a given point per unit of time; inversely related to wavelength.

funnel support A device made to support a funnel above the piece of glassware into which a chemical is transferred.

Fusarium graminearum The scientific name for a fungus that grows on barley and produces vomitoxin.

gas chromatography (GC) A system that uses gas as the mobile phase.

GC column A tube containing the stationary phase used in gas chromatography.

germination The combination of processes that occur when a seed begins to grow.

gibberellic acid A molecule that triggers enzyme production and release within barley.

glassware Laboratory equipment that includes glass beakers, flasks, rods, dishes, and pipettes.

globulins Smaller proteins that tend to be fairly soluble during the mash and get extracted into the wort.

glucose A fermentable sugar that can't be broken down into smaller sugars.

glutelins Larger proteins that tend to be less soluble during the mash.

glycolysis The metabolism of glucose into pyruvate for the cell to use as energy.

good brewing practices (GBPs) The standards in a brewery that govern quality and consistency in production.

good manufacturing practices (GMPs) Production standards that govern quality and consistency.

gradient elution or **gradient HPLC** The change in the polarity of the mobile phase during a chromatographic separation.

graduated cylinder A cylinder marked along the side with volumes.

graduated pipette A tube marked along the side with very specific volumes.

gram negative A microbe that turns red when treated with Gram's stain.

gram positive A microbe that turns purple when treated with Gram's stain.

gravimetric density meter An instrument with a hollow, flexible, U-shaped tube that measures the density of a liquid.

gravity filtering Draining a liquid using a funnel and filter paper.

green beer Beer that has finished primary fermentation but still presents off-flavors.

ground state The lowest possible energy state of a molecule.

gustation The process in which chemicals from food are captured within a taste bud pore and then absorbed into receptor cells in the taste bud.

half-life The time required for a reaction to lose 50% of its current reactants.

hazard Anything that harms the safety of the product for consumption.

hazard analysis and critical control points (HACCP) A brewery-wide quality assurance program that ensures the consistent production of a product that's safe for consumption.

Hazardous Materials Identification System A numerical rating method that incorporates colored labels to identify types of safety hazards.

head Beer foam.

headspace analysis The evaluation of an analyte by testing the vapors above the sample.

high-performance liquid chromatography (HPLC) A system that uses pressure to push a liquid through the system.

hordeins Larger proteins that tend to be less soluble during the mash.

Hordeum vulgare The scientific name for all species of barley.

Humulus lupulus The scientific name for hops.

hydrogen bonding A strong intermolecular force between molecules containing an oxygen, nitrogen, or fluorine atom and a hydrogen atom attached to one of those atoms.

hydrometer A simple instrument that can measure the density of the liquid in which it's placed based on the instrument's buoyancy.

hypothesis A proposed answer to a question that is based on experimental evidence.

in-line analysis Analysis conducted by a measuring device placed directly in the product stream.

incident intensity The concentration of light that enters the sample.

index of refraction A unitless number that quantifies the refraction of light in a substance.

infrared spectrum A spectrum generated by the amount of infrared light absorbed versus the wavelength of that light.

injection port The location that the sample is added to the chromatography instrument.

instruments Sensitive machines with precisely machined parts that measure the concentration or identity of an analyte; often include on-board computers that communicate with laptop and desktop computers dedicated to given instruments.

interferogram A plot of the intensity of the light versus the position of the moving mirror in an interferometer.

interferometer A device that creates an interference pattern in a beam of light by moving a mirror.

intermolecular force of attraction An attraction between molecules that holds them closer together.

internal standard A compound added to a sample in a known amount; used to measure the quantity of the analyte in the sample.

invertase An enzyme on the outer surface of the yeast cell that converts sucrose into glucose and fructose.

ion An atom or collection of atoms with a net charge caused by the addition or elimination of an electron.

ionic compound The combination of an anion and cation, such that the overall charge is 0.

isomers Compounds that have the same number of atoms but differ in how the atoms are attached.

isotopes Atoms that differ only in the number of neutrons in their nuclei.

joule (J) A unit of energy measured in $kg \cdot m^2 \cdot s^{-2}$.

kilning Heating to remove water from grain; the process stops germination, eliminates some enzymes, and initiates Maillard reactions that result in color and flavor enhancements.

kinematic viscosity The measure of the ratio of the internal resistance to flow versus the density.

Krebs cycle A sequence of reactions by which most living organisms produce energy when presented with aerobic conditions.

laboratory-based analysis Analysis performed in a laboratory on samples from the production floor.

lager Beer fermented with *Saccharomyces pastorianus*, a bottom-fermenting yeast; lagers are fermented at colder temperatures than ales.

Lambert absorption law Describes how objects absorb light.

laminar flow Movement of all the particles of a substance in the same parallel direction, with no diffusion lateral to the direction of flow.

lautering A step after mashing that allows the grains to settle to form a filter bed; the sweet wort is drained through this bed.

law A tested and confirmed hypothesis that predicts a principle but does not explain why.

limit dextrins Large, branching portions of the starch molecule that are not fermentable.

linear regression A mathematical treatment of a set of data in which a trendline is drawn so it's as close to each specific data point as possible.

liquid funnel A funnel with a narrow opening; typically used in transferring liquids.

London dispersion force The weakest intermolecular force that results from electron movement in a molecule.

loop A hand-held stick made of wire that has been bent into a circle at the end.

Lovibond comparator A device that uses a set of standards to determine the color of a beer by comparison with shaded panes of glass, plastic, or other material.

lupulin glands Areas in the hop cone along the strig from which oils are secreted.

macro A set of operations that the user creates and the software performs.

Maillard reactions A series of complex organic chemistry reactions that involve the combination of sugars and proteins and the loss of water.

malt Grain that has undergone the steeping, germination, and kilning process.

malt color unit (MCU) The amount of color in SRM that a given amount of malt will add to a beer.

malting A location that performs malting; also known as a malthouse.

maltose A fermentable sugar composed of 2 glucose units.

maltotriose A fermentable sugar composed of 3 glucose units.

mash tun The vessel in which mashing occurs.

mass A measure of the amount of matter.

mass number The number of protons and neutrons in the nucleus of an atom.

Material Safety Data Sheet (MSDS) An old term for Safety Data Sheet (SDS) that is still commonly used.

matrix A solution containing the analyte and other compounds that may obscure detection of the analyte.

matter Anything that occupies space and has a mass.

media The base material used to grow microbes.

membrane filtration The process of separating and concentrating microbes from a liquid using filtration.

meniscus The curve in the upper surface of a liquid in a tube.

metabolism The biological reactions in a living organism.

method The steps and processes given to an instrument to successfully report results that are relevant.

microbes Organisms so small that they can be viewed only with a microscope.

microbiology lab A lab that focuses solely on measuring and/or identifying the microbial levels in the product stream.

microscope An instrument used to view microbes and yeast.

microscope slide A glass plate on which a sample is placed for viewing under a microscope.

microvilli Tiny hair-like components on the surface of a taste bud that grab chemicals to taste.

mill A machine used to crush malt for optimal utilization of starch and enzymes within the seed.

minimal broth A liquid that lacks sufficient amounts of a single nutrient or multiple nutrients required to support the growth of microbes.

mobile phase The portion of the chromatography system that moves during an analysis.

molar absorptivity A proportionality constant that relates the ability of a compound to absorb light.

molar mass The mass of 1 mole of a substance.

molarity (M) A unit of concentration determined by the number of moles of solute (analyte) per liter of solution.

mole A number equivalent to 6.02×10^{23}.

molecule A combination of atoms that doesn't have a net charge.

monochromator A device that selects and transmits a single wavelength of electromagnetic radiation.

mouthfeel A measure of the feel of a beer in the mouth.

National Fire Protection Agency (NFPA) An international nonprofit organization devoted to eliminating risks and losses from fire and related hazards; developed a hazard identification system for emergency responders.

neutron A neutrally charged particle found in the nucleus of an atom.

Nibem method A method of beer foam analysis.

nicotinamide adenine dinucleotide (NADH) An organic molecule used to store energy in living organisms; the low-energy form is NAD+.

nonpolar A molecule in which the electrons are evenly distributed.

normality (N) A unit of concentration determined by the number of equivalents of analyte per liter of solution.

nucleus The center of an atom; made up of neutrons and protons.

Occupational Health and Safety Administration (OSHA) A U.S. federal organization that provides standards for specific work activities and workplaces (such as breweries) to ensure safe and healthy working conditions.

off-flavor A flavor in the beer that was not intended or does not fit the style of the beer.

olfactometer A detector that uses a human to smell and detect the individual components in a mixture; typically used with a chromatography system.

on-site analysis Analysis of a sample performed at the site of the sample collection; for example, using a hydrometer next to the brewhouse.

Ostwald viscometer An instrument that measures the flow of a liquid by determining the amount of time that the liquid takes to traverse a specific distance.

overall impression A summary of the beer-tasting experience as a whole.

oxidation A loss of electrons from a species.

oxidative phosphorylation Reactions that occur after the Krebs cycle to add phosphorous to the molecule; results in the production of ATP.

oxidation-reduction reaction A reaction that involves the oxidation of one compound and the reduction of another.

packaging lab A lab that handles adherence to packaging specifications, such as bottle quality, quality of the can seams, labeling, and CO_2 package levels.

pasteurization unit (PU) A measure of the effect of heat and time on the survivability of microbes.

path length The thickness of a sample where the light traverses it.

peak broadening The effect of diffusion during a chromatographic process that causes the analyte to spread out.

percent transmittance (%T) One hundred times the transmittance.

percentage alcohol by volume (% ABV) The volume of ethanol dissolved in 100 mL of beer.

periodic table An organized table of all the known elements in the universe, whether natural or produced by humans.

personal protective equipment (PPE) Apparel that is required to provide safety and health protection to an employee (such as goggles, face shields, gloves, and aprons).

petri dish A low, flat dish with a lid.

pH A measure of the concentration of protons (H+) in a solution.

photocathode A detector that converts photons into an electrical signal.

photodiode A detector that measures a change in voltage when it's struck by a photon.

photomultiplier tube A detector that cascades electrons to result in an enhanced electrical signal.

photon A discrete packet of energy.

pitching Adding yeast to the fermenter.

polar A molecule in which the electrons are unevenly distributed.

polarity A measure of the distribution of electrons within a molecule.

polyatomic ion An ion made up of multiple atoms.

polymerase chain reaction (PCR) A technique used to make copies of DNA using primers specific for a particular organism; can be used to detect the presence of an organism in a mixture.

polyphenols Compounds that are part of the grain husk and vegetative matter; one of the two precursors to haze in beer.

powder funnel A funnel with a large opening; typically used in transferring solids.

precipitation reaction A reaction that forms a solid as the product.

precision The closeness of a measurement to other measurements.

preventive maintenance The steps required by the manufacturer to ensure that an instrument continues to operate optimally.

primer A short, single-stranded version of DNA specific for a particular organism.

propagation lab A lab that deals exclusively with yeast storage, yeast health, and growing samples of pitchable yeast for the yeast propagation system.

proton A positively charged particle found in the nucleus of an atom; determines the name of the element.

Pyrex A brand of glass that can withstand large rapid temperature changes.

quality Consistency in a product; can also mean adherence to product specifications, customer satisfaction, lack of off-flavors, or another measure.

quality assurance The systematic analysis of production that utilizes quality control measures, among other things, to ensure a consistently high-quality product.

quality control A program to ensure quality; reactionary if used alone.

quality control point (QCP) A location in the production line where evaluation of a hazard that risks the quality of the product can be reduced or eliminated.

radioactive An element that's relatively unstable and undergoes reactions within its nucleus.

reducing environment An environment that lacks oxygen and as a result doesn't allow oxidation reactions to occur.

reduction A gain of electrons to a species.

refraction The change in direction of a beam of light as it passes from one medium to another.

refractometer An instrument that measures the refraction of light as it passes through a sample of the solution.

relative humidity A value reported as the percentage of water vapor in the air versus the maximum percentage of water vapor at the given temperature.

retention factor (R_f) The ratio of the time an analyte spends versus the time that an unretained compound spends in the chromatography system.

retention time (RT) The amount of time an analyte spends in the chromatography system during separation.

Reynolds number (Re) A measure of the turbulence in the flow of a liquid; can be calculated to determine if a flow is turbulent or laminar.

rhizome A perennial root-like structure; a part of the hop plant that grows year-round.

ring clamp A metal circle that can be attached to a vertical pole.

ring stand A vertical pole attached to a heavy base so that it resists tipping.

round-bottom flask A spherical flask used for heating liquids in the laboratory.

Rudin method A method of beer foam analysis.

Saccharomyces cerevisiae The scientific name for a fungus capable of converting wort into beer by fermenting sugars into ethanol and carbon dioxide.

Safety Data Sheet (SDS) A sheet containing data about a particular chemical, including its properties, handling, storage, and safety.

salts A common synonym for "ionic compounds."

sanitary Characterized by a significantly reduced number of microbes; a quantity less than found in the normal environment.

scale A piece of equipment used to measure the weight of an item.

scatter plot An x-y plot of data points.

scientific method A series of steps that scientists follow to determine the answer to a question.

scientific notation A method of representing numbers using a factor of 10.

selective broth A liquid that contains a chemical or chemicals that allow only certain microbes to grow.

sensory analysis A program that provides the data needed to assess the flavors found in a sample of beer.

sensory lab A lab that uses sensory analysis methods to evaluate off-flavors, trueness to style, and other parameters of the beers produced.

sensory overload Occurs when the body's ability to detect certain flavors is overwhelmed by overstimulation of the taste buds or olfactory system.

sensory threshold The lowest concentration at which a compound can be sensed by a particular individual.

sigma value The standard deviation of the method of analysis.

simple plot An x-y plot.

slant A gelatin-like substance that's poured into a test tube and solidified at an angle.

slope of the line The rise over the run; the change in the *y*-axis versus the corresponding change in the *x*-axis along a straight line.

solubility rules A set of common statements that help determine whether an ionic compound is soluble in water.

solute *See* analyte.

source A lightbulb or other object that generates energy in the region of the spectrum of interest.

sparging The process in which additional water is added to the mash to extract additional sugars.

spatula A stick made of glass, metal, or wood that is wide at one or both ends; used to scoop solids.

specific gravity The density of a solution divided by the density of pure water.

spectroscopy The study of how matter and light interact.

spectrum (plural: spectra) A plot of the wavelength of electromagnetic radiation versus its intensity as it interacts with a sample.

speed of light 3×10^8 m/s; the velocity of electromagnetic radiation in a vacuum.

spreadsheet A computer program that allows the entry, tabulation, and storage of data and calculations.

standard curve An *x-y* plot of data points for a series of known concentrations.

standard operating procedure (SOP) A detailed explanation of how a task is performed.

standard reference method (SRM) A numerical rating that describes the intensity of the color of a beer determined using a laboratory method of analysis.

standard samples A set of solutions with known different concentrations.

standard taper ($) Standardized dimensions of a glass joint that include the angle of the joint.

starchy endosperm The largest portion of the seed; made up of starch storage cells.

stationary phase The fixed portion of the chromatography system that doesn't move.

steep To submerge barley in water during the malting process to increase the hydration level of the barleycorn before germination.

sterile Free of microbes.

still head An h-shaped tube that allows connecting three glassware items.

stoichiometric factor The relationship between compounds in a reaction written as an equivalence.

stoichiometric ratio The relationship between compounds in a reaction.

stopper A plug inserted into a glassware item to close it; made of glass, rubber, or cork.

strig Stem of the hop cone.

strobili The flowers of hop plants.

sucrose A fermentable sugar composed of 1 glucose unit and 1 fructose unit.

sweet wort The solution containing food and nutrients required for yeast growth obtained after the mashing process has been performed; because of the high concentration of sugars (mostly maltose), the solution is sweet.

tailing The retention of an analyte behind its main mass as it moves through the chromatography system.

tare To zero the balance or scale.

target value The desired value of a specification.

taste bud A receptor organ that contributes to the sensation of tasting; thousands are located along the tongue, palate, pharynx, and larynx.

terminal velocity The fastest an object will fall through a medium based on the force of gravity.

test tube Glass shaped into a tall, slender cylinder; requires a rack to stand upright.

test tube holder A twisted wire device that can be used to hold a test tube.

test tube rack A device made to hold multiple test tubes.

testing against a standard color An evaluation to determine the intensity of a color.

testing for the presence of a color An evaluation to determine whether a color exists.

text-based reporting Using tables, words, and phrases to report the results of an analysis.

theory A tested and confirmed hypothesis that explains an answer to a question.

thermal degradation unit (TDU) A measure of the impact of heat and time on the quality of the beer.

thermometer A device used to measure the temperature of a substance.

thin-layer chromatography (TLC) The application of chromatography to a thin plate containing the stationary phase and a pool of liquid that moves upward through the plate by capillary action.

tongs A scissor-like utensil used to hold hot objects.

total quality management (TQM) A brewery-wide program that ensures quality production in which all employees constantly work to improve quality.

transmittance (T) The intensity of electromagnetic radiation that passes through a sample; related to absorbance via the equation: $A = -\log(T)$.

transmitted intensity The concentration of light that passes out of the sample.

trub A solid material made up of protein, tannins, hop material, and malt material that settles during the whirlpool.

turbulent flow Movement of the particles of a substance with variations in the direction of the flow; lateral diffusion occurs.

ultraviolet-visible spectroscopy (UV-vis) The study of how ultraviolet-visible light interacts with a solution.

vacuum adapter A bent tube fitted with a nipple, where a vacuum can be applied.

vacuum filter flask A heavy-walled flask that supports a Büchner funnel.

vacuum filtering Draining a liquid using a funnel, filter paper, and vacuum; requires using a Büchner funnel.

vibrations Movement between atoms within a molecule in which the distances between the atoms change.

viscosity The resistance of a liquid to flow.

visual perception The ability to interpret the environment based on visual stimuli; a measure of how a particular sample is observed.

volatile compound A chemical with a low boiling point.

volumetric flask A flask that is similar in appearance to a Florence flask but carefully machined and marked with a specific volume.

volumetric pipette A tube that typically has a wide portion in the middle and is carefully machined and marked with a specific volume.

vomitoxin A compound that causes vomiting in humans and can result in "gushing" of packaged beer.

warning boundary A value for a specification that indicates the process needs to be scrutinized.

wavelength A single "color" of electromagnetic radiation, or light; inversely related to frequency.

weigh paper A piece of paper used to protect the balance or scale during a measurement.

weight A measure of the effect of gravity on an object.

wetting agent A chemical that decreases the surface tension of water.

whirlpool The process of swirling the wort (usually with pumps) as it cools from boiling; this causes solid materials to gather at the center of the vessel.

wort The food- and nutrient-rich liquid that's obtained after the mash.

yeast A fungus that ferments different sugars (maltose, glucose, etc.) into ethanol and carbon dioxide during anaerobic conditions.

zinc A micronutrient needed for healthy yeast growth.

Index

Note: A bold number indicates the page on which the topic is defined as a key term.

absorbance, 131–137. *See also*
Beer-Lambert law
definition of, **131**
of infrared energy, 186–188
overtones or harmonics of, 188
relationship with concentration, 133–138
relationship with percent transmittance, 181–182
within ultraviolet-visible spectrum, 189–190, 191
y-axis, 133
absorbance values, 132–133, 134, 181
ABV. *See* alcohol by volume (ABV); percentage alcohol by volume (% ABV)
accuracy and precision, of measurement, 32–33
bullseye analogy of, 32
of colorimetry, 164–165
definition of, **32**
factors affecting, 120–124
of instrument-based analysis, 117, 120–125
with laboratory glassware, 93–94, 96, 97, 98
acetaldehyde
boiling point of, 199, 200
dipole–dipole interactions of, 199
in fermentation, 92
flavor threshold of, 21
line drawing of, 74
as off-flavor cause, 27
reduction to ethanol, 73, 75–76
shape, mass, boiling point relationship of, 200
acetic acid, 27, 154
Acetobacter, 86
acetyl CoA, 90, 91
acid-base reactions, **71**–72
acidic cleaners, 109

acidic off-flavor, 27
acidic solutions, **69**–70
acidity, effect of carbonation on, 150
acids
hazard codes for, 45
molar equivalents of, 68
pH calculation for, 69–70
spills of, 44
acrospire, **4, 78**
action boundary, 138
adenosine triphosphate (ATP), **90**, 91
adjuncts, **77**–78
aerobic conditions, 5, 75, **89**
agar, **112**. *See also* agar plates
agar plates, 237, 239–240
definition of, **237**
differential and selective, 242, 243
inoculation of, 239–240
reading of, 242
sterilization of, 112
agaropectin, 237
agarose, 237
aging of beer, as off-flavor cause, 27
albumins, **83**
alcohol(s). *See also* ethanol
antimicrobial properties of, 17
production during fermentation, 4
as sanitizer, 109
alcohol by volume (ABV), 13. *See also* percentage alcohol by volume (% ABV)
density and, 214
measurement of, 102, 117–118, 188, 189, 214
alcolyzers, 51
ale, **1**
alkalies, 44, 45
alkaline solutions, pH of, **69**, 70
allergens, in craft beer, 17
alpha acids, **81**–82

aluminum, as chromatography backing material, 203, 204
American National Standards Institute (ANSI), **42**
American Society of Brewing Chemists (ASBC), 22–25, 117
ASBC Methods of Analysis. *See ASBC Methods of Analysis*
cause-and-effect (fishbone) diagrams, 23–24, 29
Common Brewery-Related Microorganisms, 242
description of, **22**
flavor wheel, 153–154
Grow Your Own Lab guide, 46–47
Laboratory Proficiency Program, 35
Sensory Analysis Methods, 158
services provided by, 22–23
amino acids, 4, 91–92
ammonium, 64
α-amylase, 79
β-amylase, 79
anabolism, 89
anaerobic conditions, 4, 5, **90**
analytes
concentrations of, 45
conversion for measurement, 125
definition of, **31**–32, **115**
light absorbance in, 131–133
molarity determination for, 66–68
normality determination for, 68–69
separation and detection of, 45, 46
analytical laboratory, **48**
anions
definition of, **59**
formation of, 60, 64
in ionic compounds, 60–61, 62
wort content of, 4, 85
ANSI. *See* American National Standards Institute (ANSI)

343

antimicrobial properties,
  of beer, 17
Anton Paar, 189
appearance, of beer, **147**–150
Appert, Nicolas, 232
aroma, of beer, **147,** 152
  flavor perception and, 151,
    152–154
  olfactometer detection of,
    225–226
  sensory analysis of, 151, 152–154
ASBC. *See* American Society of
  Brewing Chemists
ASBC fishbone diagrams, 23–**24,**
  29
ASBC Methods of Analysis, **24**–25, 46,
  172–173, 245–331
  Beer 4A, 284–286
  Beer 4C, 287–288
  Beer 4D, 289–291
  Beer 10A, **173,** 292–293
  Beer 11C, 294–296
  Beer 13C, 297–298
  Beer 22A, 299–300
  Beer 23A, 301–302
  Beer 23C, 303–306
  Beer 25D, **173,** 307–309
  Beer 26, **173**
  Beer 38, 310–312
  Hops 6A, 260–262
  Hops 13, 263–264
  Hops 15, 265–267
  Hops 17, 268–270
  Malt 4, 247–249
  Malt 6A, 250–253
  Malt 6B, 254–257
  Malt 15B, 258–259
  Sensory Analysis 7, 323–327
  Sensory Analysis 18, 328–331
  Wort 5, 271–272
  Wort 12, 273–275
  Wort 22, 276–278
  Wort 23A, 279–280
  Wort 23C, 281–283
  Yeast 4, 313–314
  Yeast 6, 315–317
  Yeast 9, 318–319
  Yeast 11B, 320–322
aseptic sampling, **241**
Aspergillus niger, 232
atomic mass, 55
atomic mass units, 56–57
atomic number, **55,** 57, 58–59
atomic symbols, 58–59

atoms, 53–65
  with charges. *See* anions;
    cations; ions
  collections of, 55
  combinations of, 60–61
  definition of, **54**
  degrees of freedom of, 186
  structure of, 54, 57–58
  subatomic particles of, 57
audits/auditing, 22
autoclaves, **110,** 112
autopipettes, **110**–111
averages, **141**
Ayurveda medicine, 151

bacillus/bacilli, 87, 88
bacteria, 86–88
  beer-spoilage, 19–20, 27, 29–30,
    86, 111–112
  cell structure and morphology
    of, 86–88
  coliform, 19–20
  gram-negative, **242**
  gram-positive, **242**
  growth of, 235–240
  identification of, 242–244
  illness-causing, 18, 19–20
  as off-flavor cause, 27, 29–30
  sampling of, 241–242
  size of, 111–112
balances, 38, 39, 49, 50, **102**–104
bar graphs, **128**
barley, **77**–79. *See also* malt
  chitting of, 4
  conversion to malt, 78
  cultivars of, 3
  dormancy period of, 78
  six-row variety of, 77–78
  structure of, 77–79
  two-row variety of, 77
barley endosperm, 4
batches, of beer
  consistency of, 30
  recordkeeping files for, 130–131
  starting of, 4
beakers, **94,** 95
Beer, August, 131, 161
Beer Flavor Database, 26
beer journals, 146–147
Beer Judge Certification Program
  (BJCP), **146,** 150
Beer-Lambert law, 131–135
  concentration variable of, 133
  definition of, **131, 161,** 162

  linear correlations and, 135
  molar absorptivity value of, 132
  path length variable of, 133
Beer-Lambert law plot, **131.**
  See also Beer-Lambert law
beer stone. *See* calcium oxalate
Beer's law. *See* Beer-Lambert law
Bernard, Felix, 161
beta acids, **81**–82
beta-glucans, **83,** 85
bias, in research, 127, 128
bines, **80,** 81
biological hazards, 17, 18
bitter wort, **4.** *See also* wort
bitterness, of beer
  iso-alpha acids and, 81–82
  sensory analysis of, 160
BJCP. *See* Beer Judge Certification
  Program (BJCP)
blank samples, **169**–170, 171, 189
boiling, of wort, **4**
  alpha acid conversion during,
    81–82
  effect on microorganisms, 86
  sterilizing effect of, 234
boiling point
  intermolecular forces of
    attraction and, 198–199
  relationship to polarity,
    209–210
  of water, 107
bonds/bonding, 194–200
  dipoles of, 197–198
  forces of attraction in, 195–200
  hydrogen, 196, 199–200, 237
  nonpolarized, 197
  polarized, 194–195, 198
borosilicate glass, **94.**
  See also glassware
bottled beer, 17, 18, 19–20, 21
Bouguer, Pierre, 161
bracteoles, **80,** 81
bracts, **80,** 81
Brewer's Association, on number
  of breweries, 1
brewery laboratory, 31–51. *See also*
  microbiological laboratory
  budget for, 46–51
  chemicals for, 38–40
  design of, 35–36
  essential items for, 48–51
  lighting for, 124
  overview of, 31–33
  planning of, 46–51

practices for, 93–114
reporting of results. *See* reporting, of results
safety guidelines for, 40–45
space requirements for, 36–38, 119–120
supplies for, 38–40
temperature of, 121, 123
brewing process, overview of, 3–5
broken glass, in bottled beer, 17, 18, 19–20, 21
broths, for microbial growth, 114, **236**–237, 238–239
Büchner funnels, 94–**95,** 101. *See also* funnels
budget, for brewery laboratory, 46–51
 basic equipment and instrumentation, 48, 49
 under $1,000, 50–51
 under $10,000, 48–49
Bunsen burners, **97,** 112–113, 238, 239
burettes, 95, **96,** 97
burnt flavor, 78–79, 83
butanoic acid, as off-flavor cause, 27
butyric acid, 74–75

calcium
 brewing water content, 118
 as hot liquor additive, 4
 solubility of, 73
 wort content, 4, 83
calcium chloride, solubility of, 73
calcium hydroxide, solubility of, 73
calcium oxalate, formula for, 65
calcium sulfate, solubility of, 73
capillary action, 42, 204
caramel flavor, 78–79
caramelization reactions, 79
carbohydrates, 83. *See also* sugars
carbon
 hop oil content, 200–201
 isotopes of, 55–56
 molecular combinations of, 62, 63
carbonate, wort content, 4, 85
carbonation, 150
carbon dioxide
 in fermentation, 4, 5, 92
 in foam formation, 226, 227
 formula and chemical structure of, 63, 197

infrared analysis of, 189
nonpolarity of, 197
in oxidative phosphorylation, 91
in reducing environment, 17
vibrational modes of, 186, 187
carbon dioxide lines, bacterial contamination of, 29–30
carbon dioxide volume meters, 49
carbon monoxide, formula and chemical structure of, 63
Carlsberg flasks, **114,** 238, 239
carrier gases, **206**
Carter, Jimmy, 1
caryophyllene, aroma and flavor of, 82
catabolism, **89**
cations
 definition of, **59**
 formation of, 60, 64
 in ionic compounds, 60–61, 62
 wort content, 83
cause-and-effect diagrams, 23–24, 29
caustic cleaners, 109
caustic solutions, as chemical hazard, 18
CCPs. *See* critical control points
CDC. *See* Centers for Disease Control and Prevention (CDC)
cell counts, in-line analysis of, 35
Centers for Disease Control and Prevention (CDC), **40**
centipoise units, 222
centrifuges, 49
charts. *See also* tables
 cumulative sum (CUSUM), 138, 139–140
 for good brewing practices, 8
 software programs for, 128
 spreadsheet-based construction of, 138–140
chelating agents, **109**
chemical formulas, 61–62
chemical fume hoods, 37
chemical hazards, 17, 18
chemical hygiene plans, 18
chemical reactions, 71–73
 acid-base, 71–72
 in gas chromatography, 209–210
 graphic representation of, 74–76
 oxidation-reaction (redox) reactions, 73, 75–76
 precipitation reactions, 72–73

chemicals, in brewery laboratory, 38–40
 cost of, 49
 handling of, 102–108
 labeling of, 40, 41
 personal protective equipment for, 40, 42–43
 safety guidelines for, 40, 44, 102
 storage of, 37, 40
chemistry, essentials of, 53–76
 atomic symbols, 58–59
 atoms and radioactivity, 54–58
 calculation of ions in water, 66–69
 chemical formulas, 61–62
 chemical reactions, 71–73
 collections of atoms, 61–65
 definition of, **53**
 ionic compounds, 60–61
 ions, 59–60
 line drawings of molecules, 74–75
 pH, 69–70
chilling, of wort, **4–5,** 7
chitin, **88**
chitting, of barley, 4
chloride, wort content, 4, 85
chlorophenols, as off-flavor cause, 27
chromatographs/chromatography, 193–212
 columns of, 118, 193, 201, 202
 as common laboratory instruments, 45, 46, 116
 definition of, **193**
 detectors in, 201, 208–209, 225–226
 general design of, 202–212
 injection ports in, 205, 206, 207
 laboratory methods and, 201–202
 mobile and stationary phases, 193–194, 200–202
 molecular interaction prediction with, 195–200
 operation/purpose of, 118, 145, 193–194
 pros and cons of, 118
 retention time in, 203
 sensory analysis use of, 145, 154
 spectrophotometry versus, 193
 tailing in, 205
 theory underlying, 193
 types of, 45, 46, 116, 193, 202–212. *See also specific types*

chromium, wort content, 83, 85
cinnamaldehyde, 28
citric acid cycle, 89, 90–91, 92
clarification, of beer, 4
clean
    in brewery laboratory context, 109
    definition of, **108**
cleaners/cleaning solutions, 109–110
coccus, 87, 88
collagen, 237
collar, of beer, **226**. *See also* foam
color, of beer, 147–150
    deficiencies in perception of, 163–164
    effect on taste perception, 147
    intensity of, 165, 166–167, 170–171, 172, 173
    measurement of, 147–150
    presence of, 165–166, 167
color blindness, 163–164
color charts and scales, 147–148, 165, 173
colorimeters/colorimetric analysis, 161–173
    accuracy and precision of, 164–165
    ASBC Methods of Analysis for, 172–173
    as common laboratory instruments, 45, 46, 50–51
    cost of, 50–51
    development of, 161–163
    human sight and, 163–165
    measurement methods of, 165–167
    minimum and maximum concentrations in, 171–172
    operation/purpose of, 45, 46, 116
    reference charts for, 167–172
    setting up, 168–171
    spectrometers versus, 179
    standard solutions for, 168–171
    types of, 161–163, 167–168. *See also specific types*
    column ovens, **206**, 207
    columns, of gas chromatography, 193, 202, **206**–208
comparators
    homemade, 167–168
    Lovibond, 163, 164, 166–167
Compton, Arthur, 176

computers. *See also* spreadsheet programs
    dust buildup on, 124
    instrument-interactive, 45, 119–120
concentration
    definition of, **133**
    relationship to density, 213–214
    standard curve-based determination, 133–138
    of unknown samples, 136–137
condensers, 95, **96**
Conditional Formatting, **141**
conditioning tanks, 3, 5
    safety hazards in, 17
    standard operating procedures for, 30
congress wort, 160. *See also* wort
consistency
    of beer batches, 30
    importance of, 2–3
    monitoring of, 30
    role in quality maintenance, 5–6
contaminants
    in craft beer, 17
    glass and metal shards, 7, 17, 18, 234
    by microbes. *See* microbial contamination
    as off-flavor cause, 27
continuous wavelength spectrometers, **183**–184, 185. *See also* spectrometers
control points, **20**
copper
    as chemical hazard, 17
    wort content, 83, 85
    as yeast micronutrient, 92
Coriolis density meters, **218**–219. *See also* density meters/densimeters
corrective actions
    in HACCP programs, 18, 21–22
    for off-flavor control, 28–30
corrosive materials, hazard codes for, 45
cost–benefit analysis, of product analysis, 12–13
costs
    of equipment, 48
    of glassware, 51, 93, 94
    of infrared analysis instruments, 188

    of in-line analysis instruments, 35
    of instrument-based analysis, 118–119, 120
    of laboratory instruments, 46
counterions, **64**
cover slips, **111**
craft beer, allergens and contaminants in, 17
craft brewing industry, growth of, 1–2
Cranston, Alan, 1
critical control points (CCPs), 15
    actions for, 9
    definition of, **7, 20**
    employee responsibility for, 9
    examples of, 7–8
    identification of, 18, 20
    on-site analysis as, 10
    risk ratings of, 19
    upper and lower limits for, 20–21
Cryptosporidium infections, 21
cultivars, **3, 80**
culture broths, **236**. *See also* broths, for microbial growth
culture media. *See* media, for microbial growth
culture of safety, **41, 42**. *See also* safety guidelines and programs
cumulative sum (CUSUM) charts, 138, **139**–140
CUSUM charts, **139**. *See also* cumulative sum (CUSUM) charts
cuvettes, **180**
cycloheximide, 236, **241**, 243
cylinders, 95, 96, 97

D-value, 232–233
Darcy's law, 127
data logs, 34–35
decimal reduction time, 232–233
degrees Brix (°Bx), 220
degrees of freedom, **186**
degrees Plato (°P), 220, 238
density
    of beer, 117
    in-line analysis of, 218
    relationship to concentration, 213–214
    units and calculations for, 104

density meters/densimeters, 213–219
  definition of, **214**
  types of, 214–219. *See also specific types*
densitometers, 11, **214**
deoxynivalenol (DON), **78**
detectors
  diode array, 181, 184
  flame-ionization, 117–118, 201, 209
  in gas chromatography, 117–118, 208–209
  mass spectrometric, 117–118, 201, 208, 209, 211–212
  in UV-vis spectrometers, 181
detergents, **109**. *See also* cleaners/cleaning solutions
deuteranopia, 163
diacetyl
  analytical instruments for, 51
  concentration measurements of, 13, 51
  detection of, 35, 154, 173, 209
  detection threshold for, 28
  in fermentation, 92
  green beer content, 8–9
  as off-flavor cause, 27, 28, 29–30
  sampling in multiple fermenters, 11
diastatic power, **79**
differential broths, **236**. *See also* broths, for microbial growth
diffraction grating, **183**–184
dilution equation, 105–106, **168**
dimethyl ether, 63, 64
dimethyl sulfide, 27
diode arrays, **181**, 184
dioecious plants, **80**–81
dipole moments, **194**, 195, 198
dipole–dipole forces/interactions, 196, **198**–199
dissolved oxygen, 35
distillation, 101–102
  definition of, 101–102, 106
  process, 106–108
DNA, 87, 89, 243–244
DON. *See* deoxynivalenol (DON)
Drebbel, Cornelis, 230
Duboscq, Louis Jules, 161
Duboscq colorimeter, **161**, 162, 166–167. *See also* colorimeters/colorimetric analysis

duo-trio test, 158
dust, in laboratory environment, 123–124, 235
dynamic viscosity, **221,** 223, 225. *See also* viscosity

EBC. *See* European Brewery Convention (EBC)
ECDs. *See* electron capture detectors (ECDs)
Einstein, Albert, 176, 177, 181
electrical charge
  of electrons, 57
  of ions, 59–60
electrical equipment, 38, 43
electrical fields, within molecules, 194, 195–196
electrical power and services
  for laboratories, 38
  lines for, 119
  surges and spikes in, 122
electromagnetic radiation, 175
electromagnetic spectrum, 175
  definition of, **176,** 181
  molecular transitions of, 177, 178
  visible region of, 175–176, 178, 179–180
electromagnetic waves, frequency of, 176
electron capture detectors (ECDs), 201, 208–209
electronegativity, **194**–195, 197
electrons
  addition or removal of, 59–60
  definition of, **57**
  distribution within molecules, 194–195, 197
  polarity of, 194
element symbols, 58
elements, **54,** 58–59. *See also* periodic table of the elements
eluent, **204**
elute, **201**
employees. *See* staff
endosperm, starchy, **79**
energy, 54
  calculation of, 177
  cellular production of, 89, 90, 91–92
  definition of, **177**
  of photons, 176–177
Enterobacter, selective agar plates for, 243

Environmental Protection Agency (EPA), **40**
enzymes
  activation during mashing, 4
  activation in barley seeds, 4
  definition of, **79**
  inactivation during boiling, 4
  of yeast cell, 88, 92
EPA. *See* Environmental Protection Agency (EPA)
equipment
  ABSC guidelines for, 46–47
  bench space for, 37
  cleaning of, 109
  cost of, 48
  definition of, **38, 115**
  instruments versus, 45, 115
  for microbiological laboratories, 110–113, 235–236
  types of, 38, 39, 48–50. *See also specific types*
equivalences, **67**–68, 69
Eriochrome Black T, 118
Erlenmeyer flasks, 38, **95**. *See also* flasks
  in color determination tests, 167
  cost of, 51
  media preparation in, 122, 237
  sterilization of, 114
  for yeast sample growth, 114
errors, in measurement, 33
Escherichia coli, 19–20
ester flavors, 1
ethanethiol, as off-flavor cause, 27
ethanol
  boiling point of, 199, 200
  concentration measurement of, 51
  dipole–dipole interactions of, 199
  effect on refractive index, 221
  formula and chemical structure of, 63
  polarity of, 200–201
  production during fermentation, 5
  shape, mass, boiling point relationship, 200
  vibrational modes of, 186
ethyl acetate, as off-flavor cause, 27
eukaryotes, 88
European Brewery Convention (EBC), 117
  color scale of, **148,** 173

European Union, Regulation (EC) Number 852/2004, 16
Excel. *See* Microsoft Excel
excited state, **177**–179
eyes
    color perception deficiencies, 163–164
    protection for, 41, 42–43, 44
eyewash stations, 44

falling ball viscometers, **224**–225. *See also* viscometers
FAN levels. *See* food and nutrition (FAN) levels
far-infrared spectrum, 187
farnesene, aroma and flavor, 82
fats. *See* lipids
FDA. *See* Food and Drug Administration (FDA)
fermentation
    antimicrobial byproducts of, 234
    completion of, 4
    definition of, **4–5, 90**
    Pasteur's research on, 231
    process of, 4, 92
    temperature for, 1
    warm, as off-flavor cause, 27
    of wine, 231
fermenters, 3
    aseptic microbial sampling in, 241
    multiple, diacetyl sampling in, 11
FIDs. *See* flame-ionization detectors (FIDs)
filter cakes, 4
filter paper, 100–101
filtering, of liquids, 100–101
fire extinguishers, 43–44
fire safety, 40, 43–44
fishbone diagrams, 23–24, 29
fixed wavelength spectrometers, **182,** 183, 185. *See also* spectrometers
flame-ionization detectors (FIDs), 117–118, 201, 209
flammable hazards, 40, 41, 43, 45
flash chromatography, 193. *See also* chromatographs/chromatography
flasks, 95–96. *See also specific types*
flavor, of beer, 150
    aroma and, 152–155
    complexity of, 145
    consumers' perception of, 148
    definition of, **147**
    descriptors for, 153–154
    distinguished from off-flavor, 26
    effect of pasteurization temperature on, 233
    evaluation of, 145, 147. *See also* sensory analysis
    Maillard reactions and, 78–79
    relationship to aroma, 151, 152–154
    role of alpha and beta acids in, 82
flavor threshold
    definition of, **21**
    for off-flavors, 26, 27, 28
flavor wheel, 153–154
flocculation, 88
Florence flasks, **95.** *See also* flasks
flow conditioners, 223
fluorimeters, 45, 46, 116
fluorine, 194, 195
foam
    collapse of, 227
    compounds affecting, 150
    consumer's perception of beer quality and, 226
    formation of, 226–227
    measurements of, 226–227
    sensory analysis of, 150
Food and Drug Administration (FDA), 15–16, 234
food and nutrition (FAN) levels, 35
foodborne illnesses, 15–16
Food Safety and Modernization Act (FSMA), 31
food safety hazard controls, 15–16. *See also* hazard analysis and critical control points (HACCP)
Food Safety Modernization Act (FSMA), **16**
formazin turbidity units (FTUs), **173**
formulas (chemical), **61**–62
Fourier transform (FT) spectrometers, **184**–185. *See also* spectrometers
free amino nitrogen (FAN), **83**
frequency, of photon of light, **176**
fructose, 63
fructose-1,6-diphosphate, 91
fruit-like off-flavors, 27
FSMA. *See* Food Safety and Modernization Act (FSMA)
FT instruments. *See* Fourier transform (FT) spectrometers
FTUs. *See* formazin turbidity units (FTUs)
fungal contamination, 27, 78
funnel supports, **94**
funnels
    gravity filtering with, 100–101
    types of, 94–95, 101. *See also specific types*
Fusarium graminearum, **78**

gamma- (γ-) rays, 178
gas chromatographs (GCs)/gas chromatography, 205–210. *See also* chromatographs/chromatography
    ASBC Methods of Analysis using, 24
    carrier gases of, 206
    columns/column ovens of, 193, 202, **206**–208
    as common laboratory instruments, 45, 46, 116
    cost of, 51
    definition of, **203**
    detectors of, 117–118, 208–209
    diacetyl detection with, 154
    general design of, 206
    internal standards for, 209
    mobile and stationary phases of, 202, 206–207
    operation/purpose of, 45, 46, 116, 118, 205–210
    peak broadening in, 204–205
    pros and cons of, 209–210
gas chromatography-mass spectrometry (GC-MS), 116, 208, 209, 226. *See also* chromatographs/chromatography
gas lines, 119
GBPs. *See* good brewing practices (GBPs)
GC columns, **206.** *See also* gas chromatographs (GCs)/gas chromatography, columns/column ovens of
GC-MS. *See* gas chromatography-mass spectrometry (GC-MS)
GCs. *See* gas chromatographs (GCs)/gas chromatography
gelatin, 237
general relativity, 176
geraniol, 82, 201
germ theory of disease, 231

germination, 3–4, **78,** 79
gibberellic acid, **79**
glass. *See also* glassware
   as bottled beer contaminant, 7, 17, 19–20, 21, 234
   as chromatography backing material, 203, 204
glass pipettes and bulbs, 38, 50. *See also* glassware
glassware, 93–102
   breakage, 93, 94
   cleaning of, 109
   connections between, 96–97, 107
   cost of, 51, 93, 94
   definition of, **38**
   for microbiological laboratories, 109, 110–111, 112
   sterilization, 112
   storage of, 40
   "to contain" (TC), 96, 105
   "to deliver" (TD), 96, 105
   types of, 49, 94–100. *See also specific types*
globulins, **83**
gloves, safety, 41, 42–43
glucan, 88
glucan fragments, 4
β-glucanase, 79
glucose
   alpha-form, 84
   beta-form, 84
   as cell energy source, 90, 91–92
   chemical structure of, 84
   definition of, **83**
   fermentation of, 5
   formula and chemical structure of, 63
   metabolism of, 89–90, 91–92
   oxidative phosphorylation of, 91–92
   production of, 4
   vibrational modes of, 186
   wort content, 84
glutelins, **83**
glycolysis, **89,** 90, 91, 92
GMPs. *See* good manufacturing practices (GMPs)
goggles, safety, 41, 42
"go/no go" decisions, 12, 25, 31
good brewing practices (GBPs), 16, 235
   brewery laboratory and, 31
   definition, **6, 15**
   use of charting in, 8

good manufacturing practices (GMPs), **16**
Goppelsröder, Friedrich, 193
gradient elution/gradient HPLC, **210,** 211
graduated cylinders, 95, **96,** 97
graduated glassware, 97–98. *See also* glassware
graduated pipettes, **96.** *See also* pipettes
grain, moisture content of, 3
Gram, Hans Christian, 242
gram-negative microbes, **242**
gram-positive microbes, **242**
gram stainings, 242–243
graphs, software programs for, 128
gravimetric density meters, **217**–218. *See also* density meters/densimeters
gravity. *See also* specific gravity
   effect on mass, 103
   in-line analysis, 35
gravity filtering, **100**–101
gravity meters, 10
green beer, 3, **5,** 8–9
grist case, 3
ground state, **177**–178
gushing, 24, 78
gustation, **151**–152

HACCP. *See* hazard analysis and critical control points (HACCP)
half-life, **58**
hazard analysis and critical control points (HACCP), 6–9, 15–22, 31
   categories of, 17–18
   conducting an analysis, 18–20
   definition of, **6, 15**
   "go/no go" decisions and, 12, 25, 31
   history and development of, 15–16
   implementation of, 7
   limit setting for, 8–9
   principles of, 15–16
   programs. *See* hazard analysis and critical control points (HACCP) programs
hazard analysis and critical control points (HACCP) programs, 234
   corrective actions, 21–22

   critical control points identification in, 20
   hazard analysis in, 18–19
   implementation of, 17–22
   for microbial contamination, 234
   monitoring procedures in, 21
   recordkeeping procedures, 22
   steps in, 18–22
   upper and lower critical limits in, 20–21
   verification procedures in, 22, 25
Hazardous Materials Identification System, **40,** 41
hazards
   as critical control points, 20
   definition of, **15,** 17
   risk ratings for, 19–20
haze, in beer, 13, 83
head, of beer, **226.** *See also* foam
head blight, 78
headspace analysis, **203**
health hazards, 16, 41, 45, 234
heating, ventilation, and air conditioning (HVAC), effects on instruments' performance, 121
heavy metals, 18
hemocytometers, 49, 50
hexane, 201
high-gravity beers, 214
high-performance liquid chromatography (HPLC), 210–212. *See also* chromatographs/chromatography
   columns of, 210, 211
   as common laboratory instruments, 45, 46, 116
   cost of, 203
   definition of, **203**
   detectors in, 211–212
   gas chromatography versus, 211–212
   gradient elution/gradient HPLC of, 210, 211
   instrument design, 210
   internal standard for, 209
   mobile and stationary phases of, 210–211
   operation/purpose of, 210–210
   peak broadening in, 204–205
   pros and cons of, 203

Hop Flavor Database, 26
hop oils, 80, 81
 chromatographic analysis of, 200–202
 compounds in, 82
 effect on foam and foam stability, 150
 as off-flavor cause, 27
 yeast growth-inhibiting activity, 238
hop teas, 160
hops, 80–82, 81
 bacteriostatic properties of, 238
 cultivars of, 3
 cultivation and growth of, 80–81
 drying of, 81
 off-flavors, 160
 old, as off-flavor cause, 27
 sensory analysis, 160
 structure and components of, 80–81
hordeins, **83**
Hordeum distichum, 77. *See also* barley
Hordeum vulgare, **77.** *See also* barley
hot liquor, 4
hot liquor tanks, 3
hot plates, 37, 38, 49, 115
HPLC. *See* high-performance liquid chromatography (HPLC)
H. R. 1337—An Act to Amend the Internal Revenue Code of 1954, 1
humidity, 122
humulene
 aroma and flavor of, 82
 chemical structure of, 201
humulone
 chemical structure of, 81
 line drawing of, 74
Humulus lupulus, **80.** *See also* hops
HVAC systems. *See* heating, ventilation, and air conditioning (HVAC)
Hydrion indicator papers, 166
hydrochloric acid molecule, vibrational modes of, 186
hydrogen
 atoms, 55, 199–200
 hop oil content, 200–201
hydrogen bonding, 196, **199–200**, 237
hydrogen sulfide, as off-flavor cause, 27

hydrometers, 214–216. *See also* density meters/densimeters
 as common laboratory instruments, 45, 46, 49, 50, 116
 cost of, 50
 definition of, **214**
 general design of, 214–215
 maintenance of, 6
 on-site analysis using, 10
 operation/purpose of, 46, 49, 116, 215–216
hydroxide, 64, 70
hypotheses, **127**

IBUs. *See* international bitterness units (IBUs)
ICP–OES. *See* inductively coupled plasma–optical emission spectrometer (ICP–OES)
incident intensity, **132**
index of refraction, **220–221**. *See also* refraction
inductively coupled plasma–optical emission spectrometer (ICP–OES), 118, 119
infrared analysis, 186–189
infrared rays, 175–176
infrared spectrum, **186**–188
infrared wavelengths, 180–181
injection ports, **205,** 206, 207
in-line analysis, 34–35
 data logs for, 34–35
 definition of, **11**
 of density, 218
 instruments for, 35
 pros and cons of, 11, 21
instrument-based analysis
 accuracy and precision of, 120–124
 cost of, 118–119, 120
 method for, 125
 pros and cons of, 117–120
instruments, 115–126, 213–227
 analyte detection devices of, 121
 bench space for, 36
 categories of, 116–117
 cleaning of, 109, 124
 common, 45, 46
 conditions for optimal operation, 120–124
 cost of, 48, 49, 50
 definition of, **45, 115**

 demonstration ("demo") and refurbished, 119
 dust buildup on, 123–124
 equipment versus, 45, 115
 guides to, 24
 maintenance of, 118–119, 120, 124–126
 measurement errors of, 33
 for microbiological laboratories, 235–236
 standard operating procedures for, 46
 static electricity buildup on, 122–123
 types of, 45–46, 116. *See also specific types*
 validation procedures, 35
interferograms, **184,** 185
interferometers, **184,** 185
intermolecular forces of attraction, 195–200
 definition of, **195**
 intramolecular forces versus, 195–196
 of polymers, 237
 types of, 196–200
internal standards, **209**
international bitterness units (IBUs), 35, 160
International Standards Organization, standard 2200, 16
intramolecular forces of attraction, 195–196
invertase, **88**
ion chromatographs, 46
ionic compounds, 60–61
 atomic structure of, 63
 common, 62
 definition of, **60**
 polyatomic ion-containing, 64–65
 solubility of, 60, 61
 solubility rules for, 72–73
ions, 59–61
 common, 65
 definition of, **59**
 polyatomic, 64–65
 water content, 66–69, 83, 85
 wort content, 83, 85
iron
 atomic structure of, 55
 as off-flavor cause, 27
 wort content, 83, 85
 as yeast micronutrient, 92

Ishikawa (fishbone) diagrams, 23–24, 29
iso-alpha acids, 17, 81–82, 234
    antimicrobial properties of, 17
    concentration of, 191
    in foam formation, 226, 227
    formation of, 81–82
isomers, 63, **64**
isopropanol, 188
isopropyl alcohol, 235
isotopes, **55**–56, 58

Janssen, Zacharias, 229–230
Joule, James Prescott, 177
joules, **177**
Journal of the American Society of Brewing Chemists, 23

kilning, **4, 78,** 79
kilograms, as common units of measurement, 56
kilojoules, 177
kinematic viscosity, **222,** 223. *See also* viscosity
Krebs cycle, **89**–90, 90–91, 92

laboratory-based analysis. *See also* brewery laboratory
    ASBC proficiency program in, 23
    definition of, **11**
    limits of detection and linearity in, 20–21
laboratory information management systems (LIMS), 128
laboratory personnel
    dedicated, 34, 35
    HACCP implementation role, 17, 20–21
    training and qualifications, 33, 34
lactic acid bacteria, 243. *See also* Lactobacillus
Lactobacillus, 86, 88, 229
    D- and z-values for, 232, 233
    as gram-positive microbe, 243
    as off-flavor cause, 29–30
    selective agar plates for, 243
lagers, **1,** 21
Lambert, Johann Heinrich, 131, 161
Lambert absorption law, **161**
laminar flow, **222**–223
laminar flow hoods, 37, 49, 236, 238

latex allergies, 42
lautering, **4**
laws (scientific), **127**
Lemgo D-value and z-value Database for Food, 232
light
    absorbance of. *See* absorbance
    "color" of. *See* wavelengths
    as energy, 177
    as off-flavor cause, 27
    speed of, 176, 177
light-struck beer, 56, 127
lighting
    for beer sensory analysis, 147, 148, 155–156
    effect on instruments' performance, 124
limit dextrins, 4, **83,** 84
limonene, 75
LIMS. *See* laboratory information management systems (LIMS)
linalool, 82, 201
linear regression, **133**–134
line drawings, of molecules, 74–75
lipids, 88, 91–92
Lippershey, Hans, 230
liquid chromatography, 193, 202. *See also* high-performance liquid chromatography (HPLC)
liquid funnels, **94.** *See also* funnels
liquids
    distillation of, 101–102, 106–108
    filtering of, 100–101
    mass, measurement of, 104
    measurement of, 93–100
    menisci of. *See* meniscus/menisci
    volume, measurement of, 104–105
London, Fritz, 197
London dispersion forces, 196, **197**
loops, 112–113
    definition of, **112**
    sterilization of, 112–113
    use in agar plate inoculation, 239–240
Lovibond, Joseph, 147, 163
Lovibond color scale, 147–148
Lovibond comparator/tintometer, 162, **163**
low-alcohol beers, microbial contamination, 86, 234
lupulin glands, **80,** 81
lupulone, chemical structure, 81

macros, **141**–144
magnesium, wort content, 4, 83
magnetic stir bars/stirrers, 38, 39, 112, 114
magnification, 230
Maillard, Louis-Camille, 78
Maillard reactions, **78**–79, 149
malt
    definition of, **4**
    diastatic power of, 79
    flavor profiles, 149–150
    milling of, 3
    moisture content of, 4, 78
    pilsner, as off-flavor cause, 27
    specialty, 148–150, 160
    storage of, 4
    vomitoxin content, 78
malt color unit (MCU) analysis, **149**
malthouse, 3
malting, **3**–4
maltose
    chemical structure, 84
    conversion into glucose, 89
    definition of, **83**
    entry into yeast cell, 88, 89
    fermentation of, 5
    production of, 4, 79
    wort content, 84
    yeast cell's consumption of, 88
maltotriose, 4, **83**
maltster, 4, 78
mannan, 88
Martin, Archer John Porter, 193
mash, as off-flavor cause, 27
mash mixing, 4
mash-outs, efficiency of, 225
mash tun, 3, **4,** 7
mashing
    process of, 4, 83
    temperature during, 13
mass, **102**–104
mass-flow density meters, 218–219. *See also* density meters/densimeters
mass number, **55,** 58, 59
mass spectroscopic detectors, 117–118, 208, 209, 211–212
Material Safety Data Sheets (MSDSs), **40**
matrix, **32**
matter, **54**
MCU analysis. *See* malt color unit (MCU) analysis

measurement
  accuracy and precision of. *See* accuracy and precision, of measurement
  errors in, 33
  with graduated glassware, 97–98
  of mass, 56, 102–104
  of off-flavors, 26, 27, 28
  with volumetric glassware, 98–100
meat, as beer flavoring agent, 17
mechanical cleaning, 110
media, for microbial growth, 235–240
  definition of, **236**
  liquid, 236, 238–239
  preparation of, 235–237
  solid, 236, 237, 239–240
megagrams, 57
Megasphaera, 86, 243
melting point, 198–199
membrane filtration, **241**–242
meniscus/menisci, **97**, 98, 215, 216
8-Mercapto-p-menthan-3-one, as off-flavor cause, 27
metabolism, **89**, 90
metal contaminants, in bottled beer, 17, 18, 234
metal ions, sweet wort content, 4
metallic elements, 62
methanol
  dipole–dipole interactions of, 198
  hydrogen bonding in, 199–200
  in infrared analysis, 188
methods (standardized), **125.** *See also ASBC Methods of Analysis*
3-Methyl-2-butene-1-thiol, as off-flavor cause, 27
3-Methylbutanoic acid, as off-flavor cause, 27
2-Methylpropanal, as off-flavor cause, 27
microbes, 229–244
  aseptic sampling of, 241
  bacteria. *See* bacteria
  brewing process and, 234–235
  contamination with, 13, 25, 229, 234, 235. *See also* bacteria
  definition of, **229**
  early study of, 229–231
  growing of, 235–242
  identification of, 242–244
  killing of, 231–233
  membrane filtration of, 241–242
  nutritional requirements of, 236
  sanitary levels of, 110

microbial contamination, 13, 229, 234. *See also* bacteria, beer-spoilage
  via dust particles, 235
  HACCP programs for, 234
  as off-flavor cause, 25
microbiological laboratory, 235–236
  cleaning and sanitizing protocols, 108–110
  definition of, **48**
  equipment, 110–113
  glassware, 109, 110–111
  growing microbial samples, 108, 112–113, 238–240
  growing yeast samples, 114
  media for microbial growth, 236–240
  off-flavor analysis in, 29
  sterilized environment of, 108–109
microbiology, development of, 229–231
microbreweries, 1, 2
  closure rate of, 2
  flavors of beers of, 150
micrograms, 56, 57
microorganisms, types of, 86–88. *See also* bacteria; microbes
microscope slides, **111**
microscopes, **111**–112, 229–230
  cleaning of, 111, 112
  as essential laboratory equipment, 49, 50, 110
  invention of, 229–230
  slides, 111
  use of, 111–112, 242–243
Microsoft Excel, 128, 129
  digital data transfer in, 130–131
  macros and advanced functions, 141–144
  standard curve preparation with, 133–138
microvilli, **151**
microwaves, 175, 178
midrange-infrared spectrum, 187, 188
milk, as beer flavoring agent, 17
milligrams, 56, 57
milling, of malt, 3, 4
mills, **4**
minimal broths, **236.** *See also* broths, for microbial growth
mitochondria, 89, 90

mobile phase, in chromatography system, **193**–194, 200–202. *See also* chromatographs/chromatography
Mohr pipettes, 38. *See also* pipettes
moisture contents
  of barley, 78
  of grain, 3
  of hop cones, 81
  of malt, 4, 78
molar absorptivity, 131, **132,** 133
molar equivalent, 68–69
molar mass, **66**
molarity, 66–69
  definition of, **66,** 98
  in molar absorptivity calculations, 133
mold, as off-flavor cause, 27
mole (counting unit), **56**–57, 66, 69, 177
molecules, **62**–65
  degrees of freedom of, 186
  dipole moments of, 194, 195
  ground and excited states of, 177–179
  linear, 197
  line drawings of, 74–75
  nonlinear, 186
  nonpolar, 194, 195, 197
  polar, 194–195, 200
  solubility of, 63
  vibrational modes of, 186–188
monochromators, 180, **182,** 183, 184
mouthfeel, of beer, 145, **147,** 152
MSDSs. *See* Material Safety Data Sheets (MSDSs)
myrcene, 82, 201

NADH. *See* nicotinamide adenine dinucleotide (NADH)
nanograms, 57
Napoleon Bonaparte, 231
Natick Research Laboratories, 15
National Aeronautics and Space Administration, 15
National Fire Protection Agency (NFPA), **44,** 45
near-infrared rays, 178
near-infrared spectrometers, 46. *See also* spectrometers
near-infrared spectroscopy, 189
near-infrared spectrum, 187, 188, 189
neutrons, **55,** 58

Newton, Isaac, 221
NFPA. *See* National Fire Protection Agency (NFPA)
Nibem method, **227**
nicotinamide adenine dinucleotide (NADH), **90,** 91, 92
nitric acid, as acidic cleaner, 109
nitrogen, in foam formation, 226
no-hop beers, 234
nonmetallic elements, 62
nonmetals, molecular combinations of, 62–64
nonpolar molecules, **194.** *See also* polarity
normality, 66, **68**–69
nuclear magnetic resonance spectroscopy, 179
nucleus/nuclei
 of atoms, 55, 57
 of eukaryotes, 88
 definition of, **55**
 of yeast cells, 88, 89
nuts, as beer flavoring agents, 17

oats, cultivars of, 3
Obesumbacterium, 243
Occupational Health and Safety Administration (OSHA), **40**
octanoic acid, as off-flavor cause, 27
off-flavors, of beer, 25–30
 common, 26–28
 corrective actions for, 28–30
 definition of, **21**
 flavor thresholds for, 26, 27, 28
 of green beer, 5
 identification of causes, 25, 27, 28–29
 malt-related, 160
 masking of, 25
 measurement units, 26, 27, 28
 sensory/tasting panels for, 28–29. *See also* sensory analysis
olfactometers, **225**–226
olfactory system, 151, 152, 154,155, 225–226
on-site analysis, **10,** 12, 34
OpenOffice Calc, 129. *See also* spreadsheet programs
optical windows, cleaning of, 124
overall impression, as sensory input, **147**
OSHA. *See* Occupational Health and Safety Administration (OSHA)

Ostwald, Wilhelm, 224
Ostwald viscometers, **224.** *See also* viscometers
ounces, as common units of measurement, 56
oxidation, 17, **73**
oxidation-reaction (redox) reactions, **73,** 75–76
oxidative phosphorylation, **90,** 91–92
oxidizers, 45, 109
oxygen
 atomic structure of, 55
 isotopes of, 58
 low levels, as off-flavor cause, 27
 molecular combinations of, 62, 63
 nonpolarity of, 197
 as off-flavor cause, 27
oxygen meters, 49
oxygenation, of barley, 78

packaged beer, 78. *See also* bottled beer
packaging laboratory, **48**
paired-comparison test, 157, 158–159
paper chromatography, 193–194
Pasteur, Louis, 231
pasteurization, 231–233
pasteurization units (PUs), 231–**232,** 233
path length, **133**
Pauling, Linus, 194
PCR. *See* polymerase chain reaction (PCR)
peak broadening, **204**–205
Pectinatus, 86, 243
Pediococcus, 86, 88, 229
 D- and *z*-values for, 232
 selective agar plates for, 243
percentage alcohol by volume (% ABV), **32**–33. *See also* alcohol by volume (ABV)
percent transmittance, **181**–182
periodic table of the elements, 54–57, 60
 definition of, **54**
 electronegativity values, 195, 196
 metals and nonmetals on, 62
 use for molarity determination, 66
personal protective equipment (PPE), **40,** 41, 42–43

pesticides, 18
petri dishes, 110, **112**–113. *See also* agar plates
pH, **69**–70
 of beer, 17
 calculation of, 69–70
 determination of, 166
 of polyphenol solubility, 83
pH indicator papers, 70, 166
pH meters, 49, 70, 189
phenolphthalein indicators, 166, 167
phosphate, 85, 92
phosphorescence, 179
phosphoric acid, 109
photocathodes, **181**
photodiodes, **181**
photoelectric effect, 181
photomultiplier tubes, **181**
photons, **176**–177, 178, 179
physical hazards, 17, 18
physical properties analysis, instruments for, 116, 117. *See also specific instruments*
phytase, 79
Pillsbury Company, 15
pinene, 82
pipettes
 graduated, **96,** 97
 volumetric, 95, **96,** 99–100, 169
piping, size determination of, 225
pitching, **4**–5
Planck, Max, 176, 177
plastics, 53, 54
 as chromatography backing material, 203, 204
 "glassware," 94
 shards, 17, 18
plates. *See also* agar plates
 for thin-layer chromatography, 203–204, 205
Pliny the Elder, 80
plots, 138–140
polar compounds, boiling and melting points of, 198–199
polar molecules, **194**
polarity, 194–195
 definition of, **194**
 intermolecular forces of attraction and, 195–200
 relationship to boiling point, 209–210
polyatomic ions, **64.** *See also* ions

polymerase chain reaction (PCR), **243**–244
polymers, 237
polypeptides, 227
polyphenols, 4, **83,** 85, 150
polysaccharides, 88
potassium, wort content, 83
pounds, as common units of measurement, 56
powder funnels, **94.** *See also* funnels
PPE. *See* personal protective equipment (PPE)
precipitation reactions, **72**–73
precision, **32.** *See also* accuracy and precision, of measurement
preventive maintenance, **125**–126, 130
primers, DNA, **243**–244
procedures, versus methods for analysis, 125
prodelphinidin B3, wort content, 85
product analysis
　cost–benefit analysis of, 12–13
　"go/no go" decisions based on, 12
　in-line, 11, 12, 34–35
　laboratory-based, 11, 20–21, 23
　on-site, 10, 12, 34
Prohibition, 1
prokaryotes, 86
propagation laboratory, **48**
propane, 199, 200
propanol, 187, 188
proportionality constants, 131, 132, 134
protanopia, 163
protease, 79
proteins
　cellular production of, 91–92
　effect on foam and foam stability, 150
　insoluble, 4
　soluble, 4
　of yeast cells, 88
protons, 57, 58, 64
　definition of, **55**
　pH as measure of, 69–70
pumps, size determination of, 225
PUs. *See* pasteurization units
pycnometers, 10
Pyrex, **94.** *See also* glassware
pyruvate, 90, 91, 92

QCPs. *See* quality control points (QCPs)

quality, **3,** 5–13
　consistency and, 2, 3
　consumer's determination of, 5
　investments in, 12–13
　maintenance of, 5–10
　range of values for, 8–9
　unacceptable, 12
quality assurance, **9,** 12–13
quality assurance programs, 31
　methods, 9–11
　for off-flavor prevention, 25, 28, 29
　total quality management in, 6
quality control, **9**
quality control points (QCPs), **7**–9
quality control programs, 31
　for off-flavors control, 21, 28
quercetin, wort content, 85

β-radiation, 208
R-squared values, 135
radioactive, **58**
radioactive decay rate, 58
radioactive elements, 58
radioactive materials, hazard codes for, 45
radio waves, 175, 178
raw materials, as beer contaminants, 17
reactivity hazards, 41
reagents, 38, 40
　measurement and preparation of, 24
　storage of, 125
recalls, of beer, 234
recordkeeping, 140–141
　in HACCP programs, 22
　for laboratory results, 35
　in sensory analysis, 146–147
reducing environment, **17**
reduction, **73**
refraction, **219,** 220, 229
refractive index, 219, 220–221
refractometers, 219–221
　as common laboratory instruments, 45, 116
　cost of, 50
　definition of, **219**
　operation/purpose of, 45, 116, 201, 219, 220, 221
refrigerators, 48
regional breweries, 1
relative humidity, **122,** 123

reporting, of results, 127–144
　with spreadsheets. *See* spreadsheet programs
　visual methods, 140–141
restaurants, closure rates, 2
retention factor, **204**
retention time, **203**
Reynolds number, **222**–223, 225
rhizomes, **80**
ring clamps, **94,** 95
ring stands, **94,** 95, 106–107
risk assessment, 19–20
risk ratings, of hazards, 19–20
rotation mass-flow analysis, 218–219
round-bottom flasks, **96,** 106–108. *See also* flasks
Rudin method, **227**

Saccharomyces cerevisiae, 1, **88,** 232
Saccharomyces cerevisiae var. *diastaticus,* 232
Saccharomyces pastorianus, 1, 232
safety, culture of, 41, 42. *See also* safety guidelines and programs
Safety Data Sheets (SDSs), **40,** 102
safety equipment, 38
safety guidelines and programs, 18, 40–45
　for chemicals, 40, 102
　for food production. *See* hazard analysis and critical control points (HACCP) programs
　for laboratory instruments, 120
safety hazards, 18
safety showers, 44
salts, **60,** 61. *See also* sodium chloride
sample collection, 34
samples
　blank, 169–170, 171, 189
　bulk, 46
　holders for, 118
　standard, 165
　storage of, 37
sampling, aseptic, 241
sanitary (environment), **108**
sanitizers, 109, 110
scales, **103**
scatter plots, **128,** 140
Schönbein, Christian Friedrich, 193
scientific method, **127,** 128
scientific notation, **56**–57
SDSs. *See* Safety Data Sheets (SDSs)

selective broths, **236.** *See also* broths, for microbial growth
semivolatile compounds, gas chromatographic analysis, 209–210
sensors, in-line, 11
sensory analysis, 145–160
   bias in, 159
   definition of, **145**
   examples, 160
   of foam, 150
   guidelines for, 146–147, 155–157
   journals/notebooks for, 146–147
   lighting for, 147, 148, 155–156
   microbial contamination detection in, 234
   for off-flavor control, 30
   sensory thresholds and sensory overload in, 152, 154–155
   training in, 150, 156–157. *See also* sensory/tasting panels
   types of, 157, 158–159
sensory laboratory, **48**
sensory overload, 152, 154–**155**
   sensory panels/tasting panels, 155–157. *See also* sensory analysis, training in
   operation of, 29, 154, 155–157
   purposes of, 25, 29, 30, 150, 152–153
   taster characteristics and responsibilities, 28, 154–155, 157, 158–159
sensory thresholds, 152, **154**–155
shear force, 221–222, 225
shellfish, as beer flavoring agent, 17
SI system, of measurement, 56–57
sigma values, 138
silica gels, 202
simple plots, **138**
single wavelength spectrometers. *See* fixed wavelength spectrometers; spectrometers
skunk flavor, 127
slants, **112,** 237, 238, 239
slope of the line, **134**
sodium, wort content, 83
sodium chloride
   as infrared analysis chamber, 188
   molarity determination for, 67
   molar solution of, 98–99
sodium hydroxide, 109

solids
   measurement of, 97, 103–104
   solutions prepared from, 105
solubility
   of ionic compounds, 60, 61
   of molecules, 63
solubility rules, **72**–73
solutes, **31.** *See also* analytes
solutions
   color intensity of, 116
   cost of, 118
   molecular interactions in, 196–200
   preparation by dilution, 105–106
   preparation from solids, 105
   stock, dilution of, 105–106
solvents
   as beer component, 32
   effect on ionic compounds, 60
SOPs. *See* standard operating procedures (SOPs)
sources, of energy, **180**
sparging
   Darcy's law of, 127
   definition of, **4**
   excessive, as off-flavor cause, 27
   flow rates during, 225
spatulas, 38, **97**
specifications
   action boundary of, 138, 139
   cost–benefit analysis of, 12–13
   evaluation of, 10–11
   measurement of, 10–13
   relationship to quality, 5–6
   target values of, 138–140
   values outside, 141
   warning boundary of, 138, 139
specific gravity
   for alcohol by volume calculations, 117
   calculation of, 215–216
   correlation with refractive index, 220, 221
   definition of, **213,** 214
   relationship to density, 213–214
   of wort, 8, 9, 238
spectrometers, 179–185. *See also* spectrophotometers/spectrophotometric analysis
   colorimeters versus, 179
   general design of, 179–182
   operation/purpose of, 179–182
   types of, 182–185. *See also specific types*

spectrophotometers/spectrophotometric analysis, 175–191
   accuracy and precision of, 179
   ASBC Methods of Analysis for, 24
   bench space needed for, 119–120
   as common laboratory instruments, 45, 46, 49, 50–51, 116
   cost of, 50–51, 120
   chromatography versus, 175, 179, 193
   colorimetry versus, 175
   cost of, 50–51
   definition of, **175**
   lighting needed for, 124
   operation/purpose of, 45, 116, 148, 175
   types of, 45, 46, 118, 119, 131. *See also specific types*
spectroscopy, **175.** *See also* colorimeters/colorimetric analysis; spectrophotometers/spectrophotometric analysis
spectrum/spectra, **181**
speed of light, **176,** 177
spices, as beer flavoring agents, 17
splash goggles, 41, 42–43
spoilage bacteria. *See* bacteria, beer-spoilage
spreadsheet programs, 128–140
   capabilities of, 129
   chart creation with, 138–140
   definition of, **128**
   entering results in, 130–131
   functions of, 140–141
   graphing with, 128–140
   macros in, 141–144
   sources of, 128–129
   standard curve preparation with, 133–138
SRM. *See* standard reference method (SRM)
staff
   of brewery laboratories, 33. *See also* laboratory personnel
   hazard analysis roles, 18–19
   involvement in quality assurance, 6
   training of. *See* training, of staff

standard curves, 131–138
  definition of, **131**
  preparation with spreadsheets, 133–138
  R-squared values, 135
  for ultraviolet-visible spectroscopy, 190
  x- and *y*-axes of, 133–137
standard deviation, 20–21, 142–144
standard operating procedures (SOPs), 31, 34
  corrective procedures, 21–22
  definition of, **18**
  for hazard analysis, 19, 20, 21–22
  for laboratory instruments, 46
  for off-flavor control, 30
  safe practices component of, 41
  validation procedures, 35
standard reference method (SRM), **148,** 149, **165,** 173, 191
standard samples, **165**
standard solutions, for colorimetric analysis, 168–172
  blanks, 169–170
  errors in measurement of, 169
standard taper (⚡), **97**
starch
  barley endosperm content, 4, 84
  conversion into sugars, 79, 83
starchy endosperm, **79**
starting process, in brewing, 4
static electricity, buildup on instruments, 122–123
stationary phase, in chromatography system, **193**–194, 200–202. *See also* chromatographs/chromatography
steam cleaning, 109, 110
steep/steeping, **78**
sterile (environment), **108**–110
sterilization
  of agar, 112
  of fermenter sample valves, 241
  of glassware, 112
  of microbiological laboratories, 108, 110
  pasteurization, 231–233
  of wort, 234
  of yeast-growing equipment, 238, 239
still heads, 95, **96,** 108
stir plates, 112, 114

stoichiometric factor, **72**
stoichiometric ratio, **71,** 75
Stokes units, 222
stoppers, 95, **97**
storage
  of chemicals, 37, 40
  guidelines for, 40
  of malt, 4
  of samples, 37
Stout, Russian Imperial, 146
stress, effect on yeast, 23, 27
strigs, **80,** 81
strobili, **80,** 81
sucrose, **83,** 88
sugars
  concentration of, 213–214
  effect on refractive index, 221
  fermentable, 4, 83, 84, 90
  fermentation of. *See* fermentation
  unfermentable, 4, 83, 84
sulfate
  as hot liquor additive, 4
  wort content, 85
sulfuric acid, normality of solution, 68–69
suppliers/supply chains
  changes in, 30
  hazard control in, 16
sweet wort, **4.** *See also* wort
Synge, Richard Laurence Millington, 193

tables, 140–141
  cumulative sum (CUSUM), 139–140
  software-based creation of, 128
tailing, **205**
tanks, for laboratory instruments, 36
tannins. *See* polyphenols
tare/taring, **103**
target values, **138**–140
taste, of beer. *See* flavor, of beer
taste buds, **151,** 152
tasting, sensation of, 151–152
taxation, 1, 13
TDUs. *See* thermal degradation units (TDUs)
tea-like flavor, 83
temperature
  of brewery laboratories, 121, 123
  effect on density, 214
  effect on viscosity, 223

  for fermentation, 1
  for in-line analysis, 35
  during mashing, 13
  for optimal cleaner use, 109
  for optimal instrument performance, 121, 123
  of pasteurization, 231–233
temperature probes, 11
terminal velocity, **225**
terpene, 82
terpenoids, 82
test tube holders, **97**
test tube racks, **94**
test tubes, 94, 95
  for colorimetry, 168, 169, 170
  containing agar, 112, 113
  definition of, **94**
  in media preparation, 237
  racks and holders for, 50, 51, 94, 95, 97
testing against a standard color, **165.** *See also* colorimeters/colorimetric analysis, measurement methods of
testing for the presence of a color, **165.** *See also* colorimetry/colorimetric analysis, measurement methods of
tetrad test, 159
text-based reporting, **141**
theories (scientific), **127**
thermal degradation units (TDUs), 233
thermometers, **97,** 106
thin-layer chromatography (TLC), 202–205. *See also* chromatographs/chromatography
  backing materials for, 203, 294
  definition of, **202**
  operation/purpose of, 193–194, 204–205
  peak broadening in, 204–205
  stationary and mobile phases of, 204–205
threshold test, 159
tin, electronegativity values, 195
tintometers, 162, 163, 173
TLC. *See* thin-layer chromatography (TLC)
tongs, **97**
total quality management (TQM), **6**
toxic compounds, 18, 78

TQM. *See* total quality management (TQM)
training, of staff, 33
   in instrument use, 46
   in laboratory-based analysis, 11
   in product analysis, 10, 12
   in safety, 18
   in sensory analysis, 150, 157
Trans-2-nonenal, as off-flavor cause, 27, 28
transmittance, **181**
transmitted intensity, **132**
trends, charting of, 138, 139–140
triangle test, 157, 158
2,4,6-trichloroanisole, as off-flavor cause, 27
tritanopia, 163
trub, **4–5**
true-to-brand test, 157
Tsvet, Mikhail, 193
turbidimeters, 116
turbidity, measurement of, 116, 173, 191
turbulence, measurement of, 222–223
turbulent flow, **222–223**
21st Amendment, 1

U-tubes, 217, 218, 224
ultraviolet rays, 178
ultraviolet-visible (UV-vis) spectrophotometry/spectroscopy, 189–191. *See also* spectrophotometers/spectrophotometric analysis
   bench space needed for, 119, 120
   colorimetry versus, 179
   as common laboratory instruments, 44, 46, 49, 50, 116, 117
   cost of, 50–51
   definition of, **178**
   detectors in, 181
   methods of use, 189–191
   operation/purpose of, 131, 178–179
   pros and cons of, 191
   samples and standards for, 189–191
U.S. Hazardous Materials Identification System, **40**, 41
unknown samples, concentration determination of, 136–137
utility services, for laboratories, 38

UV-vis spectrophotometry. *See* ultraviolet-visible (UV-vis) spectrophotometry/spectroscopy

vacuum adapters, 95, **96,** 107
vacuum filter flasks, **95**
vacuum filtering, **101**
validation procedures, 35
Van Leeuwenhoek, Anton, 230–231
vapors, ocular exposure to, 42
varieties, of beer, 1–2
vegetable-like off-flavors, 27
ventilation systems, 36, 48
verification procedures, 22, 25
vibrations, of molecules, **186–188**
vicinal diketones, detection of, 201
viruses, illness-causing, 18
viscometers, 221–225
   as common laboratory instruments, 45, 46, 116
   operation/purpose of, 45, 46, 224–225
   types of, 224–225
viscosity, **221–222,** 223, 225
vision, color perception deficiencies, 163–164
visual evaluation, of beer. *See* appearance, of beer
visual perception, **164–165,** 175
volatile compounds
   definition of, **203**
   gas chromatographic analysis of, 203, 209–210
   ventilation system for, 48
volatile samples, 37
volume measurements
   accuracy and precision of, 93–94
   of liquids, 104–105
   for molarity determination, 69 66
volumetric flasks, **96.** *See also* flasks
volumetric glassware
   as basic lab equipment, 38, 49, 51
   types of, 96, 97
   use of, 67, 98–100, 104, 105, 106, 168, 169
volumetric pipettes, **96.** *See also* pipettes
vomitoxin, **78**
vortex mixing, 4

warning boundary, **138**
water
   atomic structure of, 63

   boiling point of, 107
   calcium content, 118
   deionized or distilled, 38
   density of, 213, 214, 215
   formula for, 63
   as hot liquor, 4
   ions in, 65, 66–69
   microbial contamination of, 21
   oxidative phosphorylation of, 91
   poor quality, as off-flavor cause, 27
   purification of, 4
   as solvent, 32
   strike, 4
   vibrational modes of, 186
water lines, 119
water-reactive materials, hazard codes for, 45
wavelengths, 175–177
   characteristics of, 175–176
   definition of, **175**
   fixed, 182
   frequency and, 176
   of infrared energy, 186–188
   measurement of, 175–176
   for refraction index measurement, 220–221
   relationship to frequency, 176–177
   of ultraviolet-visible spectrum, 189, 190
weigh paper, **103**
weight, **103**
welding, sanitary, 16
wetting agents, **109**
whirlpool/whirlpooling, of wort, **4**
wort, **83–85**
   boiling of, 234
   color evaluation of, 147–149
   component compounds of, 4, 83–85
   density of, 214, 215
   hopped, 238
   inoculation with yeast, 114
   kettle-soured, 160
   microbial contamination of, 86
   sensory analysis, 160
   sterilization of, 234
   sugar content measurement, 219–221
   types of, 3, 4, 5, 8, 10–11, 160. *See also specific types*

X-rays, 175, 178

yeast, 88–89, 229
    biological reactions, 89–90
    bottom-fermenting, 1
    budding, 88, 89
    cell structure of, 88–89
    concentration in solution, 225
    cross-contamination, 234, 235
    cultivars of, 3
    dead, as off-flavor cause, 27
    definition of, **4–5**
    in fermentation, 4
    growing techniques for, 114, 236–244
    growth and reproduction (budding), 4, 88, 89, 92
    health of, 160
    laboratory storage areas, 38
    media for, 236–240
    metabolism in, 89–92
    mixing with wort, 4
    non-*Saccharomyces*, selective agar plates for, 243
    non-yeast microbial contamination of, 236, 241–242
    overpitching of, 50
    Pasteur's experimentation with, 231
    storage of, 112
    stressors on, 23, 27
    top-fermenting, 1
    underpitching of, 50
    viability testing of, 13

z-values, 231–232
zinc, 83, 85, **92**
Zymomonas, 86